Quantum Computational Chemistry

Taku Onishi

Quantum Computational Chemistry

Modelling and Calculation for Functional Materials

Springer

Taku Onishi
CTCC, Department of Chemistry
University of Oslo
Oslo
Norway

and

Department of Applied Physics
Osaka University
Osaka
Japan

and

CUTE
Mie University
Mie
Japan

ISBN 978-981-10-5932-2 ISBN 978-981-10-5933-9 (eBook)
DOI 10.1007/978-981-10-5933-9

Library of Congress Control Number: 2017947029

Printed on acid-free paper

This Springer imprint is published by Springer Nature
The registered company is Springer Nature Singapore Pte Ltd.
The registered company address is: 152 Beach Road, #21-01/04 Gateway East, Singapore 189721, Singapore

Preface

This book is written for both theoretical and experimental scientist (chemists and physicists) to help understand chemical bonding and electronic structure, from the viewpoint of molecular orbital theory. A long time ago, quantum theory was applied to very simple atoms. To connect quantum theory with complex systems, there were many research activities in the fields of quantum chemistry and physics: the Bohr model, wave-function, Schrödinger's equation, the Hartree-Fock method, Mulliken charge density analysis, density functional theory, etc. Due to this research, we are now able to perform molecular orbital calculations from small molecules through to advanced materials including transition metals. In this book, chemical bonding and electronic structure are explained with the use of concrete calculation results, density functional theory, and coupled cluster methods.

In Part I the theoretical background of quantum chemistry is clearly explained. In Part II we introduce molecular orbital analysis of atoms and diatomic molecules via concrete calculation results. After introducing the theoretical background of inorganic chemistry in Part III, the concrete calculation results for advanced materials such as photocatalysts, secondary batteries, and fuel cells are introduced in Part IV. Finally, helium chemistry and the future of the subject are considered in Part V.

Acknowledgements

My research work has been supported by the Research Council of Norway (RCN) through CoE Grant No. 179568/V30 (CTCC) and through NOTUR Grant No. NN4654K for HPC resources. I thank Professor Trygve Helgaker for his kind research support and encouragement since my first stay in Norway.

I wrote most of the book during my research stay at the Centre for Theoretical and Computational Chemistry (CTCC), University of Oslo, Norway, and Department of Applied Physics, Graduate School of Engineering, Osaka University, Japan.

Finally, I am especially grateful for the constructive discussions I had with TO.

Quantum chemistry is not virtual but real

Research is first discovery

Kobe, Japan
May 2017

Taku Onishi

Contents

About the Author

Taku Onishi was born in Kobe, Japan. He is an international quantum chemist who graduated from the Faculty of Science, Osaka University in 1998, and gained his Ph.D. from the Department of Chemistry, Osaka University in 2003. He took up a permanent position in the Faculty of Engineering, Mie University, Japan in 2003. He has been a guest researcher for the Department of Chemistry, University of Oslo, Norway since 2010, and a guest academician at the Department of Applied Physics, Osaka University, Japan since 2016. His research is in the areas of quantum chemistry, computational chemistry, quantum physics, and material science. His scientific positions held include: a Member of Royal Society of Chemistry; Chair of the Computational Chemistry (CC) Symposium; a position on the science committee of the International Conference of Computational Methods in Sciences and Engineering (ICCMSE); the General Chair of Advanced Materials World Congress; a member of the editorial board of Cogent Chemistry as well as the Journal of Computational Methods in Sciences and Engineering (JCMSE). He has reviewed many international proceedings, books, and journals including: AIP conference proceedings, Progress in Theoretical Chemistry and Physics, Cogent Chemistry, Physical Chemistry Chemical Physics, Molecular Physics, Dalton Transaction, Chemical Physics, The Journal of Physical Chemistry Letters, Journal of Computational Chemistry, Journal of Solid State Chemistry, Solid State Ionics, Chemistry of Materials, Materials Chemistry and Physics, Chemical Engineering Journal, etc.

Part I
Theoretical Background of Quantum Chemistry

Chapter 1
Quantum Theory

Abstract By the difference of scale, matter is largely classified into solid, molecule and cluster. The basis unit of matter is atom. Atom consists of quantum particles such as electron, proton and neutron. In Bohr model, quantum effect is incorporated through the concept of matter wave. In the case of hydrogen, the orbit radius was estimated to be 0.5292 Å, corresponding to the experimental distance. In addition, the discrete energy was also reproduced. However, Bohr model was not able to be applicable to many-electron system. In order to incorporate particle-wave duality in universal manner, quantum wave-function was proposed. In wave-function theory, electron does not correspond to classical point, but spreads as wave. It is difficult to interpret wave-function itself. It is because it does not represent figure. Instead, the square of wave-function represents electron density. Wave-function can be obtained by solving the Schrödinger equation, where electron energy is given by operating wave-function with Hamiltonian. As a feature of wave-function, it is normalized and satisfies orthogonality. In quantum mechanics, one electron occupies one wave-function. It implies that one electron is not distributed to several wave-functions.

Keywords Wave-particle duality · Bohr model · Quantum wave-function · Schrödinger equation

1.1 Matter and Atom

By the difference of scale, matter is largely classified into the three: solid, molecule and cluster (nano-cluster), as shown in Fig. 1.1. Molecule and cluster exist in the basic three fundamental states: gas, liquid and solid (molecular solid). In quantum chemistry, electronic structure is normally discussed in three fundamental states. As the extreme environment, matter exists as plasma and superconducting states. In plasma state, matter is divided into positively charged ion and negatively charged electron at very high temperature. It has been considered that most of matter in space exists as plasma state. On the other hand, in superconducting state, electric resistance becomes zero at very low temperature, though matter keeps the same crystal structure.

T. Onishi, *Quantum Computational Chemistry*,
DOI 10.1007/978-981-10-5933-9_1

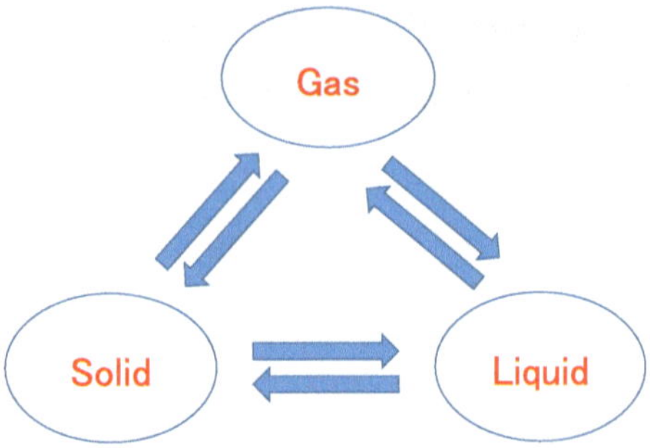

Fig. 1.1 Basic three fundamental states of matter

The basis unit of matter is atom. Atom consists of quantum particles such as electron, proton and neutron. As they belong to Fermi particle, the spin angular momentum becomes half-integer. Atom has an atomic nucleus at the centre, consisting of proton and neutron. As is well known, there are several kinds of atoms in space. The kind of atom is called element. Element is represented by atomic number (Z) that corresponds to the total number of protons. For example, the elements of $Z = 1$, 2 and 3 denote hydrogen, helium and lithium, respectively. When the same element has the different total number of neutrons, it is called isotope. As the magnitude of charge density of proton is e, the total charge density of atomic nucleus becomes $+Ze$. Z electrons are allocated around atomic nucleus. Note that the magnitude of electron charge density is $-e$.

1.2 Wave-Particle Duality

Quantum particle is defined as particle with wave-particle duality. The wave property is incorporated through the concept of matter wave.

$$\lambda = \frac{h}{mv} \tag{1.1}$$

where λ is the wave-length of matter wave: h is Plank constant; m is the mass of quantum particle; v is the velocity of quantum particle. In electron, m denotes the mass of electron, which is expressed as m_e. Though the energy of classical particle continuously changes, quantum particle has the discrete energy.

1.3 Bohr Model

Niels Bohr proposed a theoretical hydrogen model, which is well known as Bohr model, to express positions of electron and atomic nucleus, under consideration of wave-particle durability. In Bohr model, proton is located at atomic centre, and

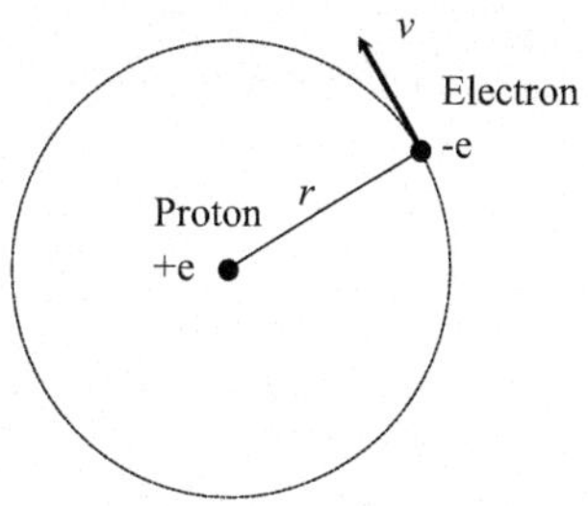

Fig. 1.2 Schematic drawing of Bohr model. Proton is located at atomic centre, and electron goes around an orbit

electron rotates around atomic centre, as shown in Fig. 1.2. The driving factor of the rotation is the Coulomb interaction (*f*) between proton and electron:

$$f = \frac{e^2}{4\pi\varepsilon_0 r^2} \tag{1.2}$$

where ε_0 is dielectric constant of vacuum; r is orbit radius. The centrifugal force, which is obtained from the classical equation of circular motion, is equal to the Coulomb interaction.

$$\frac{m_e v^2}{r} = \frac{e^2}{4\pi\varepsilon_0 r^2} \tag{1.3}$$

The equation implies that quantum effect is taken into account, by applying the concept of matter wave to electron. Electron goes around an orbit, and orbit distance is mathematically determined to be $2\pi r$. It must be also integer-multiple of wave-length of matter wave.

$$2\pi r = n\lambda (n = 1, 2, 3, \ldots) \tag{1.4}$$

By the substitution of Eq. (1.1) in Eq. (1.4), it is rewritten:

$$m_e v r = \frac{nh}{2\pi} (n = 1, 2, 3, \ldots) \tag{1.5}$$

By the substitution of Eq. (1.5) in Eq. (1.3), the orbit radius of hydrogen is obtained:

$$r = \frac{\varepsilon_0 h^2 n^2}{\pi m_e e^2} (n = 1, 2, 3, \ldots) \tag{1.6}$$

As orbit radius depends on integer (n), it is found that orbit radius is quantized. It implies that orbital radius has only discrete value. The orbit radius of $n = 1$ is called Bohr radius. The value is estimated to be 0.5292 Å.

Let us consider an electron energy. It is obtained by the summation of electron kinetic energy (KE) and potential energy (PE). In classical manner, KE is given by

$$\mathrm{KE} = \frac{m_{\mathrm{e}}v^2}{2} = \frac{e^2}{8\pi\varepsilon_0 r} \tag{1.7}$$

From Coulomb's law, PE is given by

$$\mathrm{PE} = -\frac{e^2}{4\pi\varepsilon_0 r} \tag{1.8}$$

It is defined that PE is zero, when electron is infinitely apart from proton. Hence, potential energy exhibits negative value. Finally, the total energy of electron is given by

$$\mathrm{KE} + \mathrm{PE} = -\frac{e^2}{8\pi\varepsilon_0 r} \tag{1.9}$$

By the substitution of Eq. (1.6) in Eq. (1.9), it is rewritten:

$$\mathrm{KE} + \mathrm{PE} = -\frac{m_{\mathrm{e}}e^4}{8\varepsilon_0^2 h^2 n^2}(n = 1, 2, 3, \ldots) \tag{1.10}$$

It is found that the electron energy is also quantized by the introduction of matter wave. It implies that the discrete electron energy of hydrogen is successfully reproduced in Bohr model. When n = 1, the electronic state is called ground state, exhibiting the smallest energy. When n is larger than two, the electronic state is called excited state. In general, the electron energy of excited state is larger than ground state.

1.4 Quantum Wave-Function

In Bohr model, wave property is incorporated through the concept of matter wave. Bohr model was not able to be extended to many-electron system. As the solution, quantum wave-function was proposed. Let us consider an electron isolated in space. It is mathematically represented by wave-function ($\Psi(r_1)$), which contains one radial parameter (r_1). In wave-function theory, electron corresponds to not classical

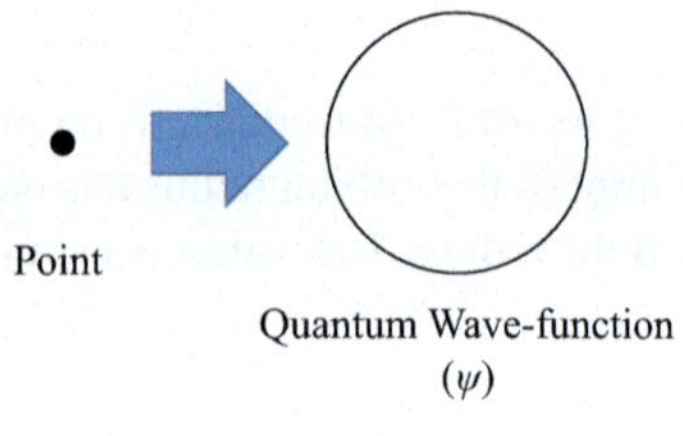

Fig. 1.3 Schematic drawing of quantum wave-function

point, but spreads as wave (see Fig. 1.3). Note that radial parameter is used for representing the spread of electron. In many-electron system, n-radial parameters are included. The wave-function is expressed as $\Psi(r_1, r_2,\ldots,r_n)$, where $r_1, r_2,\ldots,r_n$ are defined for electron 1, electron 2,…, electron n, respectively. Note that the time-independent wave-function is considered in this book, though time evolution is possible in wave-function.

1.5 Wave-Function Interpretation

One electron can be expressed as one wave-function. However, it is difficult to interpret wave-function itself. It is because it does not represent figure (line, curved surface, etc.). Instead, the square of wave-function represents electron density. It is given by,

$$|\psi|^2 = \psi^*\psi \tag{1.11}$$

Electron density within the volume element (dτ) is proportional to $|\psi|^2\mathrm{d}\tau$, where $\mathrm{d}\tau$ is equal to dxdydz (see Fig. 1.4).

For the correspondence to the real electron, the normalization is performed for the wave-function. The normalized wave-function (ψ') is expressed as

$$\psi' = N\psi \tag{1.12}$$

where N is the normalization constant. When integrating electron density within the whole space, it must represent one electron.

$$\int_{-\infty}^{\infty} \psi'^*\psi'\mathrm{d}\tau = 1 \tag{1.13}$$

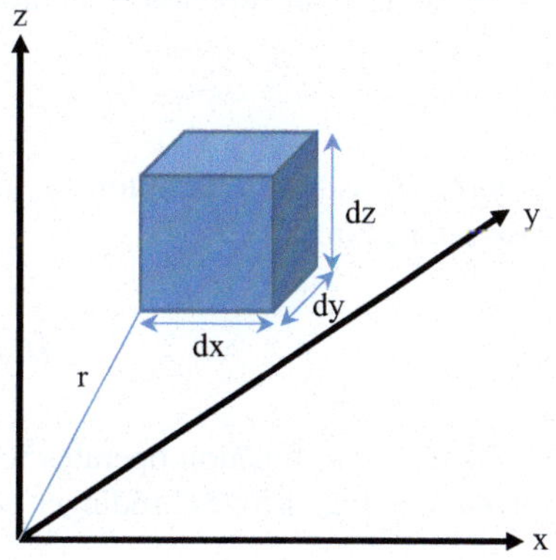

Fig. 1.4 Schematic drawing of the relationship between electron density and volume element

By substitution of Eq. (1.12) in Eq. (1.13), it is rewritten:

$$N^2 \int_{-\infty}^{\infty} \psi^* \psi \mathrm{d}\tau = 1 \quad (1.14)$$

In general, the normalized wave-function is utilized.

1.6 Schrödinger Equation

The basic equation of classical particle is motion equation. On the other hand, the basic equation of electron is Schrödinger equation, where electron is expressed as wave-function (ψ). In one-dimensional system, it is expressed as

$$-\frac{\hbar^2}{2m}\frac{\mathrm{d}^2\psi}{\mathrm{d}x^2} + V(x)\psi = E\psi \quad (1.15)$$

where $V(x)$ denotes the potential energy at x; E is the total energy; $\hbar$ is defined as

$$\hbar = \frac{h}{2\pi} \quad (1.16)$$

Extending to three-dimensional system, it is expressed as

$$-\frac{\hbar^2}{2m}\nabla^2\psi + V\psi = E\psi \quad (1.17)$$

where ∇^2 is defined as

$$\nabla^2 = \frac{\partial^2}{\partial x^2} + \frac{\partial^2}{\partial y^2} + \frac{\partial^2}{\partial z^2} \quad (1.18)$$

In the general expression, Schrödinger equation is expressed as

$$\hat{H}\psi = E\psi \quad (1.19)$$

where $\hat{H}$ is the Hamiltonian operator, which mathematically operates to wave-function.

$$\hat{H} = -\frac{\hbar^2}{2m}\nabla^2 + \hat{V} \quad (1.20)$$

When wave-function operates with the Hamiltonian operator, the total energy is given (see Fig. 1.5). Schrödinger equation is eigenvalue equation, where E and ψ

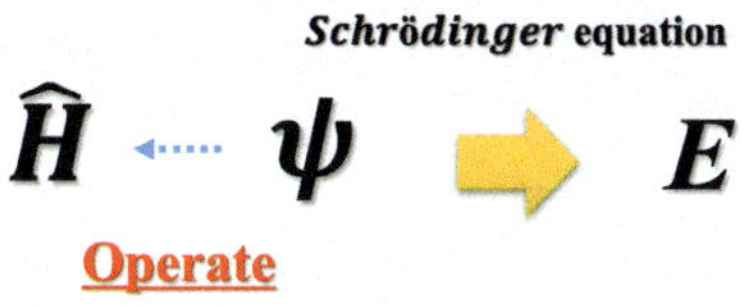

Fig. 1.5 Schematic drawing of Schrödinger equation

denote eigenvalue and eigenfunction, respectively. It implies that one eigenvalue is given for one wave-function.

Let us consider two different wave-functions. The wave-functions (ψ_i and ψ_j) satisfy the following equations:

$$\hat{H}\psi_i = E_i\psi_i \tag{1.21}$$

$$\hat{\mathrm{H}}\psi_j = E_j\psi_j \tag{1.22}$$

where E_i and E_j are eigenvalues for ψ_i and ψ_j, respectively. Integrating Eq. (1.21) within the whole space, combined with the product of ψ_i^* on the left side,

$$\int \psi_j^*\hat{\mathrm{H}}\psi_i \mathrm{d}\tau = E_i \int \psi_j^*\psi_i \mathrm{d}\tau \tag{1.23}$$

Integrating Eq. (1.22) within the whole space, combined with the product of ψ_i^* on the left side,

$$\int \psi_i^*\hat{H}\psi_j \mathrm{d}\tau = E_j \int \psi_i^*\psi_j \mathrm{d}\tau \tag{1.24}$$

The complex conjugate of Eq. (1.23) becomes as

$$\left(\int \psi_j^*\hat{H}\psi_i \mathrm{d}tau\right)^* = E_i \int \psi_i^*\psi_j \mathrm{d}\tau \tag{1.25}$$

In general, Hermitian operator satisfies the following relationship:

$$\int \psi_i^*\hat{H}\psi_j \mathrm{d}\tau = \left(\int \psi_j^*\hat{H}\psi_i \mathrm{d}\tau\right)^* \tag{1.26}$$

By the substitution of Eqs. (1.24)–(1.26),

$$(E_i - E_j)\int \psi_i^*\psi_j \mathrm{d}\tau = 0 \tag{1.27}$$

When $E_i \neq E_j$ is satisfied,

$$\int \psi_i^* \psi_j \mathrm{d}\tau = 0 \tag{1.29}$$

It implies that eigenfunctions with different eigenvalues are orthogonal.

1.7 Quantum Tiger

Quantum electron is represented as quantum wave-function, which is obtained from Schrödinger equation. For the easy understanding, it is, here, assumed that one tiger represents one quantum electron, and one box represents one wave-function. When there are two boxes (box A and box B), where is tiger is staying? (See Fig. 1.6). In conventional world, tiger stays in box A or B, without changing its figure. It means that the density of tiger must be 0 or 100% in one box. If the density is between 0 and 100%, tiger must be separated into two pieces. It does not occur.

If tiger is separated into two pieces, it means that one electron is delocalized over two wave-functions. In fact, the ith quantum electron has specific energy (E_i) and exists in one wave-function (ψ_i) (see Fig. 1.7). The ith electron cannot be allocated in the different wave-function, due to the orthogonality between wave-functions with different energies. In degenerated case, although the wave-functions are different, they have the same energy. However, the ith electron is allocated into one wave-function.

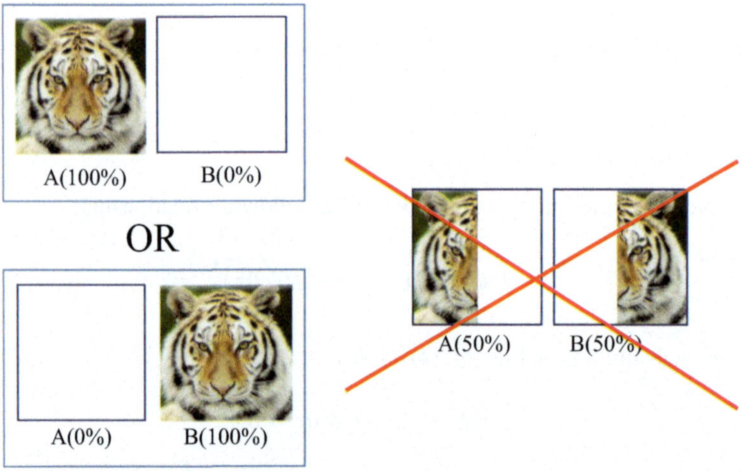

Fig. 1.6 Schematic figure of quantum tiger in two boxes

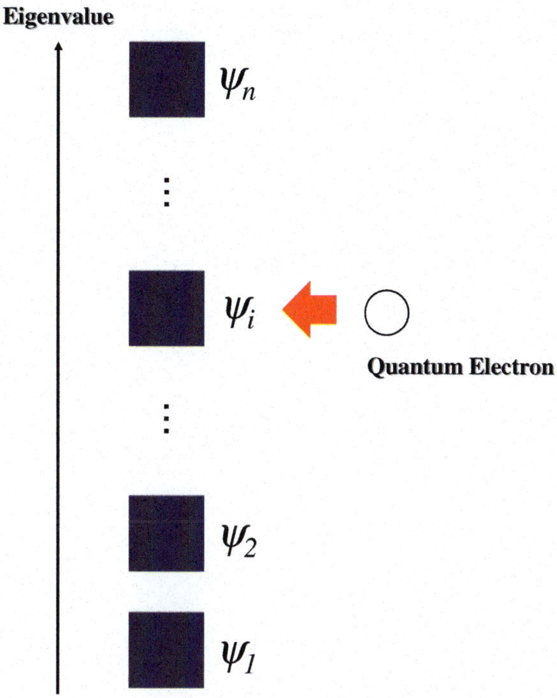

Fig. 1.7 Schematic drawing of the relationship between wave-function and eigenvalue

Further Readings

1. Atkins P, de Paula J (2006) Physical chemistry 8th edn, Chapter 8 (in Japanese)
2. Atkins P, de Paula J, Friedman R (2009) Quanta, matter, and change a molecular approach to physical chemistry, Chapter 1 (in Japanese)
3. Barrow GM (1999) Physical chemistry 6th edn, Chapter 9 (in Japanese)

Chapter 2
Atomic Orbital

Abstract In one-electron atom such as hydrogen atom and hydrogenic atom, the exact solution of Schrödinger equation can be obtained. The wave-function, which stands for atomic orbital, is separated into the two radial and angular wave-functions. Radial wave-function contains two quantum numbers such as principal quantum number and orbital angular momentum quantum number. The former and latter denote shell and subshell, respectively. Due to the relationship between two quantum numbers, 2p and 3d orbitals have the three and five orbitals. Angular wave-function expresses electron spread by using two angular parameters. The wave-function cannot be directly plotted into three-dimensional space. Instead, it is possible to visualize electron density, which is given by the square of wave-function. In many-electron atom, the effect of spin cannot be negligible. Spin has two quantum numbers of total spin angular momentum and spin angular momentum along the standard direction. When the latter quantum number is +1/2 or −1/2, it is called α or β spins, respectively. To incorporate electron spin in wave-function, spin function is introduced. Spin orbital is expressed by the product between spatial orbital and spin function. To satisfy inversion principle, the total wave-function is represented by Slater determinant. Finally, building-up principle is also explained.

Keywords Hydrogenic atom · Radial wave-function · Angular wave-function · Electron spin · Slater determinant

2.1 Hydrogenic Atom

2.1.1 Schrödinger Equation

As explained in Chap. 1, in quantum manner, electron is represented by wave-function. The wave-function standing for an electron is called orbital. In atom and molecule, it is called atomic orbital (AO) and molecular orbital (MO), respectively. Let us explain atomic orbitals of hydrogenic atom, where one electron

T. Onishi, *Quantum Computational Chemistry*,
DOI 10.1007/978-981-10-5933-9_2

exists around atomic nucleus with nuclear charge Ze. Note that Z is positive integer. Coulomb potential energy (V) between atomic nucleus and electron is expressed as

$$V = -\frac{Ze^2}{4\pi\varepsilon_e r} \tag{2.1}$$

where e, ε_0 and r denote charge, vacuum permittivity and electron-atom distance, respectively. The Hamiltonian of hydrogenic atom is given by

$$H = -\frac{\hbar^2}{2m_e}\nabla^2 - \frac{\hbar^2}{2m_N}\nabla^2 - \frac{Ze^2}{4\pi\varepsilon_e r} \tag{2.2}$$

where $\hbar = h/2\pi, m_e$ is the mass of electron, and m_N is the mass of nucleus. The first, second and third terms denote kinetic energy of electron, kinetic energy of atomic nucleus and Coulomb potential energy, respectively. The Schrödinger equation for hydrogenic atom is expressed as

$$\left(-\frac{\hbar^2}{2m_e}\nabla^2 - \frac{\hbar^2}{2m_N}\nabla^2 - \frac{Ze^2}{4\pi\varepsilon_e r}\right)\Psi = E\Psi \tag{2.3}$$

where Ψ and E denote the wave-function of an electron and the total energy, respectively. The wave-function can be separated into two parts by three variables such as radial (r) and two angular (θ, ϕ) components (see Fig. 2.1).

$$\Psi(r, \theta, \phi) = R(r)Y(\theta, \phi) \tag{2.4}$$

When the reduced mass (μ) is defined as

$$\mu = \frac{m_e m_N}{m_e + m_N} \tag{2.5}$$

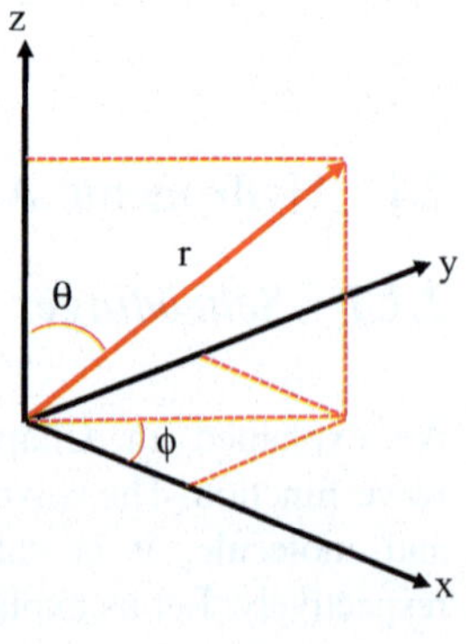

Fig. 2.1 Relationship between Cartesian coordinates and polar coordinates

the μ value is approximately similar to the electron mass.

$$\frac{1}{\mu} = \frac{1}{m_{\mathrm{e}}} + \frac{1}{m_{\mathrm{N}}} \approx \frac{1}{m_{\mathrm{e}}} \tag{2.6}$$

It is because the nucleus mass is much larger than electron mass. By the substitution of Eq. (2.6), Eq. (2.3) is written as

$$\left(-\frac{\hbar^2}{2\mu}\nabla^2 - \frac{Ze^2}{4\pi\varepsilon_{\mathrm{e}}r}\right)\Psi = E\Psi \tag{2.7}$$

In spherical polar coordinates, ∇^2 is defined as

$$\nabla^2 = \frac{\partial^2}{\partial r^2} + \frac{2}{r}\frac{\partial}{\partial r} + \frac{1}{r^2}\Lambda^2 \tag{2.8}$$

where Λ^2 is defined as

$$\Lambda^2 = \frac{1}{\sin^2\theta}\frac{\partial^2}{\partial\phi^2} + \frac{1}{\sin\theta}\frac{\partial}{\partial\theta}\sin\theta\frac{\partial}{\partial\theta} \tag{2.9}$$

The Schrödinger equation is rewritten as

$$-\frac{\hbar^2}{2\mu}\left(\frac{\partial^2}{\partial r^2} + \frac{2}{r}\frac{\eth}{\partial r} + \frac{1}{r^2}\Lambda^2\right)RY - \frac{Ze^2}{4\pi\varepsilon_{\mathrm{e}}r}RY = ERY \tag{2.10}$$

Finally, it is rewritten as

$$-\frac{\hbar^2}{2\mu R}\left(\mathrm{r}^2\frac{\mathrm{d}^2R}{\mathrm{d}r^2} + 2r\frac{\mathrm{d}R}{\mathrm{d}r}\right) + \frac{Ze^2}{4\pi\varepsilon_{\mathrm{e}}}r^2 - \frac{\hbar^2}{2\mu\Theta\Phi}\Lambda^2Y = Er^2 \tag{2.11}$$

Equation (2.11) can be separated into two equations.

$$-\frac{\hbar^2}{2\mu\Theta\Phi}\Lambda^2Y = \text{constant} \tag{2.12}$$

$$-\frac{\hbar^2}{2\mu R}\left(\frac{\mathrm{d}^2R}{\mathrm{d}r^2} + \frac{2}{r^2}\frac{\mathrm{d}R}{\mathrm{d}r}\right) + \frac{Ze^2}{4\pi\varepsilon_{\mathrm{e}}}r^2 - Er^2 = -\text{constant} \tag{2.13}$$

The parameters of Eq. (2.12) are two angular components (ϕ and θ). On the other hand, r is the sole parameter in Eq. (2.13).

Table 2.1 Radial wave-functions of hydrogenic atom

n	l	$R(r)$
1	0	$R_{1s} = 2\left(\frac{Z}{a_0}\right)^{3/2} e^{-\rho/2}$
2	0	$R_{2s} = \frac{1}{2\sqrt{2}}\left(\frac{Z}{a_0}\right)^{3/2}(2-\rho)e^{-\rho/2}$
2	1	$R_{2p} = \frac{1}{2\sqrt{6}}\left(\frac{Z}{a_0}\right)^{3/2}\rho e^{-\rho/2}$
3	0	$R_{3s} = \frac{1}{9\sqrt{3}}\left(\frac{Z}{a_0}\right)^{3/2}(6-6\rho+\rho^2)e^{-\rho/2}$
3	1	$R_{3p} = \frac{1}{9\sqrt{6}}\left(\frac{Z}{a_0}\right)^{3/2}(4-\rho)\rho e^{-\rho/2}$
3	2	$R_{3d} = \frac{1}{9\sqrt{30}}\left(\frac{Z}{a_0}\right)^{3/2}\rho^2 e^{-\rho/2}$

2.1.2 *Radial Wave-Function*

We go through the detailed mathematical process to solve the radial equation. Table 2.1 shows the radial wave-functions of hydrogenic atom. The wave-functions are written in terms of dimensionless quantity (ρ).

$$\rho = \frac{Zr}{a_0} \tag{2.14}$$

where a_0 is Bohr radius.

$$a_0 = \frac{4\pi\varepsilon_0\hbar^2}{m_e e^2} \tag{2.15}$$

In radial wave-function, two quantum numbers are defined. One is the principal quantum number (n), corresponding to a shell. For example, electrons with $n = 2$ belong to the L shell. The other is orbital angular momentum quantum number (l), corresponding to a subshell. Two quantum numbers satisfies the following condition.

$$l = 0, 1, 2, 3, \ldots, (n-1) \tag{2.16}$$

Table 2.2 shows the relationship between quantum numbers, shell and subshell in hydrogenic atom. When $n = 1$, there is only one s-type subshell ($l = 0$). Quantum numbers of $n = 1$ and $l = 0$ stand for 1s atomic orbital. When $n = 2$, there are s-type ($l = 0$) and p-type ($l = 1$) subshells. Quantum numbers of $n = 2$ and $l = 0$ stand for 2s atomic orbital, and $n = 2$ and $l = 1$ stand for 2p atomic orbital. Figure 2.2 shows the variation of $R/(Zr/a_0)^{3/2}$ value, changing Zr/a_0 value. In 2s, 3s and 3p AOs, positive and negative $R/(Zr/a_0)^{3/2}$ values are given. The total energy (E) is given by

Table 2.2 Relationship between quantum numbers, shell and subshell in hydrogenic atom

n	Shell	l	Subshell	Atomic orbital
1	K	0	s	1s
2	L	0	s	2s
2	L	1	p	2p
3	M	0	s	3s
3	M	1	p	3p
3	M	2	d	3d

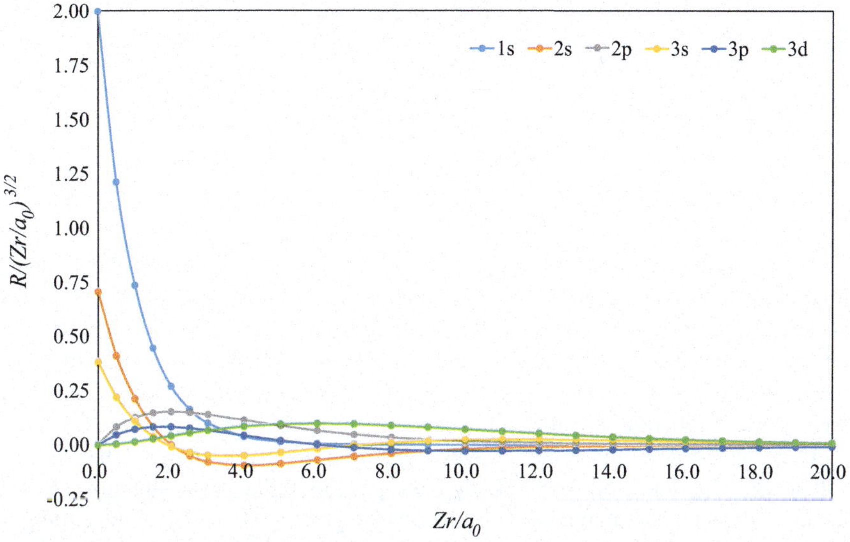

Fig. 2.2 Variation of $R/(Zr/a_0)^{3/2}$ value, changing Zr/a_0 value

$$E = -\frac{\hbar^2}{2m_e a_0^2}\frac{Z^2}{n^2} \tag{2.17}$$

E depends only on principal quantum number. It is why 2s and 2p atomic orbitals of hydrogenic atom are degenerated.

2.1.3 Angular Wave-Function

The sign of the total wave-function is determined by the signs of the radial and angular wave-functions. The angular wave-functions are written in terms of angular components (θ and ϕ). We go through the detailed mathematical process to solve the angular equation. Table 2.3 shows the angular wave-functions of hydrogenic atom.

Table 2.3 Angular wave-functions of hydrogenic atom

l	$Y(\theta, \phi)$
0	$\frac{1}{2\sqrt{\pi}}$
1	$\frac{1}{2}\sqrt{\frac{3}{\pi}} \cos \theta$
1	$\frac{1}{2}\sqrt{\frac{3}{\pi}} \sin \theta \cos \phi$
1	$\frac{1}{2}\sqrt{\frac{3}{\pi}} \sin a \sin \phi$
2	$\frac{1}{4}\sqrt{\frac{5}{\pi}}(3\cos^2\theta - 1)$
2	$\frac{1}{2}\sqrt{\frac{15}{\pi}} \sin \theta \cos \theta \cos \phi$
2	$\frac{1}{2}\sqrt{\frac{15}{\pi}} \sin \theta \cos \theta \sin \phi$
2	$\frac{1}{4}\sqrt{\frac{15}{\pi}} \sin^2\theta \cos 2\phi$
2	$\frac{1}{4}\sqrt{\frac{15}{\pi}} \sin^2\theta \sin 2\phi$

In 1s, 2s and 3s AOs, one angular wave-function has no angular parameter. It implies that the AOs spread uniformly to all directions. The sign of wave-function is determined by the radial wave-function. In 1s AO, the sign of wave-function becomes positive. On the other hand, in 2s and 3s AOs, the sign of radial wave-functions is positive or negative, depending on a radius. It implies that the sign of wave-function is changeable.

In 2p and 3p AOs, three angular wave-functions are given. In $n = 2$, the AOs are called $2p_x$, $2p_y$ and $2p_z$ AOs. Though the sign of radial wave-function (R_{2p}) is positive, the signs of angular wave-functions are positive or negative, depending on angular parameters. Hence, the sign of wave-function is changeable in 2p AOs.

In 3d AOs, five angular wave-functions are given. The AOs are called $3d_{xy}$, $3d_{yz}$, $3d_{xz}$, $3d_{x^2-y^2}$, $3d_{3z^2-r^2}$ AOs. Though the sign of radial wave-function (R_{3d}) is positive, the signs of angular wave-functions are positive or negative, depending on angular parameters. Hence, the sign of wave-function is changeable in 3d AOs.

The positive and negative signs in the wave-function represent the qualitative difference of wave-function. In electron–electron interaction, the difference has an important role.

2.1.4 Visualization of Hydrogenic Atomic Orbital

In hydrogenic atom, one electron spreads as one wave-function. The wave-function of the ground state consists of R_{1s} and $Y(= 1/2\sqrt{\pi})$. It is because minimum total energy is given when $n = 1$. However, the wave-function (Ψ) cannot be directly plotted into three-dimensional space. Instead, it is possible to visualize electron density, which is given by the square of wave-function ($|\Psi^2|$). Electron density in an finite volume (dτ) is given by

$$|\Psi^2|\mathrm{d}\tau \tag{2.18}$$

Electron density is normalized in three-dimensional space.

$$\int |\Psi^2|\mathrm{d}\tau = 1.00 \tag{2.19}$$

In general, the atomic orbital envelope diagrams are drawn based on the contours, within which the values of electron density is 0.95. Figure 2.3 depicts the atomic orbital envelope diagram of hydrogenic atom. Note that electron density is

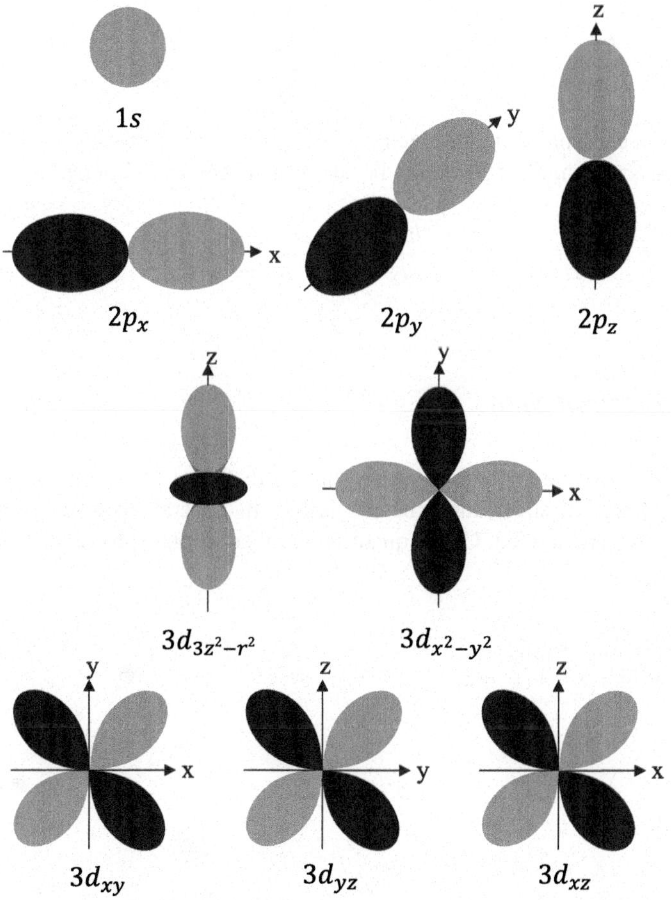

Fig. 2.3 Atomic orbital envelope diagrams of hydrogenic atom

dense around the centre, though the radial wave-function spreads in a large distance. The positive and negative signs of wave-functions are discriminated by colour difference. In this book, grey- and blue-coloured lobes represent the positive and negative signs of wave-function, respectively.

2.2 Many-Electron Atom

2.2.1 Schrödinger Equation

First, we consider helium atom as the simple example of two-electron atom (see Fig. 2.4). Two electrons are labelled as electron 1 and electron 2. Each electron has both electron–atomic nucleus interaction and electron–electron interaction. The Hamiltonian of the Schrödinger equation is expressed by

$$H = -\frac{\hbar^2}{2m_e}\nabla_1^2 - \frac{\hbar^2}{2m_e}\nabla_2^2 - \frac{\hbar^2}{2m_N}\nabla^2 - \frac{2e^2}{4\pi\varepsilon_e r_1} - \frac{2e^2}{4\pi\varepsilon_e r_2} + \frac{e^2}{4\pi\varepsilon_e r_{12}} \quad (2.20)$$

In many-electron atom (n-electron system), all electron–atomic nucleus and electron–electron interactions must be included in the Hamiltonian.

$$H = -\frac{\hbar^2}{2m_e}\sum_i^n \nabla_i^2 - \frac{\hbar^2}{2m_N}\nabla^2 - \frac{Ze^2}{4\pi\varepsilon_e}\sum_i^n \frac{1}{r_i} + \frac{e^2}{4\pi\varepsilon_e}\sum_{i<j}^n \frac{1}{r_{ij}} \quad (2.21)$$

2.2.2 Electron Spin

In many-electron atom, which means that more than two electrons exist in one atom, the effect of spin cannot be negligible. In general, two quantum numbers related to spin are defined. One is quantum number of total spin angular momentum

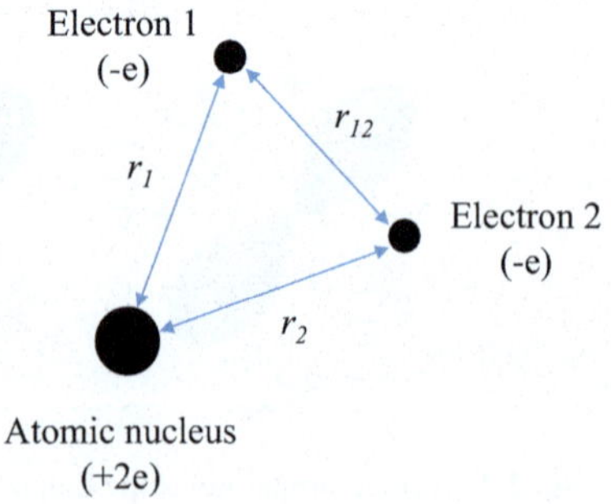

Fig. 2.4 Schematic drawing of helium atom

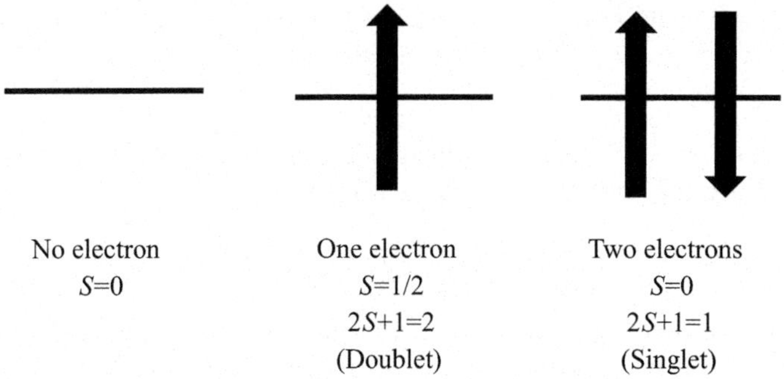

Fig. 2.5 Electron allocation in one atomic orbital

(S). The other is quantum number of spin angular momentum along the standard direction (m_s). When m_s =+1/2, electron has α spin. On the other hand, when $m_s = -1/2$, electron has β spin. Figure 2.5 depicts the schematic drawing of electron allocation in one atomic orbital. In one electron case, one electron occupies one atomic orbital. The spin multiplicity, which is defined as $(2S + 1)$, becomes two. It is called doublet spin state. When two electrons occupy one atomic orbital, two spins are paired, due to Pauli exclusion principle. Two electrons with paired spins have zero resultant spin angular momentum. Hence, the spin multiplicity becomes one. It is called singlet spin state.

2.2.3 *Spin Orbital*

To incorporate electron spin in wave-function, two kinds of spin functions such as $\alpha(\omega)$ and $\beta(\omega)$ are introduced. ω is a parameter of spin coordinates. Two spin functions are normalized.

$$\int \alpha^*(\omega)\alpha(\omega)\mathrm{d}\omega = \int \beta^*(\omega)\beta(\omega)\mathrm{d}\omega = 1 \tag{2.22}$$

Using the bra and ket symbols, they are rewritten:

$$\langle \alpha(\omega)|\alpha(\omega)\rangle = \langle \beta(\omega)|\beta(\omega)\rangle = 1 \tag{2.23}$$

Two spin functions satisfy an orthogonality.

$$\int \alpha^*(\omega)\beta(\omega)\mathrm{d}\omega = \int \beta^*(\omega)\alpha(\omega)\mathrm{d}\omega = 0 \tag{2.24}$$

Using the bra and ket symbols, they are rewritten:

$$\langle\alpha(\omega)|\beta(\omega)\rangle = \langle\beta(\omega)|\alpha(\omega)\rangle = 1 \tag{2.25}$$

The spin orbital (χ) is defined by the product between spatial orbital (ψ) and spin function. When one electron has α spin, it is expressed as

$$\chi(x) = \psi(r)\alpha(\omega) \tag{2.26}$$

where χ denotes both space and spin coordinates. On the other hand, one electron has β spin, and it is expressed as

$$\chi(x) = \psi(r)\beta(\omega) \tag{2.27}$$

Note that it is assumed that α and β spins are allowed in one spatial orbital. The different spatial orbitals are also orthonormal.

$$\int \psi_i^*(r)\psi_j(r)\mathrm{d}r = \delta_{ij} \tag{2.28}$$

where δ_{ij} is called Kronecker delta.

$$\delta_{ij} = \begin{cases} 1(i = j) \\ 0(i \neq j) \end{cases} \tag{2.29}$$

As the result, the different spin orbitals are orthonormal.

$$\int \chi_i^*(r)\chi_j(r)\mathrm{d}r = \delta_{ij} \tag{2.30}$$

2.2.4 *Total Wave-Function*

By using Hartree product, the total wave-function (Φ) of n-electron system is expressed as the product of all spin orbitals.

$$\Phi^{\mathrm{HP}}(x_1, x_2, \ldots, x_n) = \chi_1(x_1) \cdot \chi_2(x_2) \ldots \chi_n(x_n) \tag{2.31}$$

If there is no electron–electron interaction, Φ^{HP} is the eigenfunction of the Schrödinger equation. However, if there is an electron–electron interaction, Hartree product is different from the exact total wave-function.

Electron belongs to fermion, which is a quantum particle with half-integer quantum number for spin angular momentum. Fermion must satisfy "inverse principle" that the total wave-function changes the sign, when the labels of any two identical fermions are exchanged.

$$\Phi(x_1,\ldots,x_i,\ldots,x_j,\ldots x_n) = -\Phi(x_1,\ldots,x_j,\ldots,x_i,\ldots x_n) \tag{2.32}$$

In order to satisfy inverse principle, by using Slater determination, the total wave-function is expressed as

$$\Phi(\chi_1,\chi_2,\cdots\chi_n) = (n!)^{-\frac{1}{2}} \begin{bmatrix} \chi_1(x_1) & \chi_2(x_1) & \cdots & \chi_n(x_1) \\ \chi_1(x_2) & \chi_2(x_2) & & \chi_n(x_2) \\ \vdots & & \ddots & \vdots \\ \chi_1(x_n) & \chi_2(x_n) & \cdots & \chi_n(x_n) \end{bmatrix} \tag{2.33}$$

The convenient representation of Slater determination is expressed as

$$|\chi_1(x_1)\chi_2(x_2)\ldots\chi_n(x_n)\rangle \tag{2.34}$$

Let us consider the simple example of many-electron atom. Helium atom has two electrons with paired spins. They are allocated into the same atomic orbital. The spin orbitals are given by

$$\chi_1(x_1) = \psi_1(r_1)\alpha(\omega_1) \tag{2.35}$$

$$\chi_2(x_2) = \psi_1(r_2)\beta(\omega_2) \tag{2.36}$$

Slater determination is rewritten as

$$\Phi(\chi_1,\chi_2) = |\chi_1(x_1)\chi_2(x_2)\rangle = (2!)^{-\frac{1}{2}} \begin{bmatrix} \chi_1(x_1) & \chi_2(x_1) \\ \chi_1(x_2) & \chi_2(x_2) \end{bmatrix} \tag{2.37}$$

$$= (2!)^{-\frac{1}{2}} \begin{bmatrix} \psi_1(r_1)\alpha(\omega_1) & \psi_1(r_1)\beta(\omega_1) \\ \psi_1(r_2)\alpha(\omega_2) & \psi_1(r_2)\beta(\omega_2) \end{bmatrix} \tag{2.38}$$

$$= \frac{\psi_1(r_1)\psi_1(r_2)}{\sqrt{2}} \{\alpha(\omega_1)\cdot\beta(\omega_2) - \beta(\omega_1)\cdot\alpha(\omega_2)\} \tag{2.39}$$

However, the Schrödinger equation for many-electron atom cannot be analytically solved. Many calculation methods have been developed to solve it approximately with high precision.

2.2.5 Building-Up Rule

Building-up principle is the plausible and empirical rule to predict the ground-state electron configuration of a single atom. It is a starting point before actual calculation. It is based on two types of allocations for one AO: (1) one electron with α spin and (2) two electrons with paired spins. The order of occupation is as follows.

$$(1)1s, (2)2s, (3)2p, (4)3s, (5)3p, (6)3d\ldots \tag{2.40}$$

In hydrogen atom, which is one-electron system, one electron occupies 1s orbital with α spin.

$$\mathrm{H} : 1s^1 \tag{2.41}$$

In helium atom, which is two-electron system, two electrons occupy 1s orbital with paired spins.

$$\mathrm{He} : 1s^2 \tag{2.42}$$

In lithium atom, which is three-electron system, two electrons occupy 1s orbital with paired spins, and one electron occupies 2s orbital with α spin.

$$\mathrm{Li} : 1s^2 2s^1 \tag{2.43}$$

Table 2.4 summarizes the number of electrons in atomic orbitals, based on building-up principle. Neutral carbon, oxygen and nitrogen have six, seven and eight electrons. Their electron configuration is expressed as follows.

$$\mathrm{C} : 1s^2 2s^2 2p^2 \tag{2.44}$$

$$\mathrm{O} : 1s^2 2s^2 2p^3 \tag{2.45}$$

$$\mathrm{N} : 1s^2 2s^2 2p^4 \tag{2.46}$$

Table 2.4 Number of electrons in atomic orbitals, based on building-up principle

n		l	Atomic orbital	Number of electrons	
1	K	0	1s	2	(1α, 1β)
2	L	0	2s	2	(1α, 1β)
		1	2p	6	(3α, 3β)
3	M	0	3s	2	(1α, 1β)
		1	3p	6	(3α, 3β)
		2	3d	10	(5α, 5β)

Further Readings

1. Atkins P, de Paula J (2006) Physical chemistry 8th edn, Chapters 9 and 10 (in Japanese)
2. Atkins P, de Paula J, Friedman R (2009) Quanta, matter, and change a molecular approach to physical chemistry, Chapter 4 (in Japanese)
3. Barrow GM (1999) Physical chemistry 6th edn, Chapter 10 (in Japanese)

Chapter 3
Hartree-Fock Method

Abstract In many-electron system, it is impossible to obtain the exact solution of the Schrodinger equation by using the present mathematical approach. Hartree-Fock method was developed to solve approximately the time-independent Schrödinger equation. In Born–Oppenheimer approximation, atomic nucleus is regarded as stationary point, in comparison with electron. The total energy of many-electron system can be represented by using one-electron and two-electron operators. Schrödinger equation can be mathematically transformed to one-electron Hartree-Fock equation by minimizing the total energy. The eigenvalue and wave-function denote orbital energy and molecular orbital (atomic orbital). In closed shell system, there is one restriction that α-spin and β-spin electrons are paired in the same spatial orbital. Hartree-Fock in the closed shell system is called restricted Hartree-Fock (RHF). On the other hand, in open shell system, the spatial orbital of α electron is independent from β electron. Hartree-Fock in open shell system is called unrestricted Hartree-Fock (UHF). By using orbital energy rule, the stability of molecular orbital (atomic orbital) can be discussed from orbital energy.

Keywords Born–Oppenheimer approximation · Hartree-Fock method · Closed shell · Open shell · Orbital energy rule

3.1 Born–Oppenheimer Approximation

In Chap. 2, it was explained that electron spreads as wave-function within atom. In comparison with electron, atomic nucleus may be regarded as stationary point. In Born–Oppenheimer approximation, kinetic energy of atomic nucleus is neglected in the Hamiltonian. The Hamiltonian for n-electron atom is given by

$$H^{\text{Atom}} = -\frac{\hbar^2}{2m_\text{e}}\sum_{i=1}^{n}\nabla_i^2 - \frac{Ze^2}{4\pi\varepsilon_\text{e}}\sum_{i=1}^{n}\frac{1}{r_i} + \frac{e^2}{4\pi\varepsilon_\text{e}}\sum_{i=1}^{n}\sum_{j=1,j\neq i}^{n}\frac{1}{r_{ij}} \tag{3.1}$$

T. Onishi, *Quantum Computational Chemistry*,
DOI 10.1007/978-981-10-5933-9_3

where Z, e, r_i and r_{ij} denote atomic number, an electronic charge, the atomic nucleus-electron distance and the electron–electron distance, respectively. The first, second and third terms denote the kinetic energy of electrons, the Coulomb interaction energy between nucleus and electrons and the Coulomb repulsion energy between electrons, respectively. Atomic orbital (AO) is given as the solution (wave-function) of the Schrödinger equation.

In n-electron molecule consisting of m-atom, the Hamiltonian is given by

$$H^{\text{Molecule}} = -\frac{\hbar^2}{2m_e}\sum_{i=1}^{n}\nabla_i^2 - \frac{e^2}{4\pi\varepsilon_e}\sum_{i=1}^{n}\sum_{j=1}^{m}\frac{Z_j}{r_i} + \frac{e^2}{4\pi\varepsilon_e}\sum_{i=1}^{n}\sum_{j=1, j\neq i}^{n}\frac{1}{r_{ij}} \tag{3.2}$$

where Z_j denotes jth atomic number. Note that molecular orbital (MO) is given as the solution (wave-function) of the Schrödinger equation.

3.2 Total Energy of n-Electron Atom

Multiplied of Φ^* on the left of Schrödinger equation, then integrated both sides,

$$\int \Phi^* H \Phi \mathrm{d}x_1 \mathrm{d}x_2 \cdots \mathrm{d}x_n = \mathrm{E}\int \Phi^* \Phi \mathrm{d}x_1 \mathrm{d}x_2 \cdots \mathrm{d}x_n \tag{3.3}$$

By the using bra and ket symbols, it is rewritten:

$$\langle \Phi | H | \Phi \rangle = E \langle \Phi | \Phi \rangle \tag{3.4}$$

By the normalization of wave-function, Eq. (3.4) is rewritten as

$$E = \langle \Phi | H | \Phi \rangle \tag{3.5}$$

By using Slater determination, the wave-function of n-electron atom is expressed as

$$\Phi^{\text{Atom}} = |\chi_1(x_1)\chi_2(x_2)\cdots\chi_n(x_n)\rangle \tag{3.6}$$

where χ_i denotes the ith spin orbital. Under Born–Oppenheimer approximation, the Hamiltonian for the ith component is given by

$$H_i^{\text{Atom}} = -\frac{\hbar^2}{2m_e}\nabla_i^2 - \frac{Ze^2}{4\pi\varepsilon_e r_i} + \frac{e^2}{4\pi\varepsilon_e}\sum_{j>i}^{n}\frac{1}{r_{ij}} \tag{3.7}$$

The one-electron operator (h_i) is defined as

$$h_i^{\mathrm{Atom}} = -\frac{\hbar^2}{2m_{\mathrm{e}}}\nabla_i^2 - \frac{Ze^2}{4\pi\varepsilon_{\mathrm{e}} r_i} \tag{3.8}$$

The notation of one-electron integral is

$$\int \chi_i^*(x_i) h_i^{\mathrm{Atom}} \chi_i(x_i) \mathrm{d}x_i = \langle \chi_i(x_i) | h_i^{\mathrm{Atom}} | \chi_i(x_i) \rangle \tag{3.9}$$

By using Eq. (3.9), the total energy related to one-electron operator is expressed as

$$\begin{aligned} &\frac{1}{n}\langle \chi_1(x_1)|h_1^{\mathrm{Atom}}|\chi_1(x_1)\rangle + \frac{1}{n}\langle \chi_2(x_1)|h_2^{\mathrm{Atom}}|\chi_2(x_1)\rangle + \cdots \\ &\quad + \frac{1}{n}\langle \chi_n(x_1)|h_n^{\mathrm{Atom}}|\chi_n(x_1)\rangle + \frac{1}{n}\langle \chi_1(x_2)|h_1^{\mathrm{Atom}}|\chi_1(x_2)\rangle \\ &\quad + \frac{1}{n}\langle \chi_2(x_2)|h_2^{\mathrm{Atom}}|\chi_2(x_2)\rangle + \cdots + \frac{1}{n}\langle \chi_n(x_2)|h_n^{\mathrm{Atom}}|\chi_n(x_2)\rangle + \cdots \\ &\quad + \frac{1}{n}\langle \chi_1(x_n)|h_1^{\mathrm{Atom}}|\chi_1(x_n)\rangle + \frac{1}{n}\langle \chi_2(x_n)|h_2^{\mathrm{Atom}}|\chi_2(x_n)\rangle + \cdots \\ &\quad + \frac{1}{n}\langle \chi_n(x_n)|h_n^{\mathrm{Atom}}|\chi_n(x_n)\rangle \end{aligned} \tag{3.10}$$

By using a sigma symbol, Eq. (3.10) is rewritten as

$$\sum_{i=1}^{n} \langle \chi_i(x_i) | h_i^{\mathrm{Atom}} | \chi_i(x_i) \rangle \tag{3.11}$$

Two-electron operator (r_{ij}^{-1}) is defined as

$$r_{ij}^{-1} = \frac{e^2}{4\pi\varepsilon_{\mathrm{e}} r_{ij}} \tag{3.12}$$

The notation of two-electron integrals is

$$\int \chi_i^*(x_i)\chi_j^*(x_j) r_{ij}^{-1} \chi_i(x_i)\chi_j(x_j) \mathrm{d}x_i \mathrm{d}x_j = \langle \chi_i(x_i)\chi_j(x_j) | \chi_i(x_i)\chi_j(x_j) \rangle \tag{3.13}$$

The total energy related to two-electron operator can be separated into two parts. By using Eq. (3.13), the first part is expressed as

$$\begin{aligned}&\langle\chi_1(x_1)\chi_2(x_2)|\chi_1(x_1)\chi_2(x_2)\rangle+\langle\chi_1(x_1)\chi_3(x_2)|\chi_1(x_1)\chi_3(x_2)\rangle\\&\quad+\cdots+\langle\chi_1(x_1)\chi_n(x_2)|\chi_1(x_1)\chi_n(x_2)\rangle+\langle\chi_2(x_1)\chi_3(x_2)|\chi_2(x_1)\chi_3(x_2)\rangle\\&\quad+\cdots+\langle\chi_2(x_1)\chi_n(x_2)|\chi_2(x_1)\chi_n(x_2)\rangle+\cdots\\&\quad+\langle\chi_{n-1}(x_1)\chi_n(x_2)|\chi_{n-1}(x_1)\chi_n(x_2)\rangle\end{aligned}\tag{3.14}$$

By using a sigma symbol, Eq. (3.14) is rewritten as

$$\sum_{i=1}^{n}\sum_{j>i}^{n}\langle\chi_i(x_i)\chi_j(x_j)|\chi_i(x_i)\chi_j(x_j)\rangle\equiv\sum_{i=1}^{n}\sum_{j>i}^{n}J_{ij}\tag{3.15}$$

where J_{ij} is called Coulomb integral. Note that r_{ij} is not defined when i is equal to j. By using Eq. (3.13), the second part is expressed as

$$\begin{aligned}&\langle\chi_1(x_1)\chi_2(x_2)|\chi_2(x_1)\chi_1(x_2)\rangle+\langle\chi_1(x_1)\chi_3(x_2)|\chi_3(x_1)\chi_1(x_2)\rangle\\&\quad+\cdots+\langle\chi_1(x_1)\chi_n(x_2)|\chi_n(x_1)\chi_1(x_2)\rangle+\langle\chi_2(x_1)\chi_3(x_2)|\chi_3(x_1)\chi_2(x_2)\rangle\\&\quad+\cdots+\langle\chi_2(x_1)\chi_n(x_2)|\chi_n(x_1)\chi_2(x_2)\rangle+\cdots+\langle\chi_{n-1}(x_1)\chi_n(x_2)|\chi_n(x_1)\chi_{n-1}(x_2)\rangle\end{aligned}\tag{3.16}$$

By using a sigma symbol, Eq. (3.16) is rewritten as

$$\sum_{i=1}^{n}\sum_{j>i}^{n}\langle\chi_i(x_i)\chi_j(x_j)|\chi_j(x_i)\chi_i(x_j)\rangle\equiv\sum_{i=1}^{n}\sum_{j>i}^{n}K_{ij}\tag{3.17}$$

where K_{ij} is called exchange integral. It is because two electrons i and j are exchanged between two spin orbitals in the right ket symbol. Finally, the total energy of n-electron atom is rewritten as

$$\begin{aligned}E^{\mathrm{Atom}}&=\sum_{i=1}^{n}\langle\chi_i(x_i)|h_i^{\mathrm{Atom}}|\chi_i(x_i)\rangle+\sum_{i=1}^{n}\sum_{j>i}^{n}\{\langle\chi_i(x_i)\chi_j(x_j)|\chi_i(x_i)\chi_j(x_j)\rangle\\&\quad-\langle\chi_i(x_i)\chi_j(x_j)|\chi_j(x_i)\chi_i(x_j)\rangle\}\\&=\sum_{i=1}^{n}\langle\chi_i(x_i)|h_i^{\mathrm{Atom}}|\chi_i(x_i)\rangle+\sum_{i=1}^{n}\sum_{j>i}^{n}(J_{ij}-K_{ij})\end{aligned}\tag{3.18}$$

3.3 Total Energy of n-Electron Molecule

Here, n-electron molecule consisting of m-atom is considered. The wave-function of n-electron molecule is

$$\Phi^{\text{Molecule}} = |\chi_1(x_1)\chi_2(x_2)\cdots\chi_n(x_n)\rangle \tag{3.19}$$

where χ_i denotes the ith spin orbital. Under Born–Oppenheimer approximation, the Hamiltonian for the ith component is given by

$$H_i^{\text{Molecule}} = -\frac{\hbar^2}{2m_e}\nabla_i^2 - \frac{e^2}{4\pi\varepsilon_e r_i}\sum_{j=1}^{m} Z_j + \frac{e^2}{4\pi\varepsilon_e}\sum_{j>i}^{n}\frac{1}{r_{ij}} \tag{3.20}$$

The one-electron operator (h_i) is defined as

$$h_i^{\text{Molecule}} = -\frac{\hbar^2}{2m_e}\nabla_i^2 - \frac{e^2}{4\pi\varepsilon_e r_i}\sum_{j=1}^{m} Z_j \tag{3.21}$$

The total energy for the molecule is obtained in the same manner:

$$E^{\text{Molecule}} = \sum_{i=1}^{n}\langle\chi_i(x_i)|h_i^{\text{Molecule}}|\chi_i(x_i)\rangle + \sum_{i=1}^{n}\sum_{j>i}^{n}(J_{ij} - K_{ij}) \tag{3.22}$$

Note that Eqs. (3.20)–(3.22) are for n-electron atom, when m = 1. Hence, they are also used for n-electron atom.

3.4 Hartree-Fock Equation

Hartree-Fock method is regarded as starting point in *ab initio* calculation. Though the accurate electron–electron interactions are not reproduced due to average approximation, it provides the qualitatively correct results. The present precise calculation methods have been theoretically constructed based on the revision of Hartree-Fock method.

The n-electron Schrödinger equation is mathematically transformed to one-electron Hartree-Fock equation by minimizing the total energy of Schrödinger equation.

$$f_i\chi_i(x_i) = \varepsilon_i\chi_i(x_i) \tag{3.23}$$

where f_i denotes Fock operator; ε_i is an eigenvalue, which denotes orbital energy. In atom and molecule, atomic orbital (AO) and molecular orbital (MO) are given as a solution, respectively. Fock operator, which is one-electron operator for a spin orbital, is defined as

$$f_i = h_i + \sum_{j\neq i}^{n}\{J_j - K_j\} \tag{3.24}$$

where h_i denotes kinetic energy and Coulomb potential energy between atomic nucleus and electrons for the ith electron; J_j and K_j are Coulomb operator and exchange operator between the ith and jth electrons, respectively. Note that h_i, J_j and K_j are given by

$$h_i = -\frac{\hbar^2}{2m_e}\nabla_i^2 - \frac{e^2}{4\pi\varepsilon_e r_i}\sum_{j=1}^{m} Z_j \tag{3.25}$$

$$J_j\chi_i(x_i) = \int \chi_j^*(x_j) r_{ij}^{-1}\chi_j(x_j)d\chi_j \cdot \chi_i(x_i) \tag{3.26}$$

$$K_j\chi_i(x_i) = \int \chi_j^*(x_j) r_{ij}^{-1}\chi_i(x_j)d\chi_j \cdot \chi_j(x_i) \tag{3.27}$$

The ith orbital energy (ε_i) satisfies the following equation.

$$\varepsilon_i = \langle \chi_i(x_i)|f_i|\chi_i(x_i)\rangle \tag{3.28}$$

By substituting Eqs. (3.24), (3.26) and (3.27), Eq. (3.28) is rewritten as

$$\begin{aligned} \varepsilon_i &= \langle \chi_i(x_i)|h_i|\chi_i(x_i)\rangle + \sum_{j\neq i}^{n}\left\{\langle \chi_i(x_i)|J_j|\chi_i(x_i)\rangle - \langle \chi_i(x_i)|K_j|\chi_i(x_i)\rangle\right\} \\ &= \langle \chi_i(x_i)|h_i|\chi_i(x_i)\rangle + \sum_{j=1}^{n}\left\{\langle \chi_i(x_i)\chi_j(x_j)|\chi_i(x_i)\chi_j(x_j)\rangle\right. \\ &\quad \left.-\langle \chi_i(x_i)\chi_j(x_j)|\chi_j(x_i)\chi_i(x_j)\rangle\right\} \end{aligned} \tag{3.29}$$

Note that the second and third terms are cancelled out, when i is equal to j.

3.5 Closed Shell System

The total wave-function of closed $2n$-electron system is expressed as

$$\Phi = |\chi_1(x_1)\chi_2(x_2)\cdots\chi_{2n-1}(x_{2n-1})\chi_{2n}(x_{2n})\rangle \tag{3.30}$$

where χ_i is the ith spin orbital. As α and β electrons are paired at the same spatial orbital in closed shell system (see Fig. 3.1), the total wave-function is rewritten as

$$\Phi = |\psi_1(r_1)\alpha(\omega_1)\psi_1(r_2)\beta(\omega_2)\cdots\psi_n(r_{2n})\alpha(\omega_{2n})\psi_n(r_{2n})\beta(\omega_{2n})\rangle \tag{3.31}$$

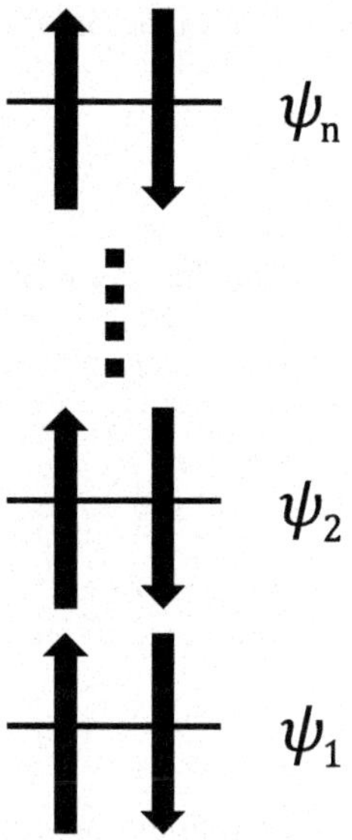

Fig. 3.1 Electron configuration of closed shell system. In the same spatial orbital, α and β spins are paired

Let us consider the total energy of $2n$-electron system. The first term of Eq. (3.22) is rewritten as

$$\sum_{i=1}^{2n}\langle\chi_i(x_i)|h_i|\chi_i(x_i)\rangle = \sum_{i=1}^{n}\langle\psi_i(r_i)\alpha(\omega_i)|h_i|\psi_i(r_i)\alpha(\omega_i)\rangle + \sum_{i=1}^{n}\langle\psi_i(r_i)\beta(\omega_i)|h_i|\psi_i(r_i)\beta(\omega_i)\rangle \tag{3.32}$$

In addition, due to the orthonormality of spin functions, it is rewritten:

$$\sum_{i=1}^{2n}\langle\chi_i(x_i)|h_i|\chi_i(x_i)\rangle = 2\sum_{i=1}^{n}\langle\psi_i(r_i)|h_i|\psi_i(r_i)\rangle \tag{3.33}$$

The second term of Eq. (3.22) is rewritten as

$$\begin{aligned}&\frac{1}{2}\sum_{i=1}^{n}\sum_{j=1}^{n}\left\langle\psi_i(r_i)\alpha(\omega_i)\psi_j(r_j)\alpha(\omega_j)\right|\left.\psi_i(r_i)\alpha(\omega_i)\psi_j(r_j)\alpha(\omega_j)\right\rangle\\&+\frac{1}{2}\sum_{i=1}^{n}\sum_{j=1}^{n}\left\langle\psi_i(r_i)\alpha(\omega_i)\psi_j(r_j)\beta(\omega_j)\right|\left.\psi_i(r_i)\alpha(\omega_i)\psi_j(r_j)\beta(\omega_j)\right\rangle\\&+\frac{1}{2}\sum_{i=1}^{n}\sum_{j=1}^{n}\left\langle\psi_i(r_i)\beta(\omega_i)\psi_j(r_j)\alpha(\omega_j)\right|\left.\psi_i(r_i)\beta(\omega_i)\psi_j(r_j)\alpha(\omega_j)\right\rangle\\&+\frac{1}{2}\sum_{i=1}^{n}\sum_{j=1}^{n}\left\langle\psi_i(r_i)\beta(\omega_i)\psi_j(r_j)\beta(\omega_j)\right|\left.\psi_i(r_i)\beta(\omega_i)\psi_j(r_j)\beta(\omega_j)\right\rangle\end{aligned} \tag{3.34}$$

Due to the orthonormality of spin functions, it is rewritten as

$$\sum_{i=1}^{n}\sum_{j=1}^{n}2\langle\psi_i(r_i)\psi_j(r_j)|\psi_i(r_i)\psi_j(r_j)\rangle=\sum_{i=1}^{n}\sum_{j=1}^{n}2J_{ij} \tag{3.35}$$

The third term of Eq. (3.22) is rewritten as

$$\begin{aligned}
&-\tfrac{1}{2}\sum_{i=1}^{n}\sum_{j=1}^{n}\langle\psi_i(r_i)\alpha(\omega_i)\psi_j(r_j)\alpha(\omega_j)|\psi_j(r_i)\alpha(\omega_i)\psi_i(r_j)\alpha(\omega_j)\rangle\\
&\quad-\tfrac{1}{2}\sum_{i=1}^{n}\sum_{j=1}^{n}\langle\psi_i(r_i)\alpha(\omega_i)\psi_j(r_j)\beta(\omega_j)|\psi_j(r_i)\beta(\omega_i)\psi_i(r_j)\alpha(\omega_j)\rangle\\
&\quad-\tfrac{1}{2}\sum_{i=1}^{n}\sum_{j=1}^{n}\langle\psi_i(r_i)\beta(\omega_i)\psi_j(r_j)\alpha(\omega_j)|\psi_j(r_i)\alpha(\omega_i)\psi_i(r_j)\beta(\omega_j)\rangle\\
&\quad-\tfrac{1}{2}\sum_{i=1}^{n}\sum_{j=1}^{n}\langle\psi_i(r_i)\beta(\omega_i)\psi_j(r_j)\beta(\omega_j)|\psi_j(r_i)\beta(\omega_i)\psi_i(r_j)\beta(\omega_j)\rangle
\end{aligned} \tag{3.36}$$

Due to the orthogonality of spatial orbitals, it is rewritten as

$$-\sum_{i=1}^{n}\sum_{j=1}^{n}\langle\psi_i(r_i)\psi_j(r_j)|\psi_j(r_i)\psi_i(r_j)\rangle=-\sum_{i=1}^{n}\sum_{j=1}^{n}K_{ij} \tag{3.37}$$

Finally, the total energy of the $2n$-electron closed shell system is rewritten:

$$E=2\sum_{i=1}^{n}\langle\psi_i(r_i)|h_i|\psi_i(r_i)\rangle+\sum_{i=1}^{n}\sum_{j=1}^{n}(2J_{ij}-K_{ij}) \tag{3.38}$$

The ith orbital energy is rewritten in the same manner:

$$\varepsilon_i=\langle\psi_i(r_i)|h_i|\psi_i(r_i)\rangle+\sum_{j=1}^{n}(2J_{ij}-K_{ij}) \tag{3.39}$$

In closed shell system, there is one restriction that α-spin and β-spin electrons are paired in the same spatial orbital. Hartree-Fock in the closed shell system is called restricted Hartree-Fock (RHF).

3.6 Open Shell System

The total wave-function of n-electron open shell system is expressed as

$$\Phi=|\chi_1(x_1)\chi_2(x_2)\cdots\chi_n(x_n)\rangle \tag{3.40}$$

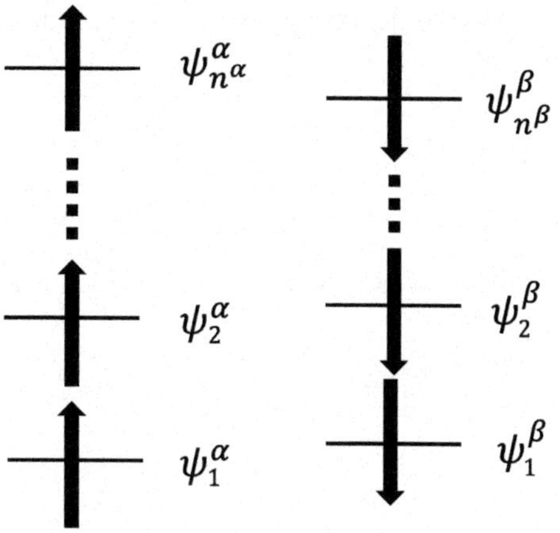

Fig. 3.2 Electron configuration of open shell system

where χ_i is the ith spin orbital. As shown in Fig. 3.2, each electron has the specific spatial orbital. Note that spatial orbital is theoretically discriminated by electron spin. The numbers of α and β electrons are denoted as n^{α} and n^{β}, respectively.

$$n = n^{\alpha} + n^{\beta} \tag{3.41}$$

By using spatial orbital and spin function, spin orbitals of α spin are expressed as

$$\psi_1^{\alpha}(r_1^{\alpha})\alpha(\omega_1^{\alpha}), \psi_2^{\alpha}(r_2^{\alpha})\alpha(\omega_2^{\alpha}), \cdots, \psi_{n^{\alpha}}^{\alpha}(r_{n^{\alpha}}^{\alpha})\alpha(\omega_{n^{\alpha}}^{\alpha}) \tag{3.42}$$

In the same manner, spin orbitals of β spin are expressed as

$$\psi_1^{\beta}\left(r_1^{\beta}\right)\beta\left(\omega_1^{\beta}\right), \psi_2^{\beta}\left(r_2^{\beta}\right)\beta\left(\omega_2^{\beta}\right), \cdots, \psi_{n^{\beta}}^{\beta}\left(r_{n^{\beta}}^{\beta}\right)\beta\left(\omega_{n^{\beta}}^{\beta}\right) \tag{3.43}$$

Note that position and spin coordinates are separately defined in α and β spins. When n^{α} is larger than n^{β}, Eq. (3.40) is rewritten as

$$\Phi = \left|\psi_1^{\alpha}(r_1^{\alpha})\alpha(\omega_1^{\alpha})\psi_1^{\beta}\left(r_1^{\beta}\right)\beta\left(\omega_1^{\beta}\right)\cdots\psi_{n^{\alpha}}^{\alpha}(r_{n^{\alpha}}^{\alpha})\right\rangle \tag{3.44}$$

The first term of Eq. (3.22) is rewritten as

$$\begin{aligned}\sum_{i=1}^{n}\langle\chi_i(x_i)|h_i|\chi_i(x_i)\rangle &= \sum_{i=1}^{n^{\alpha}}\langle\psi_i^{\alpha}(r_i^{\alpha})\alpha(\omega_i^{\alpha})|h_i|\psi_i^{\alpha}(r_i^{\alpha})\alpha(\omega_i^{\alpha})\rangle \\ &+ \sum_{i=1}^{n^{\beta}}\left\langle\psi_i^{\beta}\left(r_i^{\beta}\right)\beta\left(\omega_i^{\beta}\right)\middle|h_i\middle|\psi_i^{\beta}\left(r_i^{\beta}\right)\beta\left(\omega_i^{\beta}\right)\right\rangle\end{aligned} \tag{3.45}$$

Due to the orthogonality of spin functions, it is rewritten as

$$\sum_{i=1}^{n}\langle\chi_i(x_i)|h_i|\chi_i(x_i)\rangle = \sum_{i=1}^{n^\alpha}\langle\psi_i^\alpha(r_i^\alpha)|h_i|\psi_i^\alpha(r_i^\alpha)\rangle + \sum_{i=1}^{n^\beta}\left\langle\psi_i^\beta\left(r_i^\beta\right)\middle|h_i\middle|\psi_i^\beta\left(r_i^\beta\right)\right\rangle \tag{3.46}$$

The second term of Eq. (3.22) is rewritten as

$$\begin{aligned}
&\tfrac{1}{2}\sum_{i=1}^{n^\alpha}\sum_{j=1}^{n^\alpha}\left\langle\psi_i^\alpha(r_i^\alpha)\alpha(\omega_i^\alpha)\psi_j^\alpha\left(r_j^\alpha\right)\alpha\left(\omega_j^\alpha\right)|\psi_i^\alpha(r_i^\alpha)\alpha(\omega_i^\alpha)\psi_j^\alpha\left(r_j^\alpha\right)\alpha\left(\omega_j^\alpha\right)\right\rangle\\
&\quad+\tfrac{1}{2}\sum_{i=1}^{n^\alpha}\sum_{j=1}^{n^\beta}\left\langle\psi_i^\alpha(r_i^\alpha)\alpha(\omega_i^\alpha)\psi_j^\beta\left(r_j^\beta\right)\beta\left(\omega_j^\beta\right)|\psi_i^\alpha(r_i^\alpha)\alpha(\omega_i^\alpha)\psi_j^\beta\left(r_j^\beta\right)\beta\left(\omega_j^\beta\right)\right\rangle\\
&\quad+\tfrac{1}{2}\sum_{i=1}^{n^\beta}\sum_{j=1}^{n^\alpha}\left\langle\psi_i^\beta\left(r_i^\beta\right)\beta\left(\omega_i^\beta\right)\psi_j^\alpha\left(r_j^\alpha\right)\alpha\left(\omega_j^\alpha\right)|\psi_i^\beta\left(r_i^\beta\right)\beta\left(\omega_i^\beta\right)\psi_j^\alpha\left(r_j^\alpha\right)\alpha\left(\omega_j^\alpha\right)\right\rangle\\
&\quad+\tfrac{1}{2}\sum_{i=1}^{n^\beta}\sum_{j=1}^{n^\beta}\left\langle\psi_i^\beta\left(r_i^\beta\right)\beta\left(\omega_i^\beta\right)\psi_j^\beta\left(r_j^\beta\right)\beta\left(\omega_j^\beta\right)|\psi_i^\beta\left(r_i^\beta\right)\beta\left(\omega_i^\beta\right)\psi_j^\beta\left(r_j^\beta\right)\beta\left(\omega_j^\beta\right)\right\rangle
\end{aligned} \tag{3.47}$$

Due to the orthogonality of spin functions, it is rewritten as

$$\begin{aligned}
&\tfrac{1}{2}\sum_{i=1}^{n^\alpha}\sum_{j=1}^{n^\alpha}\left\langle\psi_i^\alpha(r_i^\alpha)\psi_j^\alpha\left(r_j^\alpha\right)|\psi_i^\alpha(r_i^\alpha)\psi_j^\alpha\left(r_j^\alpha\right)\right\rangle\\
&\quad+\tfrac{1}{2}\sum_{i=1}^{n^\alpha}\sum_{j=1}^{n^\beta}\left\langle\psi_i^\alpha(r_i^\alpha)\psi_j^\beta\left(r_j^\beta\right)|\psi_i^\alpha(r_i^\alpha)\psi_j^\beta\left(r_j^\beta\right)\right\rangle\\
&\quad+\tfrac{1}{2}\sum_{i=1}^{n^\beta}\sum_{j=1}^{n^\alpha}\left\langle\psi_i^\beta\left(r_i^\beta\right)\psi_j^\alpha\left(r_j^\alpha\right)|\psi_i^\beta\left(r_i^\beta\right)\psi_j^\alpha\left(r_j^\alpha\right)\right\rangle\\
&\quad+\tfrac{1}{2}\sum_{i=1}^{n^\beta}\sum_{j=1}^{n^\beta}\left\langle\psi_i^\beta\left(r_i^\beta\right)\psi_j^\beta\left(r_j^\beta\right)|\psi_i^\beta\left(r_i^\beta\right)\psi_j^\beta\left(r_j^\beta\right)\right\rangle
\end{aligned} \tag{3.48}$$

The third term of Eq. (3.22) is rewritten as

$$\begin{aligned}
&-\tfrac{1}{2}\sum_{i=1}^{n^\alpha}\sum_{j=1}^{n^\alpha}\left\langle\psi_i^\alpha(r_i^\alpha)\alpha(\omega_i^\alpha)\psi_j^\alpha\left(r_j^\alpha\right)\alpha\left(\omega_j^\alpha\right)|\psi_j^\alpha(r_i^\alpha)\alpha(\omega_i^\alpha)\psi_i^\alpha\left(r_j^\alpha\right)\alpha\left(\omega_j^\alpha\right)\right\rangle\\
&\quad-\tfrac{1}{2}\sum_{i=1}^{n^\alpha}\sum_{j=1}^{n^\beta}\left\langle\psi_i^\alpha(r_i^\alpha)\alpha(\omega_i^\alpha)\psi_j^\beta\left(r_j^\beta\right)\beta\left(\omega_j^\beta\right)|\psi_j^\alpha(r_i^\alpha)\beta(\omega_i^\alpha)\psi_i^\beta\left(r_j^\beta\right)\alpha\left(\omega_j^\beta\right)\right\rangle\\
&\quad-\tfrac{1}{2}\sum_{i=1}^{n^\beta}\sum_{j=1}^{n^\alpha}\left\langle\psi_i^\beta\left(r_i^\beta\right)\beta\left(\omega_i^\beta\right)\psi_j^\alpha\left(r_j^\alpha\right)\alpha\left(\omega_j^\alpha\right)|\psi_j^\beta\left(r_i^\beta\right)\alpha\left(\omega_i^\beta\right)\psi_i^\alpha\left(r_j^\alpha\right)\beta\left(\omega_j^\alpha\right)\right\rangle\\
&\quad-\tfrac{1}{2}\sum_{i=1}^{n^\beta}\sum_{j=1}^{n^\beta}\left\langle\psi_i^\beta\left(r_i^\beta\right)\beta\left(\omega_i^\beta\right)\psi_j^\beta\left(r_j^\beta\right)\beta\left(\omega_j^\beta\right)|\psi_j^\beta\left(r_i^\beta\right)\beta\left(\omega_i^\beta\right)\psi_i^\beta\left(r_j^\beta\right)\beta\left(\omega_j^\beta\right)\right\rangle
\end{aligned} \tag{3.49}$$

Due to the orthogonality of spin functions, it is rewritten as

$$
\begin{aligned}
&-\frac{1}{2}\sum_{i=1}^{n^\alpha}\sum_{j=1}^{n^\alpha}\left\langle \psi_i^\alpha(r_i^\alpha)\psi_j^\alpha\left(r_j^\alpha\right)|\psi_j^\alpha(r_i^\alpha)\psi_i^\alpha\left(r_j^\alpha\right)\right\rangle \\
&-\frac{1}{2}\sum_{i=1}^{n^\beta}\sum_{j=1}^{n^\beta}\left\langle \psi_i^\beta\left(r_i^\beta\right)\psi_j^\beta\left(r_j^\beta\right)|\psi_j^\beta\left(r_i^\beta\right)\psi_i^\beta\left(r_j^\beta\right)\right\rangle
\end{aligned}
\tag{3.50}
$$

Finally, by using notations of Coulomb and exchange integrals, the total energy of the n-electro closed shell system is given by

$$
\begin{aligned}
E = &\sum_{i=1}^{n^\alpha}\langle \psi_i^\alpha(r_i^\alpha)|h_i|\psi_i^\alpha(r_i^\alpha)\rangle + \sum_{i=1}^{n^\beta}\left\langle \psi_i^\beta\left(r_i^\beta\right)\middle|h_i\middle|\psi_i^\beta\left(r_i^\beta\right)\right\rangle \\
&+\frac{1}{2}\sum_{i=1}^{n^\alpha}\sum_{j=1}^{n^\alpha}\left(J_{ij}^{\alpha\alpha}-K_{ij}^{\alpha\alpha}\right)+\frac{1}{2}\sum_{i=1}^{n^\beta}\sum_{j=1}^{n^\beta}\left(J_{ij}^{\beta\beta}-K_{ij}^{\beta\beta}\right)+\sum_{i=1}^{n^\alpha}\sum_{j=1}^{n^\beta}J_{ij}^{\alpha\beta}
\end{aligned}
\tag{3.51}
$$

The ith orbital energy of α atomic orbital is rewritten in the same manner:

$$
\varepsilon_i^\alpha = \langle \psi_i^\alpha(r_i^\alpha)|h_i|\psi_i^\alpha(r_i^\alpha)\rangle + \sum_{j=1}^{n^\alpha}\left(J_{ij}^{\alpha\alpha}-K_{ij}^{\alpha\alpha}\right)+\sum_{j=1}^{n^\beta}J_{ij}^{\alpha\beta}
\tag{3.52}
$$

The ith orbital energy of β atomic orbital is rewritten in the same manner:

$$
\varepsilon_i^\beta = \left\langle \psi_i^\beta\left(r_i^\beta\right)\middle|h_i\middle|\psi_i^\beta\left(r_i^\beta\right)\right\rangle + \sum_{j=1}^{n^\beta}\left(J_{ij}^{\beta\beta}-K_{ij}^{\beta\beta}\right)+\sum_{j=1}^{n^\alpha}J_{ij}^{\alpha\beta}
\tag{3.53}
$$

In open shell system, the spatial orbital of α electron is independent from β electron. Hartree-Fock in open shell system is called unrestricted Hartree-Fock (UHF).

3.7 Orbital Energy Rule

After solving Hartree-Fock equation, orbital energy is given as an eigenvalue. Note that it is different from total energy. As it is difficult to consider the chemical meaning, orbital energy difference between n-electron system and $(n - 1)$-electron system $(E^n - E^{n-1})$ is considered.

$$
\begin{aligned}
E^n - E^{n-1} &= \langle \chi_n(x_n)|h_n|\chi_n(x_n)\rangle \\
&\quad + \frac{1}{2}\sum_{i=1}^{n-1}\{\langle \chi_i(x_i)\chi_n(x_n)|\chi_i(x_i)\chi_n(x_n)\rangle - \langle \chi_i(x_i)\chi_n(x_n)|\chi_n(x_i)\chi_i(x_n)\rangle\} \\
&\quad + \frac{1}{2}\sum_{j=1}^{n-1}\{\langle \chi_n(x_n)\chi_j(x_j)|\chi_n(x_n)\chi_j(x_j)\rangle - \langle \chi_n(x_n)\chi_j(x_j)|\chi_j(x_n)\chi_n(x_j)\rangle\} \\
&\quad + \frac{1}{2}\langle \chi_n(x_n)\chi_n(x_n)|\chi_n(x_n)\chi_n(x_n)\rangle - \frac{1}{2}\langle \chi_n(x_n)\chi_n(x_n)|\chi_n(x_n)\chi_n(x_n)\rangle
\end{aligned}
\tag{3.54}
$$

The second term is equivalent to the third term, and the fourth term can be included in the sigma symbol. It is rewritten:

$$
\begin{aligned}
E^n - E^{n-1} &= \langle \chi_n(x_n)|h_n|\chi_n(x_n)\rangle \\
&\quad + \sum_{j=1}^{n}\{\langle \chi_n(x_n)\chi_j(x_j)|\chi_n(x_n)\chi_j(x_j)\rangle - \langle \chi_n(x_n)\chi_j(x_j)|\chi_j(x_n)\chi_n(x_j)\rangle\}
\end{aligned}
\tag{3.55}
$$

From Eq. (3.29), it is found that the value corresponds to the n-th orbital energy (ε_n).

$$
E^n - E^{n-1} = \varepsilon_n \tag{3.56}
$$

Here is assumed that spatial orbitals are the same in both systems. The selection of excluded electron is arbitrary. Though spatial orbitals may be slightly different between n-electron and $(n-1)$-electron system of the same molecule or atom, it is useful for qualitative discuss to use the relationship between total and orbital energies.

It is normally considered that total energy of n-electron system is smaller than $(n-1)$-electron system. It is because the effect of Coulomb interaction is larger in n-electron system. However, when the effects of kinetic energy and electron–electron repulsion are larger, the total energy of n-electron system is larger than $(n-1)$-electron system. As the result, the positive orbital energy is given. The orbital energy rule can be constructed as follows.

Orbital energy rule

(1) When $E^n < E^{n-1}$, the negative n-th orbital energy (ε_n) is given.
The n-th orbital is stabilized
(2) When $E^n > E^{n-1}$, the positive n-th orbital energy (ε_n) is given.
The n-th orbital is destabilized
(3) When $E^n = E^{n-1}$, the n-th orbital energy is zero.

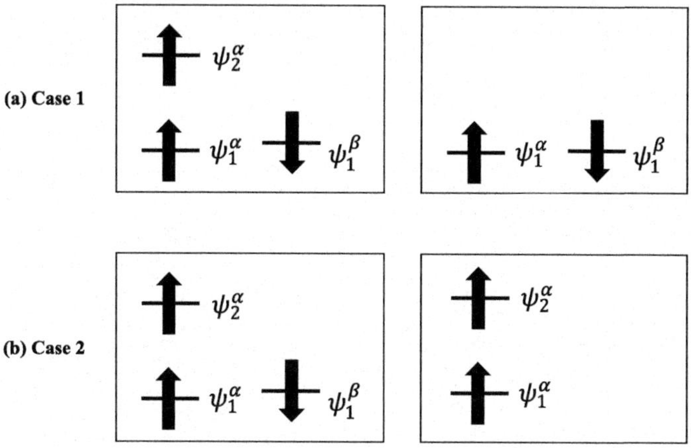

Fig. 3.3 Schematic drawing of the three-electron system and the corresponding two-electron system, when one electron is excluded

In three-electron system, two cases are considered as the pattern of electron exclusion. In case 1 (see Fig. 3.3a), the orbital energy of MO2α (ε_2^α) can be estimated from Eq. (3.56). In fact, orbital energy of MO1α is slightly different from MO1β, though the same orbital energy is given in MO1α and MO1β of two-electron closed shell system. Hence, Eq. (3.56) may not give the precise n-th orbital energy. However, the stability of n-th orbital could be qualitatively discussed. In case 2 (see Fig. 3.3b), the orbital energy of MO1β (ε_1^β) can be estimated from Eq. (3.56). However, the same situation occurs.

Further Reading

1. Szabo A, Ostlund NS (1996) Modern quantum chemistry: introduction to advanced electronic structure theory. Dover Publications Inc., New York, pp 108–230

Chapter 4
Basis Function

Abstract In many-electron system, Hartree-Fock equation has no analytical solution. To overcome the inconvenience, the introduction of basis functions to spatial orbital was considered. By the introduction of basis functions, Hartree-Fock equation can be expressed as matrix equation. In Hartree-Fock matrix equation, the problem is converted to obtain expansion coefficients and orbital energies numerically by self-consistent-field (SCF) calculation. A set of basis functions per atom is called "basis set". Initial atomic orbital is defined from designated basis set. Note that basis set must be beforehand designated per atom. The real atomic orbital and molecular orbital are represented by the combination of initial atomic orbitals. Virtual orbital is produced by the introduction of basis set. Due to inadequacy of theoretical definition, virtual orbital is often meaningless. Basis set is expressed by Gaussian basis function, due to mathematical advantages. However, Gaussian basis function has two disadvantages of a poor representation of radial wave-function near atomic nucleus, and a rapid decrease in the amplitude of the wave-function. In order to improve them, contraction is performed. In order to express flexibility of outer shell electron, split-valence basis function is applied. Polarization basis function and diffuse basis function are applied for further improvement. Several useful basis sets are introduced: minimal basis set, 6-31G basis set and correlation-consistent basis set. Finally, our empirical recommendation for basis set selection is introduced.

Keywords Hartree-Fock matrix equation · Basis set · Initial atomic orbital · Polarization basis function · Split-valence basis function · Diffuse basis function

4.1 Hartree-Fock Matrix Equation

4.1.1 Closed Shell System

In closed shell system, due to the orthogonality of spin functions, the Hartree-Fock equation for the *i*th molecular orbital is written as

T. Onishi, *Quantum Computational Chemistry*,
DOI 10.1007/978-981-10-5933-9_4

$$f_i \psi_i(r_i) = \varepsilon_i \psi_i(r_i) \tag{4.1}$$

where f_i denotes Fock operator; ε_i is eigenvalue, which denotes orbital energy; ψ_i denotes the wave-function of the ith molecular orbital. However, it is impossible to obtain eigenvalue and wave-function analytically. As one of the solutions, a set of basis functions, which is called basis set, is introduced to the wave-function of spatial orbital.

$$\psi_i(r_i) = \sum_{\lambda=1}^{N^\lambda} C_{\lambda i} \phi_\lambda \tag{4.2}$$

where N^λ is a number of basis functions that is normally larger than the number of electrons. $C_{\lambda i}$ is an unknown expansion coefficient, and ϕ_λ is a defined basis function. Note that basis set is designated per atom. By introducing basis set, Hartree-Fock equation is rewritten as

$$f_i \sum_{\lambda=1}^{N^\lambda} C_{\lambda i} \phi_\lambda = \varepsilon_i \sum_{\lambda=1}^{N^\lambda} C_{\lambda i} \phi_\lambda \tag{4.3}$$

Multiplied ϕ_γ^* on the left, and then integrated both sides,

$$\sum_{\lambda=1}^{N^\lambda} C_{\lambda i} \langle \phi_\gamma | f_i | \phi_\lambda \rangle = \varepsilon_i \sum_{\lambda=1}^{N^\lambda} C_{\lambda i} \langle \phi_\gamma | \phi_\lambda \rangle \tag{4.4}$$

Fock matrix ($F_{\gamma\lambda}$) and overlap matrix ($S_{\gamma\lambda}$) are defined as follows.

$$F_{\gamma\lambda} = \langle \phi_\gamma | f_i | \phi_\lambda \rangle \tag{4.5}$$

$$S_{\gamma\lambda} = \langle \phi_\gamma | \phi_\lambda \rangle \tag{4.6}$$

By using the notations of Eqs. (4.5) and (4.6), the Hartree-Fock equation is rewritten as

$$\sum_{\lambda=1}^{N^\lambda} F_{\gamma\lambda} C_{\lambda i} = \varepsilon_i \sum_{\lambda=1}^{N^\lambda} S_{\gamma\lambda} C_{\lambda i} \tag{4.7}$$

Finally, it can be written as matrix equation.

$$\mathbf{FC} = \mathbf{SC\varepsilon} \tag{4.8}$$

where **C** and **ε** are given by

$$\mathbf{C} = \begin{pmatrix} C_{11} & C_{12} & \cdots & C_{1N^\lambda} \\ C_{21} & C_{22} & \cdots & C_{2N^\lambda} \\ \cdot & \cdot & \cdots & \cdot \\ \cdot & \cdot & \cdots & \cdot \\ C_{N^\lambda 1} & C_{N^\lambda 2} & \cdots & C_{N^\lambda N^\lambda} \end{pmatrix} \tag{4.9}$$

$$\boldsymbol{\varepsilon} = \begin{pmatrix} \varepsilon_1 & 0 & \cdots & 0 \\ 0 & \varepsilon_2 & \cdots & 0 \\ \cdot & \cdot & \cdots & \cdot \\ \cdot & \cdot & \cdots & \cdot \\ 0 & 0 & \cdots & \varepsilon_{N^\lambda} \end{pmatrix} \tag{4.10}$$

The problem is converted to obtain expansion coefficients and orbital energies numerically by self-consistent-field (SCF) calculation. By using arbitrary set of expansion coefficients, the new set of expansion coefficients are obtained (cycle). The cycle process is continued until the convergence criterion is satisfied. For example, energy difference after cycle process is enough small, etc. Finally, converged expansion coefficients and orbital energies are given.

4.1.2 Open Shell System

In open shell system, two Hartree-Fock equations are considered for α and β spatial orbitals. The Fock operator for α spatial orbital is written as

$$f_i^\alpha = h_i + \sum_{j \neq i}^{n^\alpha} \left\{ J_j^\alpha - K_j^\alpha \right\} + \sum_{j \neq i}^{n^\beta} \left\{ J_j^\beta - K_j^\beta \right\} \tag{4.11}$$

where J_j^x and K_j^x denote Coulomb and exchange operators for x spin ($x = \alpha$ or β), respectively. The exchange operator of β spatial orbital will be cancelled out, due to the orthogonality of spin functions. Finally, it is rewritten as

$$f_i^\alpha = h_i + \sum_{j \neq i}^{n^\alpha} \left\{ J_j^\alpha - K_j^\alpha \right\} + \sum_{j \neq i}^{n^\beta} J_j^\beta \tag{4.12}$$

In the same manner, the Fock operator for β spatial orbital is written as

$$f_i^\beta = h_i + \sum_{j \neq i}^{n^\beta} \left\{ J_j^\beta - K_j^\beta \right\} + \sum_{j \neq i}^{n^\alpha} J_j^\alpha \tag{4.13}$$

Two Hartree-Fock equations for α and β spatial orbitals are written as

$$f_i^\alpha \psi_i^\alpha(r_i^\alpha) = \varepsilon_i^\alpha \psi_i^\alpha(r_i^\alpha) \tag{4.14}$$

$$f_i^\beta \psi_i^\beta(r_i^\beta) = \varepsilon_i^\beta \psi_i^\beta(r_i^\beta) \tag{4.15}$$

The same basis sets are introduced in both α and β spatial orbitals.

$$\psi_i^\alpha(r_i) = \sum_{\lambda=1}^{N^\lambda} C_{\lambda i}^\alpha \phi_\lambda \tag{4.16}$$

$$\psi_i^\beta(r_i) = \sum_{\lambda=1}^{N^\lambda} C_{\lambda i}^\beta \phi_\lambda \tag{4.17}$$

where N^λ is the number of basis functions; C_λ^α and C_λ^β are unknown expansion coefficients for the α and β spatial orbitals, respectively; ϕ_λ is a defined basis function. Note that expansion coefficients of α spatial orbitals are generally different from β spatial orbitals. By substitution of Eqs. (4.16) and (4.17), two Hartree-Fock equations (Eqs. 4.14 and 4.15) are rewritten as

$$f_i^\alpha \sum_{\lambda=1}^{N^\lambda} C_{\lambda i}^\alpha \phi_\lambda = \varepsilon_i^\alpha \sum_{\lambda=1}^{N^\lambda} C_{\lambda i}^\alpha \phi_\lambda \tag{4.18}$$

$$f_i^\beta \sum_{\lambda=1}^{N^\lambda} C_{\lambda i}^\beta \phi_\lambda = \varepsilon_i^\beta \sum_{\lambda=1}^{N^\lambda} C_{\lambda i}^\beta \phi_\lambda \tag{4.19}$$

Multiplied $\phi_\gamma{}^*$ on the left, and then integrated both sides,

$$\sum_{\lambda=1}^{N^\lambda} C_{\lambda i}^\alpha \langle \phi_\gamma | f_i^\alpha | \phi_\lambda \rangle = \varepsilon_i^\alpha \sum_{\lambda=1}^{N^\lambda} C_{\lambda i}^\alpha \langle \phi_\gamma | \phi_\lambda \rangle \tag{4.20}$$

$$\sum_{\lambda=1}^{N^\lambda} C_{\lambda i}^\beta \left\langle \phi_\gamma | f_i^\beta | \phi_\lambda \right\rangle = \varepsilon_i^\beta \sum_{\lambda=1}^{N^\lambda} C_{\lambda i}^\beta \langle \phi_\gamma | \phi_\lambda \rangle \tag{4.21}$$

Fock matrices ($F_{\gamma\lambda}^\alpha$, $F_{\gamma\lambda}^\beta$) for α and β spins, and overlap matrix ($S_{\gamma\lambda}$) are defined as follows.

$$F_{\gamma\lambda}^\alpha = \langle \phi_\gamma | f_i^\alpha | \phi_\lambda \rangle \tag{4.22}$$

$$F_{\gamma\lambda}^\beta = \left\langle \phi_\gamma | f_i^\beta | \phi_\lambda \right\rangle \tag{4.23}$$

$$S_{\gamma\lambda} = \langle \phi_\gamma | \phi_\lambda \rangle \tag{4.24}$$

By using the notations of Eqs. (4.22)–(4.24), two Hartree-Fock equations for α and β spatial orbitals are rewritten as

$$\sum_{\lambda=1}^{N^\lambda} F^{\alpha}_{\gamma\lambda} C^{\alpha}_{\lambda i} = \varepsilon^{\alpha}_{i} \sum_{\lambda=1}^{N^\lambda} S_{\gamma\lambda} C^{\alpha}_{\lambda i} \tag{4.25}$$

$$\sum_{\lambda=1}^{N^\lambda} F^{\beta}_{\gamma\lambda} C^{\beta}_{\lambda i} = \varepsilon^{\beta}_{i} \sum_{\lambda=1}^{N^\lambda} S_{\gamma\lambda} C^{\beta}_{\lambda i} \tag{4.26}$$

Finally, they can be written as matrix equations.

$$\mathbf{F}^{\alpha}\mathbf{C}^{\alpha} = \mathbf{S}\mathbf{C}^{\alpha}\boldsymbol{\varepsilon}^{\alpha} \tag{4.27}$$

$$\mathbf{F}^{\beta}\mathbf{C}^{\beta} = \mathbf{S}\mathbf{C}^{\beta}\boldsymbol{\varepsilon}^{\beta} \tag{4.28}$$

where $\mathbf{C}^{\alpha}$, $\boldsymbol{\varepsilon}^{\alpha}$, $\mathbf{C}^{\beta}$ and $\boldsymbol{\varepsilon}^{\beta}$ are given by

$$\mathbf{C}^{\alpha} = \begin{pmatrix} C^{\alpha}_{11} & C^{\alpha}_{12} & \cdots & C^{\alpha}_{1N^\lambda} \\ C^{\alpha}_{21} & C^{\alpha}_{22} & \cdots & C^{\alpha}_{2N^\lambda} \\ \cdot & \cdot & \cdots & \cdot \\ \cdot & \cdot & \cdots & \cdot \\ C^{\alpha}_{N^\lambda 1} & C^{\alpha}_{N^\lambda 2} & \cdots & C^{\alpha}_{N^\lambda N^\lambda} \end{pmatrix} \tag{4.29}$$

$$\boldsymbol{\varepsilon}^{\alpha} = \begin{pmatrix} \varepsilon^{\alpha}_{1} & 0 & \cdots & 0 \\ 0 & \varepsilon^{\alpha}_{2} & \cdots & 0 \\ \cdot & \cdot & \cdots & \cdot \\ \cdot & \cdot & \cdots & \cdot \\ 0 & 0 & \cdots & \varepsilon^{\alpha}_{N^\lambda} \end{pmatrix} \tag{4.30}$$

$$\mathbf{C}^{\beta} = \begin{pmatrix} C^{\beta}_{11} & C^{\beta}_{12} & \cdots & C^{\beta}_{1N^\lambda} \\ C^{\beta}_{21} & C^{\beta}_{22} & \cdots & C^{\beta}_{2N^\lambda} \\ \cdot & \cdot & \cdots & \cdot \\ \cdot & \cdot & \cdots & \cdot \\ C^{\beta}_{N^\lambda 1} & C^{\beta}_{N^\lambda 2} & \cdots & C^{\beta}_{N^\lambda N^\lambda} \end{pmatrix} \tag{4.31}$$

$$\boldsymbol{\varepsilon}^{\beta} = \begin{pmatrix} \varepsilon^{\beta}_{1} & 0 & \cdots & 0 \\ 0 & \varepsilon^{\beta}_{2} & \cdots & 0 \\ \cdot & \cdot & \cdots & \cdot \\ \cdot & \cdot & \cdots & \cdot \\ 0 & 0 & \cdots & \varepsilon^{\beta}_{N^\lambda} \end{pmatrix} \tag{4.32}$$

The problem is converted to obtain expansion coefficients and orbital energies of α and β spatial orbitals numerically by SCF calculation. Arbitrary set of expansion coefficients must be prepared for both α and β spatial orbitals. Finally, converged expansion coefficients and orbital energies of α and β spatial orbitals are given, when convergence criterion is satisfied.

4.2 Initial Atomic Orbital

Atomic orbital (AO) and molecular orbital (MO) are the solution of Hartree-Fock equation for atom and molecule, respectively. The concept of initial atomic orbital (IAO) is very useful to analyse obtained AOs and MOs. In each atom, IAO is defined by one basis function or combination of basis functions. In this book, IAO is just called "orbital". Note that IAO is an artificially defined orbital. AOs and MOs are represented by the combination of IAOs. In many cases, AO corresponds to IAO. However, AO is sometimes represented by the combination of IAOs. It is called hybridization. MO related to outer shell electrons is often represented by the combination of IAOs of the different atoms. It is called orbital overlap.

4.3 Virtual Orbital

In Hartree-Fock matrix equation, the number of the produced AOs and MOs correspond to the sum of all basis functions (see Eqs. 4.10, 4.30 and 4.32). For example, in hydrogen molecule, two MOs are produced, if one basis function is defined for hydrogen 1s orbital. At ground state, two electrons occupy one MO, and the other MO is unoccupied. The unoccupied AO and MO are often called "virtual orbital".

Readers may consider that virtual orbital is related to excited electronic structure. However, the obtained virtual orbital often does not correspond to excited electronic structure. It is because there is no universal method to estimate the interaction between virtual orbitals, as no electron is allocated in virtual orbital. Hence, the present virtual orbital, which is obtained from the present calculation, is often meaningless. We have to pay attention to examine virtual orbital.

4.4 Gaussian Basis Function

There are two types of basis functions: Slater-type and Gauss-type (Gaussian) basis functions. Slater-type basis function resembles the wave-function of hydrogenic atom. However, it suffers from obtaining analytical solution for two-electron integral.

On the other hand, Gaussian basis function (ϕ^{Gauss}) can overcome the problem. It is written in terms of Cartesian coordinates.

$$\phi^{\text{Gauss}} = Nx^i y^j z^k \exp(-\alpha r^2) \tag{4.33}$$

The origin of coordinates is atomic nucleus. N is normalization constant. The $x^i y^j z^k$ part stands for angular component. The i, j and k values are non-negative integers. The sum of these values determines the types of orbitals. When $i + j + k = 0$, s orbital is expressed, due to no existence of x, y and z parameters. When $i + j + k = 1$, three p_x orbital ($i = 1$), p_y orbital ($j = 1$) and p_z orbital ($k = 1$) are expressed. When $i + j + k = 2$, six types of orbitals are considered. However, only five d orbitals are allowed in hydrogenic atom. When $i = j = 1$, $j = k = 1$ and $i = k = 1$, d_{xy}, d_{yz}, and d_{xz} orbitals are expressed, respectively. Two $d_{x^2-y^2}$ orbital and $d_{3z^2-r^2}$ orbital cannot be expressed in the manner. Instead, $d_{x^2-y^2}$ orbital is expressed by the hybridization between d_{x^2} orbital ($i = 2$) and d_{y^2} orbital ($j = 2$). $d_{3z^2-r^2}$ orbital is expressed by the hybridization between d_{x^2} orbital ($i = 2$), d_{y^2} orbital ($j = 2$) and d_{z^2} orbital ($k = 2$).

4.5 Contraction

Gaussian basis function is utilized from the viewpoint of analytical advantage. However, they have two disadvantages. One is a poor representation of radial wave-function near atomic nucleus. The other is a rapid decrease in the amplitude of the wave-function. For example, hydrogen 1s orbital has a cusp around atomic nucleus. As shown in Fig. 4.1, though a cusp is reproduced well in Slater-type basis function, the figure of Gaussian basis function is smooth around atomic nucleus. In order to improve radial wave-function, one orbital is expressed by the linear combination of several basis functions. It is called a contracted Gaussian basis function.

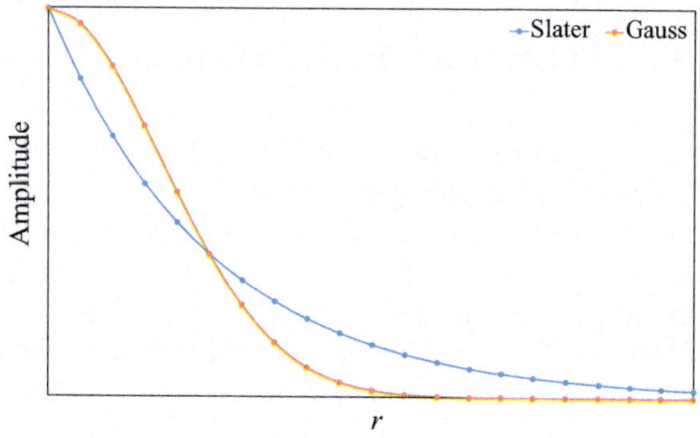

Fig. 4.1 Radial wave-functions of Slater-type and Gaussian basis functions

$$\sum_{\mu=1}^{L} d_\mu \phi_a^{\text{Gauss}} \tag{4.34}$$

L is the number of Gaussian basis functions in the linear combination. ϕ_a^{Gauss} is the primitive Gaussian basis function. d_μ is the contracted coefficient. Note that the contracted Gaussian basis function stands for one IAO. The analytical solution of two-electron integrals is also obtained when using the contracted Gaussian basis function.

4.6 Split-Valence Basis Function

In comparison with inner shell electron, outer shell electron is more interactive. To express own flexibility, IAO of outer shell electron is represented by multi-basis functions. In double-zeta split-valence basis, one IAO is represented by two Gaussian basis functions. In triple-zeta split-valence basis, one IAO is represented by three Gaussian basis functions. On the other hand, IAO of inner shell electron is represented by one Gaussian basis function.

Let us consider hydrogen atom. In double-zeta split-valence basis, hydrogen 1s IAO is represented by two Gaussian basis functions.

$$c_{\text{H}(1\text{s}')}\phi_{\text{H}(1\text{s}')} + c_{\text{H}(1\text{s}'')}\phi_{\text{H}(1\text{s}'')} \tag{4.35}$$

where $\phi_{\text{H}(1\text{s}')}$ and $\phi_{\text{H}(1\text{s}'')}$ denote two Gaussian basis functions, and $c_{\text{H}(1\text{s}')}$ and $c_{\text{H}(1\text{s}'')}$ denote the coefficients. In triple-zeta split-valence basis, hydrogen 1s IAO is represented by three Gaussian basis functions.

$$c_{\text{H}(1\text{s}')}\phi_{\text{H}(1\text{s}')} + c_{\text{H}(1\text{s}'')}\phi_{\text{H}(1\text{s}'')} + c_{\text{H}(1\text{s}''')}\phi_{\text{H}(1\text{s}''')} \tag{4.36}$$

where $\phi_{\text{H}(1\text{s}')}$, $\phi_{\text{H}(1\text{s}'')}$ and $\phi_{\text{H}(1\text{s}''')}$ denote Gaussian basis functions, and $c_{H(1s')}$, $c_{\text{H}(1\text{s}''')}$ and $c_{\text{H}(1\text{s}''')}$ denote the coefficients.

4.7 Polarization Basis Function

Own orbital flexibility can be enhanced by the introduction of split-valence basis function. When covalent bonding is formed between different orbitals, more complicated covalent bonding may be formed. For the correction, polarization basis function is introduced. There is no clear rule in the combination of polarization basis functions. In many cases, the basis function of p orbital is combined to s orbital, and the basis function of d orbital is combined to p orbital. Hence, in principal, polarization basis function does not stand for IAO.

4.8 Diffuse Basis Function

In highest spin state, electrons are allocated in more outer shell orbital where electron is unoccupied in lowest spin state. The extra basis function, which is called diffuse basis function, is added to represent excited electron configuration. In this sense, it can be regarded as "excited electron configuration basis function". For example, the ground state of helium atom has the singlet electron configuration, where two electrons occupy one helium 1s orbital. On the other hand, in triplet helium atom, though one electron is allocated in helium 1s orbital, the other is allocated in helium 2s orbital. The extra basis function must be included to represent 2s orbital. In this case, diffuse basis function stands for IAO. The other role of disuse basis function is the correction of polarization basis function. For example, in aug-cc-pVXZ basis set, diffuse basis function is used for the correction.

In summary, there are two roles in diffuse basis functions: (1) representation of excited electron configuration and (2) correction of polarization. Though the former stands for IAO, the latter is used only for the correction. We must distinguish the difference in molecular orbital analysis. Polarization and diffuse basis functions are principally defined in theoretical manner. However, there is no guarantee that they keep principal role in practical calculation.

4.9 Useful Basis Set

Basis set is a set of basis functions that is defined for each atom. We have to select the best basis set to reproduce a scientifically reasonable electronic structure. It is because there is no single and universal basis set that is applicable under all circumstances. In this chapter, several practical basis sets are introduced.

4.9.1 Minimal Basis Set

In the minimal basis set, occupied IAO are only expressed by Gaussian basis functions. Minimal basis set is often called MINI or MINI basis set. Let us explain MINI basis set for neutral copper. The electron configuration of copper atom is

$$\text{Cu: } 1s^2 2s^2 2p^6 3s^2 3d^{10} 4s^1 \tag{4.37}$$

At least, 1s, 2s, 2p, 3s, 3p, 3d and 4s orbitals must be represented by Gaussian basis functions. The general notation of basis set is as follows.

$$\text{Basis set}\left(N_{1s}.N_{2s}.N_{3s}.N_{4s}\cdots/N_{2p}.N_{3p}\cdots/N_{3d}\cdots\right) \tag{4.38}$$

Table 4.1 Contracted coefficients (d) and exponential coefficients (α) of MINI (5.3.3.3/5.3/5) for neutral copper

	1s orbital	2s orbital	3s orbital	4s orbital
d_1	−0.0051311	−0.1089833	0.2242654	−0.0971173
d_2	−0.0389436	0.6381907	−0.7327660	0.5610408
d_3	−0.1761209	0.4362349	−0.4010780	0.5192031
d_4	−0.4682401			
d_5	−0.4507014			
α_1	32311.084	161.71783	13.738109	0.92052275
α_2	4841.4341	18.731951	2.2080203	0.10255637
α_3	1094.8876	7.7018109	0.84846612	0.03649045
α_4	307.74535			
α_5	94.865639			

	2p orbital	3p orbital	3d orbital
d_1	0.0095141	0.3410642	0.0348038
d_2	0.0704695	0.5491335	0.1757100
d_3	0.2663558	0.2331493	0.3897658
d_4	0.5105298		0.4580844
d_5	0.3239964		0.3141941
α_1	963.25905	5.1070835	45.307828
α_2	227.39750	1.9450324	12.636091
α_3	72.327649	0.71388491	4.2082300
α_4	26.200292		1.3630734
α_5	9.7923323		0.37550107

where N_X (X = 1s, 2s, 3s, 4s, 2p, 3p, 3d, etc.) denotes the number of primitive Gaussian basis functions for each IAO. In MINI(5.3.3.3/5.3/5) basis set for neutral copper, copper 1s, 2s, 3s and 4s IAOs are represented by a contracted Gaussian basis function with five, three, three and three primitive Gaussian basis functions, respectively; copper 2p and 3p orbitals are represented by a contracted Gaussian basis function with five and three primitive Gaussian basis functions, respectively; copper 3d orbital is represented by a contracted Gaussian basis function with five primitive Gaussian basis functions. The exponential coefficients (α) and contracted coefficients (c) of MINI(5.3.3.3/5.3/5) for neutral copper are shown in Table 4.1.

Three types of IAOs (p_x, p_y and p_z orbitals) exist in both 2p and 3p orbitals. Though basis functions of $2p_x$, $2p_y$ and $2p_z$ orbitals have the same exponential coefficients and contracted coefficients, they have the different the $x^i y^j z^k$ term in Eq. 4.33. It implies that they have the same radial wave-function, but the angular wave-function is different. Six types of IAOs (d_x^2, d_y^2, d_z^2, d_{xy}, d_{yz}, d_{xz} orbitals) exist in 3d orbital. Though basis functions of $3d_x^2$, $3d_y^2$, $3d_z^2$, $3d_{xy}$, $3d_{yz}$, $3d_{xz}$ orbitals have the same exponential coefficients and contracted coefficients, they have the different the $x^i y^j z^k$ term. In real, $3d_{z^3-r^2}$ and $3d_{x^2-y^2}$ orbitals are represented by the hybridization between basis functions of $3d_{x^2}$, $3d_{y^2}$ and $3d_{z^2}$ orbitals. Though the basis set is optimized based on Cu: $1s^22s^22p^63s^23d^{10}4s^1$, it may work well in copper cation such as Cu^{2+}: $1s^22s^22p^63s^23d^9$.

In MINI(5.3.3.2.1/5.3/4.1) basis set, split-valence basis functions is combined. 4s and 3d IAOs are represented by two Gaussian basis functions. Polarization basis function and diffuse basis function can be added in MINI basis set. MINI basis set reproduces well the electron configuration of transition metal 3d electron. It is because five 3d orbitals have the more flexibility, in comparison with 1s, 2s and three 2p orbitals.

4.9.2 6-31G Basis Set

6-31G basis set, which belongs to double-zeta split-valence basis, was developed by Pople and coworkers. It has been recognized that it reproduces well electronic structure, combined with Hartree-Fock and density functional theory (DFT) methods. IAO of inner shell electron is represented by a contracted Gaussian basis function, which contains six primitive Gaussian basis functions. IAO of outer shell orbital is split into two Gaussian basis functions. One is a contracted Gaussian basis function, which contains three primitive Gaussian basis functions. The other is a single Gaussian basis function. Polarization basis function is added to 6-31G except for hydrogen. It is denoted as 6-31G*. 6-311G basis set belongs to triple-zeta split-valence basis. IAO of outer shell electron is split into three Gaussian basis functions. One is a contracted Gaussian basis function with three primate Gaussian basis functions. The others are a single Gaussian basis function.

Hydrogen

One electron occupies 1s IAO, and there is no inner shell electron. In 6-31G basis set, 1s IAO is represented by two Gaussian basis functions.

$$c_{\mathrm{H}(1s')}\phi_{\mathrm{H}(1s')} + c_{\mathrm{H}(1s'')}\phi_{\mathrm{H}(1s''')} \tag{4.39}$$

where $\phi_{\mathrm{H}(1s')}$ and $\phi_{\mathrm{H}(1s'')}$ denote two Gaussian basis functions for 1s IAO; $c_{\mathrm{H}(1s')}$ and $c_{\mathrm{H}(1s'')}$ denote the coefficients. In hydrogen and helium, no polarization basis function is added in 6-31G*, but p-type polarization basis function is added in 6-31G**.

Carbon

Two electrons occupy 1s IAO as inner shell electron, and it is treated that four electrons occupy 2s and 2p IAOs as outer shell electron. In 6-31G basis set, 1s IAO is represented by Gaussian basis function.

$$c_{\mathrm{C}(1s)}\phi_{\mathrm{C}(1s)} \tag{4.40}$$

where $\phi_{\mathrm{C}(1s)}$ denotes Gaussian basis function of 1s IAO; $c_{\mathrm{C}(1s)}$ denote the coefficient. On the other hand, 2s and 2p IAOs are represented by two Gaussian basis functions.

$$c_{C(2s')}\phi_{C(2s')} + c_{C(2s'')}\phi_{H(2s'')} \tag{4.41}$$

$$c_{C(2p'_x)}\phi_{C(2p'_x)} + c_{C(2p''_x)}\phi_{C(2p''_x)} \tag{4.42}$$

$$c_{C(2p'_y)}\phi_{C(2p'_y)} + c_{C(2p''_y)}\phi_{C(2p''_y)} \tag{4.43}$$

$$c_{C(2p'_z)}\phi_{C(2p'_z)} + c_{C(2p''_z)}\phi_{C(2p''_z)} \tag{4.44}$$

where $\phi_{C(2s')}$ and $\phi_{C(2s'')}$ denote two Gaussian basis functions of 2s IAO; $\phi_{C(2p'_x)}$ and $\phi_{(2p''_x)}$ denote two Gaussian basis functions of $2p_x$ IAO; $\phi_{C(2p'_x)}$ and $\phi_{(2p''_x)}$ denote two Gaussian basis functions of $2p_y$ IAO; $\phi_{C(2p'_z)}$ and $\phi_{(2p''_z)}$ denote two Gaussian basis functions of $2p_z$ IAO; $c_{C(2s')}, c_{C(2s'')}, c_{(2p'_x)}, c_{(2p''_x)}, c_{(2p''_y)}, c_{(2p''_y)}c_{(2p'_z)}$ and $c_{(2p''_z)}$ denote the coefficients. Though basis functions of $2p_x$, $2p_y$ and $2p_z$ IAOs have the same exponential coefficients and contracted coefficients, they have the different radial wave-function, due to the different the $x^iy^jz^k$ terms. Note that the difference of the $x^iy^jz^k$ terms is automatically recognized in many calculation program. In carbon, d-type polarization basis function is added in 6-31G* and 6-31G** basis sets.

Table 4.2 summarizes the initial atomic orbitals and polarization basis functions of first-row atoms (H, He), second-row atoms (Li, Be, B, C, N, O, F, Ne) and the

Table 4.2 Initial atomic orbitals and polarization basis functions of first-row atoms (H, He), second-row atoms (Li, Be, B, C, N, O, F, Ne) and the third-row atoms (Na, Mg, Al, Si, P, S, Cl, Ar) in 6-31G, 6-31G* and 6-31G** basis sets

Basis set	Row	Initial atomic orbital		Polarization
		Inner shell electron	Outer shell electron	
6-31G	First	×	1s orbital	×
	Second	1s orbital	2s orbital	×
			2p orbital	
	Third	1s orbital	3s orbital	×
		2s orbital	3p orbital	
		2p orbital		
6-31G*	First		1s orbital	×
	Second	1s orbital	2s orbital	d-type
			2p orbital	
	Third	1s orbital	3s orbital	d-type
		2s orbital	3p orbital	
		2p orbital		
6-31G**	First		1s orbital	p-type
	Second	1s orbital	2s orbital	d-type
			2p orbital	
	Third	1s orbital	3s orbital	d-type
		2s orbital	3p orbital	
		2p orbital		

third-row atoms (Na, Mg, Al, Si, P, S, Cl, Ar) in 6-31G, 6-31G* and 6-31G** basis sets. In the first-row atoms, no polarization basis function is added in 6-31G*, though p-type polarization basis function is added in 6-31G** basis set. Note that 6-31G* is equivalent to 6-31G in hydrogen and helium. In second-row and third-row atoms, d-type polarization basis function is added in 6-31G* and 6-31G** basis sets.

The general notations of 6-31G basis set for the first-row, second-row and third-row atoms are 6-31G (3.1), 6-31G (6.3.1/3.1) and 6-31G (6.6.3.1/6.3.1), respectively. The notations of 6-31G* for the first-row, second-row and third-row atoms are 6-31G* (3.1), 6-31G* (6.3.1/3.1/1) and 6-31G* (6.6.3.1/6.3.1/1), respectively. The notations of 6-31G** for the first-row, second-row and third-row atoms are 6-31G** (3.1/1), 6-31G** (6.3.1/3.1/1) and 6-31G** (6.6.3.1/6.3.1/1), respectively.

4.9.3 Correlation-Consistent Basis Sets

4.9.3.1 cc-PVXZ Basis Set

Correlation-consistent basis sets were developed by Dunning and coworkers, from the viewpoint of the improvement of electron correlation energy. Recently, it has been widely utilized, combined with accurate calculation methods beyond Hartree-Fock method. The general notation of correlation-consistent basis set is cc-pVXZ, which implies "correlation-consistent polarized valence X-zeta basis set" (X − D (double-zeta), T (triple zeta), Q (quadruple zeta), etc.).

Hydrogen
One electron occupies 1s IAO, and there is no inner shell electron. In cc-pVDZ basis set, 1s IAO is represented by two Gaussian basis functions.

$$c_{\mathrm{H}(1s')}\phi_{\mathrm{H}(1s')} + c_{\mathrm{H}(1s'')}\phi_{\mathrm{H}(1s'')} \tag{4.45}$$

where $\phi_{\mathrm{H}(1s')}$ and $\phi_{\mathrm{H}(1s'')}$ denote two Gaussian basis functions for 1s IAO; $c_{\mathrm{H}(1s')}$ and $c_{\mathrm{H}(1s'')}$ denote the coefficients. One p-type polarization basis function is added. The notation of cc-pVDZ basis set for hydrogen is cc-pVDZ (3.1/1).

Carbon
Two electrons occupy 1s IAO as inner shell electron, and it is treated that four electron occupy 2s and 2p IAOs as outer shell electron. In cc-pVDZ basis set, 1s IAO is represented by Gaussian basis function. 2s and 2p IAOs are represented by two Gaussian basis functions.

$$c_{C(2s')}\phi_{C(2s')} + c_{C(2s'')}\phi_{C(2s'')} \tag{4.46}$$

$$c_{C(2p'_x)}\phi_{C(2p'_x)} + c_{C(2p''_x)}\phi_{C(2p''_x)} \tag{4.47}$$

$$c_{C(2p'_y)}\phi_{C(2p'_y)} + c_{C(2p''_y)}\phi_{C(2p''_y)} \tag{4.48}$$

$$c_{C(2p'_z)}\phi_{C(2p'_z)} + c_{C(2p''_z)}\phi_{C(2p''_z)} \tag{4.49}$$

where $\phi_{C(2s')}$ and $\phi_{C(2s'')}$ denote two Gaussian basis functions of 2s IAO; $\phi_{C(2p'_x)}$ and $\phi_{(2p''_x)}$ denote two Gaussian basis functions of $2p_x$ IAO; $\phi_{C(2p'_y)}$ and $\phi_{(2p''_y)}$ denote two Gaussian basis functions of $2p_y$ IAO; $\phi_{C(2p'_z)}$ and $\phi_{(2p''_z)}$ denote two Gaussian basis functions of $2p_z$ IAO; $c_{C(2s')}$, $c_{C(2s'')}$, $c_{(2p'_x)}$, $c_{(2p''_x)}$, $c_{(2p'_y)}$, $c_{(2p''_y)}$, $c_{(2p'_z)}$ and $c_{(2p''_z)}$ denote the coefficients. Though basis functions of $2p_x$, $2p_y$ and $2p_z$ IAOs have the same exponential coefficients and contracted coefficients, they have the different radial wave-function, due to the different the $x^iy^jz^k$ terms. One d-type polarization basis function is also added. The notation of cc-pVDZ basis set for carbon is cc-pVDZ (8.8.1/3.1/1).

Silicon

In cc-pVDZ basis set, as 1s, 2s and 2p electrons belong to inner shell, 1s, 2s and 2p IAOs are represented by Gaussian basis function. On the other hand, as it is treated that 3s and 3p electrons belong to outer shell, 3s and 3p IAOs are represented by two Gaussian basis functions.

$$c_{Si(3s')}\phi_{Si(3s')} + c_{Si(3s'')}\phi_{Si(3s'')} \tag{4.50}$$

$$c_{Si(3p'_x)}\phi_{Si(3p'_x)} + c_{Si(3p''_x)}\phi_{Si(3p''_x)} \tag{4.51}$$

$$c_{Si(3p'_y)}\phi_{Si(3p'_y)} + c_{Si(3p''_y)}\phi_{Si(3p''_y)} \tag{4.52}$$

$$c_{Si(3p'_z)}\phi_{Si(3p'_z)} + c_{Si(3p''_z)}\phi_{Si(3p''_z)} \tag{4.53}$$

where $\phi_{Si(3s')}$ and $\phi_{Si(3s'')}$ denote two Gaussian basis functions of 3s IAO; $\phi_{Si(3p'_x)}$ and $\phi_{Si(3p''_x)}$ denote two Gaussian basis functions of $3p_x$ IAO; $\phi_{Si(3p'_y)}$ and $\phi_{Si(3p''_y)}$ denote two Gaussian basis functions of $3p_y$ IAO; $\phi_{Si(3p'_z)}$ and $\phi_{Si(3p''_z)}$ denote two Gaussian basis functions of $3p_z$ IAO; $c_{Si(3s')}$, $c_{Si(3s'')}$, $c_{Si(3p'_x)}$, $c_{Si(3p''_x)}$, $c_{Si(3p'_y)}$, $c_{Si(3p''_y)}$, $c_{Si(3p'_z)}$ and $c_{Si(3p''_z)}$ denote the coefficients. Though basis functions of $3p_x$, $3p_y$ and $3p_z$ IAOs have the same exponential coefficients and contracted coefficients, they have the different radial wave-function, due to the different the $x^iy^jz^k$ terms. One d-type polarization basis function is also added. The notation of cc-pVDZ basis set for silicon is cc-pVDZ (11.11.11.1/7.7.1/1).

4.9.3.2 aug-cc-pVXZ basis set

To represent excited electron configuration or perform the correction of polarization basis function, diffuse basis function is added to cc-pVXZ basis set. The general notation of correlation-consistent basis set with diffuse basis function is aug-cc-pVXZ, which implies “augmented correlation-consistent polarized valence X-zeta basis set” (X = D (double-zeta), T (triple-zeta), Q (quadruple-zeta), etc.).

Hydrogen
In aug-cc-pVDZ basis set, 1s IAO is represented by two Gaussian basis functions, as same as cc-pVDZ basis set. Based on cc-pVDZ basis set, s-type and p-type diffuse basis functions are added. The notation of aug-cc-pVDZ basis set for hydrogen is aug-cc-pVDZ (3.1.1/1.1).

Carbon
In aug-cc-pVDZ basis set, s-type, p-type and d-type diffuse basis functions are added, based on cc-pVDZ basis set. The notation of aug-cc-pVDZ basis set for carbon is aug-cc-pVDZ (8.8.1.1/3.1.1/1.1).

Silicon
In aug-cc-pVDZ basis set, s-type, p-type and d-type diffuse basis functions are added, based on cc-pVDZ basis set. The notation of silicon aug-cc-pVDZ basis set is aug-cc-pVDZ (11.11.11.1.1/7.7.1.1/1.1).

4.9.4 Basis Set Selection

No single and universal basis set has been developed yet. Basis set selection contains very arbitrary factors. There is no guarantee that correct electronic structure is reproduced, even if larger basis set is applied in practical calculation. It is expected that smaller eigenvalue may be given, due to mathematical advantage such as higher flexibility through contraction and diffuse and polarization basis functions. However, there is a possibility that electron may be allocated mainly in diffuse and polarization basis functions. We have to pay attention that mathematically smallest eigenvalue is not always equivalent to a real minimum total energy. Benchmarking of basis set is important.

Table 4.3 summarizes our empirical recommendation of basis set selection. For small molecule and conventional organic molecule, the use of 6-31G* basis set combined with DFT method, which is denoted as 6-31G*/DFT, has an advantage in computational cost and gives the scientifically reasonable electronic structure.

For small molecule and conventional organic molecule, the use of correlation-consistent basis set combined with coupled cluster method, which is denoted as aug-cc-pVXZ/coupled cluster, makes it possible to perform the very accurate quantitative discussion.

Table 4.3 Basis set selection

System	Calculation method	Basis sets
Small molecule Organic molecule	Hartree-Fock	6-31G*
	Coupled cluster	cc-pVDZ aug-cc-pVDZ cc-pVTZ aug-cc-pVTZ
	DFT	6-31G*
Transition metal compounds	Hartree-Fock	MINI (transition metal) 6-31G* (other atoms)
	DFT	MINI (transition metal) 6-31G* (other atoms)

For transition metal, MINI basis set combined with DFT method gives the scientifically reasonable electronic structure. Normally, 3d orbital participates in chemical bonding. It has own flexibility, due to the existence of five type orbitals. In transition metal compounds, 6-31G* basis set is normally utilized for other atoms except for transition metal.

Basis function has two scientific meaning such as the expression of IAO and the correction of chemical bonding. In atom, the definition of IAO normally corresponds to the real AO. On the other hand, in molecule, complex chemical bonding is formed between IAOs, diffuse basis function and polarization basis function. There is a possibility that electron is allocated in diffuse basis function. In practical calculation result, it is important to check whether roles of basis functions are changed or not. Basis set selection is one of the important factors for scientifically reasonable calculation.

Three main factors

(1) **Basis set**
(2) **Combination of basis set and calculation method**
(3) **Modelling**

If scientifically reasonable model is not constructed, benchmarking is meaningless. In Chap. 6, calculation methods beyond Hartree-Fock method such as coupled cluster and DFT are introduced. In Chap. 9, how to construct model is introduced.

Further Readings

1. Szabo A, Ostlund NS (1996) Modern quantum chemistry: introduction to advanced electronic structure theory. Dover Publications Inc., New York, pp 108–230
2. Helgaker T, Jørgensen P, Olsen J (2001) Molecular electronic-structure theory. Wiley, West Sussex, pp 287–335
3. Jensen F, Introduction to computational chemistry, 2nd edn. Wiley, West Sussex, pp 192–231

4. Davidson ER, Feller D (1986) Chem Rev 86:681–696
5. Huzinaga S, Andzelm J, Radzio-Andzelm E, Sakai Y, Tatewaki H, Klobukowski M (1984) Gaussian basis sets for molecular calculations. Elsevier, Amsterdam
6. Hariharan PC, Pople JA (1973) Theoret Chim Acta 28:213–222
7. Francl MM, Pietro WJ, Hehre WJ, Binkley JS, Gordon MS, DeFrees DJ, Pople JA (1982) J Chem Phys 77(7):3654–3665
8. Rassolov VA, Pople JA, Ratner MA, Windus TL (1998) J Chem Phys 109(4):1223–1229
9. Rassolov VA, Ratner MA, Pople JA, Redfern PC, Curtiss LA (2001) J Compt Chem 22 (9):976–984
10. Dunning TH Jr (1989) J Chem Phys 90(2):1007–1023
11. Woon DE, Dunning TH Jr (1994) J Chem Phys 100(4):2975–2988
12. Woon DE, Dunning TH Jr (1994) J Chem Phys 98(2):1358–1371
13. Dunning TH Jr, Peterson KA, Woon DE (1999) Encyclopedia of computational chemistry, pp 88–115

Chapter 5
Orbital Analysis

Abstract In Hartree-Fock equation, the obtained wave-function represents atomic orbitals in atom and molecular orbitals in molecule. The details of chemical bonding and charge density can be investigated from orbital analysis. Chemical bonding rule is very useful to specify chemical bonding character. Outer shell electron is shared between different atoms in covalent bonding, though different atoms are bound through Coulomb interaction in ionic bonding. Hence, chemical bonding character can be specified by checking whether orbital overlap exists or not in molecular orbital including outer shell electrons. Mulliken charge density is a useful index to estimate a net electron distribution. As orbital overlap is equally divided into different atoms, it may cause an error. However, it has been widely accepted that Mulliken charge density is applicable for a quantitative discussion. In wave-function, spin-orbital interaction is taken into account, through the product between spatial orbital and spin function. The communication relation exists between Hamiltonian and spin operator. Natural orbital is completely different from molecular orbital. The discrete orbital energy disappears, and α and β spin functions are mixed.

Keywords Chemical bonding rule · Population analysis · Mulliken charge density · Spin-orbital interaction · Natural orbital

5.1 Chemical Bonding Rule

By solving Hartree-Fock matrix equation, atomic orbital (AO) or molecular orbital (MO) coefficients are obtained. Chemical bonding can be understood, based on the interaction between initial atomic orbitals (IAOs). In molecule and solid, chemical bonding is largely divided into covalent bonding and ionic bonding. In covalent bonding, outer shell electron is shared between different atoms. On the other hand, in ionic bonding, different atoms are bound through Coulomb interaction. Hence, checking MOs including outer shell electrons, chemical bonding character can be specified.

T. Onishi, *Quantum Computational Chemistry*,
DOI 10.1007/978-981-10-5933-9_5

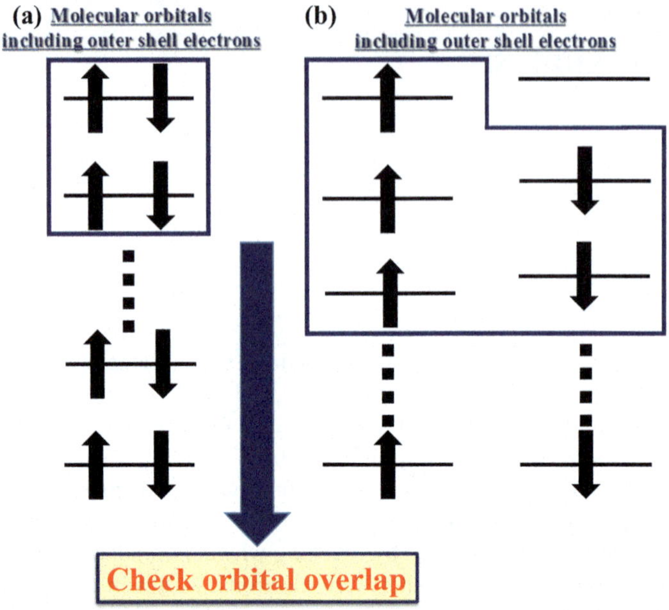

Fig. 5.1 Schematic drawing of chemical bonding rule: **a** open shell system, **b** closed shell system

Chemical bonding rule

For molecular orbitals including outer shell electrons, check whether orbital overlap exists or not.

- **With orbital overlap: Covalent.**
- **Without orbital overlap: Ionic.**

Notation

(1) Ionic bonding coexists in covalent bonding.
(2) In open shell system, outer shell electrons are often allocated in not only unpaired α MOs but also paired α and β MOs (see Fig. 5.1).
(3) MOs including outer shell electrons must be determined from obtained MO coefficients.
(4) The difference between orbital hybridization and orbital overlap: orbital hybridization occurs within atom; orbital overlap occurs between different atoms.

5.2 Mulliken Population Analysis

5.2.1 Charge Density Function

Before starting electron distribution into each atom, the total charge density function is defined for atom or molecule. In closed shell 2n-electron system, it is defined as

$$\rho(r) = 2\sum_{i=1}^{n} \psi_i(r)^* \psi_i(r) \tag{5.1}$$

where $\psi_i(r)$ is the i-th AO in atom or MO in molecule. The integration of the total charge density function corresponds to the total number of electrons in atom or molecule:

$$\int \rho(r)\mathrm{d}r = 2\sum_{i=1}^{n} \int \psi_i(r)^* \psi_i(r)\mathrm{d}r = 2n \tag{5.2}$$

On the other hand, in open shell system, the total α charge density function is defined as

$$\rho^{\alpha}(r) = \sum_{i=1}^{n^{\alpha}} \psi_i^{\alpha}(r)^* \psi_i^{\alpha}(r) \tag{5.3}$$

where n^{α} is the total number of α electrons; $\psi_i^{\alpha}(r)$ is the i-th α AO in atom or MO in molecule. The integration of the total α charge density function corresponds to the total number of α electrons in atom or molecule:

$$\int \rho^{\alpha}(r)\mathrm{d}r = \sum_{i=1}^{n^{\alpha}} \int \psi_i^{\alpha}(r)^* \psi_i^{\alpha}(r)\mathrm{d}r = n^{\alpha} \tag{5.4}$$

On the other hand, the total β charge density function is defined as

$$\rho^{\beta}(r) = \sum_{i=1}^{n^{\beta}} \psi_i^{\beta}(r)^* \psi_i^{\beta}(r) \tag{5.5}$$

where n^{β} is the total number of β electrons; $\psi_i^{\beta}(r)$ is the i-th β AO in atom or MO in molecule. The integration of the total β charge density function corresponds to the total number of β electrons in atom or molecule:

$$\int \rho^{\beta}(r)\mathrm{d}r = \sum_{i=1}^{n^{\beta}} \int \psi_i^{\beta}(r)^* \psi_i^{\beta}(r)\mathrm{d}r = n^{\beta} \tag{5.6}$$

As α and β orbitals are separated in open shell system, the total number of all electrons is given by the summation of Eqs. (5.4) and (5.6).

$$\int \{\rho^{\alpha}(r) + \rho^{\beta}(r)\}\mathrm{d}r = n^{\alpha} + n^{\beta} \tag{5.7}$$

5.2.2 *Mulliken Charge Density*

5.2.2.1 Two-Electron System

Mulliken explored how to estimate charge density of each atom in molecule, from obtained molecular orbitals. Let us consider two-electron system with singlet spin state, where two electrons are allocated in two atoms. Atom 1 and atom 2 have own atomic orbital. Atomic orbitals for atom 1 and atom 2 are denoted as ψ_1 and ψ_2, respectively. The wave-function of molecular orbital is approximately represented by the combination of ψ_1 and ψ_2.

$$\Psi(r_1, r_2) = c_1\psi_1(r_1) + c_2\psi_2(r_2) \tag{5.8}$$

where c_1 and c_2 denote a coefficient; r_1 and r_2 denote coordinate variable. The charge density function of the two-electron system is given by

$$\rho(r_1, r_2) = \Psi^*(r_1, r_2)\Psi(r_1, r_2) \tag{5.9}$$

By substitution of Eq. (5.8),

$$\rho(r_1, r_2) = (c_1)^2\{\psi_1(r_1)\}^2 + c_1^*c_2\psi_1(r_1)^*\psi_2(r_2) + c_1c_2^*\psi_1(r_1)\psi_2(r_2)^* + (c_2)^2\{\psi_2(r_2)\}^2 \tag{5.10}$$

As coefficients are real, it is rewritten as

$$\rho(r_1, r_2) = (c_1)^2\{\psi_1(r_1)\}^2 + 2c_1c_2\psi_1(r_1)\psi_2(r_2) + (c_2)^2\{\psi_2(r_2)\}^2 \tag{5.11}$$

Though the first and third terms correspond to charge density functions in pure atom 1 and atom 2, respectively, the second term is related to both atoms.

In Mulliken manner, it is equally divided into both atoms. The charge density function of atom 1 is expressed as

$$\rho(r_1, r_2)^{\text{atom1}} = (c_1)^2\{\psi_1(r_1)\}^2 + c_1c_2\psi_1(r_1)\psi_2(r_2) \tag{5.12}$$

The integration of the function gives the net number of electrons distributed to atom 1.

$$\int \rho(r_1, r_2)^{\text{atom1}} \mathrm{d}r_1\mathrm{d}r_2 = (c_1)^2 \int \{\psi_1(r_1)\}^2 \mathrm{d}r_1 + c_1c_2 \int \psi_1(r_1)\psi_2(r_2)\mathrm{d}r_1\mathrm{d}r_2 \tag{5.13}$$

The value is called Mulliken charge density. The overlap integral between two orbitals is defined as

$$S_{ij} = \int \psi_i(r_i)\psi_j(r_j)\mathrm{d}r_i\mathrm{d}r_j \tag{5.14}$$

By using the notation, it is rewritten as

$$\int \rho(r_1, r_2)^{\text{atom1}} \mathrm{d}r_1\mathrm{d}r_2 = (c_1)^2 S_{11} + c_1c_2S_{12} \tag{5.15}$$

On the other hand, the charge density function of atom 2 is expressed as

$$\rho(r_1, r_2)^{\text{atom2}} = c_1c_2\psi_1(r_1)\psi_2(r_2) + (c_2)^2\{\psi_2(r_2)\}^2 \tag{5.16}$$

The integration of the function gives Mulliken charge density of atom 2:

$$\int \rho(r_1, r_2)^{\text{atom2}} \mathrm{d}r_1\mathrm{d}r_2 = c_1c_2 \int \psi_1(r_1)\psi_2(r_2)\mathrm{d}r_1\mathrm{d}r_2 + (c_2)^2 \int \{\psi_2(r_2)\}^2 \mathrm{d}r_2 \tag{5.17}$$

By using the notation of overlap integral, it is rewritten as

$$\int \rho(r_1, r_2)^{\text{atom2}} \mathrm{d}r_1\mathrm{d}r_2 = c_1c_2S_{12} + (c_2)^2 S_{22} \tag{5.18}$$

If atom 1 and atom 2 are the same, the division will give the best approximation. It is due to molecular symmetry of the system, for example H_2 molecule, N_2 molecule, O_2 molecule, etc.

5.2.2.2 General System

Let us generalize Mulliken charge density for open shell system (m-atom; n^{α}-α electron; n^{β}-β electron). By substitution of basis function (Eq. 4.2) in the total α charge density function (Eq. 5.3),

$$\rho^{\alpha}(r) = \sum_{i=1}^{n^{\alpha}} \sum_{k=1}^{\lambda} c_k^{\alpha}(i)^* \phi_k^{\alpha}(i)^* \sum_{l=1}^{\lambda} c_l^{\alpha}(i) \phi_l^{\alpha}(i) \tag{5.19}$$

where λ is the number of basis functions. The integration of the total α charge density function gives the total number of α electrons.

$$n^{\alpha} = \int \rho^{\alpha}(r)\mathrm{d}r = \sum_{i=1}^{n^{\alpha}} \sum_{k=1}^{\lambda} \sum_{l=1}^{\lambda} c_k^{\alpha}(i)^* c_l^{\alpha}(i) \int \phi_k^{\alpha}(i)^* \phi_l^{\alpha}(i)\mathrm{d}r \tag{5.20}$$

By using the notation of overlap integral, it is rewritten as

$$n^{\alpha} = \int \rho^{\alpha}(r)\mathrm{d}r = \sum_{i=1}^{n^{\alpha}} \sum_{k=1}^{\lambda} \sum_{l=1}^{\lambda} P_{kl}^{\alpha}(i) S_{kl}^{\alpha}(i) \tag{5.21}$$

where P_{kl}^{α} is defined as

$$P_{kl}^{\alpha}(i) = c_k^{\alpha}(i)^* c_l^{\alpha}(i) \tag{5.22}$$

If the wave-function of atomic orbital for atom 1 consists of one basis function (ϕ_1), the terms related to atom 1 are

$$\begin{aligned} &P_{11}^{\alpha}(1)S_{11}^{\alpha}(1) + \{P_{12}^{\alpha}(1)S_{12}^{\alpha}(1) + P_{13}^{\alpha}(1)S_{13}^{\alpha}(1) + \cdots + P_{1\lambda}^{\alpha}(1)S_{1\lambda}^{\alpha}(1)\} \\ &\quad + \{P_{21}^{\alpha}(1)S_{21}^{\alpha}(1) + P_{31}^{\alpha}(1)S_{31}^{\alpha}(1) + \cdots + P_{\lambda 1}^{\alpha}(1)S_{\lambda 1}^{\alpha}(1)\} \end{aligned} \tag{5.23}$$

Though the first term belongs only to atom 1, other terms must be half-divided. Finally, Mulliken α charge density of atom 1 is given by

$$P_{11}^{\alpha}(1)S_{11}^{\alpha}(1) + \{P_{12}^{\alpha}(1)S_{12}^{\alpha}(1) + P_{13}^{\alpha}(1)S_{13}^{\alpha}(1) + \cdots + P_{1\lambda}^{\alpha}(1)S_{1\lambda}^{\alpha}(1)\} \tag{5.24}$$

If the wave-function of atomic orbital for atom 1 consists of two basis function (ϕ_1, ϕ_2), the terms related to atom 1 are

$$\begin{aligned} &P_{11}^{\alpha}(1)S_{11}^{\alpha}(1) + P_{12}^{\alpha}(1)S_{12}^{\alpha}(1) + P_{21}^{\alpha}(1)S_{21}^{\alpha}(1) + P_{22}^{\alpha}(1)S_{22}^{\alpha}(1) \\ &\quad + \{P_{13}^{\alpha}(1)S_{13}^{\alpha}(1) + \cdots + P_{1\lambda}^{\alpha}(1)S_{1\lambda}^{\alpha}(1)\} + \{P_{31}^{\alpha}(1)S_{31}^{\alpha}(1) + \cdots + P_{\lambda 1}^{\alpha}(1)S_{\lambda 1}^{\alpha}(1)\} \\ &\quad + \{P_{23}^{\alpha}(1)S_{23}^{\alpha}(1) + \cdots + P_{2\lambda}^{\alpha}(1)S_{2\lambda}^{\alpha}(1)\} + \{P_{32}^{\alpha}(1)S_{32}^{\alpha}(1) + \cdots + P_{\lambda 2}^{\alpha}(1)S_{\lambda 2}^{\alpha}(1)\} \end{aligned} \tag{5.25}$$

Except for the first, second, third and fourth terms, other terms must be half-divided. Mulliken α charge density is given by

$$\begin{aligned} &P_{11}^{\alpha}(1)PS_{11}^{\alpha}(1)+P_{12}^{\alpha}(1)S_{12}^{\alpha}(1)+P_{21}^{\alpha}(1)S_{21}^{\alpha}(1)+P_{22}^{\alpha}(1)S_{22}^{\alpha}(1)\\ &\quad+\{P_{13}^{\alpha}(1)S_{13}^{\alpha}(1)+\cdots+P_{1\lambda}^{\alpha}(1)S_{1\lambda}^{\alpha}(1)\}\\ &\quad+\{P_{23}^{\alpha}(1)S_{23}^{\alpha}(1)+\cdots+P_{2\lambda}^{\alpha}(1)S_{2\lambda}^{\alpha}(1)\} \end{aligned} \tag{5.26}$$

In the same manner, β Mulliken charge density can be estimated. When obtaining Mulliken charge density for specific atom, it must be checked which basis functions belong to which atom. It is summarized how to obtain Mulliken charge density as follows:

How to estimate Mulliken charge density

1. **Check which basis functions belong to which atom.**
2. **Sum PS terms consisting of considering atom.**
3. **Sum PS terms consisting of considering and other atoms, and then divide them in half.**
4. **Sum 2 and 3 = Mulliken charge density for considering atom.**

5.2.3 Summary

In Mulliken population analysis, PS term is equally divided in half. When considering orbital overlap between different orbitals or different atoms, half-division may cause an error. It is because spread of orbital is not correctly represented. If PS term is correctly distributed to each atom, precise charge density can be obtained. In addition, Mulliken charge density depends on basis set. However, it has been widely accepted that Mulliken charge density is very useful and applicable for a quantitative discussion. More precise division, combined with precise basis set, is much expected.

5.3 Spin-Orbital Interaction

5.3.1 Spin Angular Momentum

Spin angular momentum (s) of electron has two quantum numbers. One is for total spin angular momentum (s), and the other is z-component of total spin angular momentum (s_z). Note that the selection of the direction may be arbitrary, but the z direction is normally chosen. The wave-function of spin angular momentum (ϕ_{spin}) satisfies the following quantum equations.

$$s^2\phi_{\text{spin}} = s(s+1)\phi_{\text{spin}} \tag{5.27}$$

$$s_z\phi_{\text{spin}} = s_z\phi_{\text{spin}} \tag{5.28}$$

where s^2 and s_z denote operators of spin angular momentum and its z-component, respectively.

In the spin angular momentum of electron, α-type and β-type exist. The α-type and β-type wave-functions are denoted as $\phi^{\alpha}_{\text{spin}}$ and $\phi^{\beta}_{\text{spin}}$, respectively. $\phi^{\alpha}_{\text{spin}}$ and $\phi^{\beta}_{\text{spin}}$ have the same s value (1/2). They are distinguished by the different s_z values: 1/2 ($\phi^{\alpha}_{\text{spin}}$), −1/2 ($\phi^{\beta}_{\text{spin}}$). By substitution of s value, Eq. (5.27) is rewritten as

$$s^2\phi^{\alpha}_{\text{spin}} = \frac{3}{4}\phi^{\alpha}_{\text{spin}} \tag{5.29}$$

$$s^2\phi^{\beta}_{\text{spin}} = \frac{3}{4}\phi^{\beta}_{\text{spin}} \tag{5.30}$$

By substitution of s_z values, Eq. (5.27) is rewritten as

$$s_z\phi^{\alpha}_{\text{spin}} = \frac{1}{2}\phi^{\alpha}_{\text{spin}} \tag{5.31}$$

$$s_z\phi^{\alpha}_{\text{spin}} = -\frac{1}{2}\phi^{\alpha}_{\text{spin}} \tag{5.32}$$

Note that $\phi^{\alpha}_{\text{spin}}$ and $\phi^{\beta}_{\text{spin}}$ are not eigenfunctions of both s_x and s_y operators. Instead, two ladder operators (s_+, s_-) are introduced. By using ladder operators, s^2 operator can be rewritten as

$$s^2 = s_+s_- - s_z + s_z^2 \tag{5.33}$$

Ladder operators satisfy the following equations.

$$s_+\phi^{\alpha}_{\text{spin}} = 0 \tag{5.34}$$

$$s_+\phi^{\beta}_{\text{spin}} = \phi^{\alpha}_{\text{spin}} \tag{5.35}$$

$$s_-\phi^{\alpha}_{\text{spin}} = \phi^{\beta}_{\text{spin}} \tag{5.36}$$

$$s_-\phi^{\beta}_{\text{spin}} = 0 \tag{5.37}$$

5.3.2 Total Spin Angular Momentum

In *n*-electron system, the total spin angular momentum corresponds to the summation of spin angular momentum.

$$\boldsymbol{S} = \sum_{i=1}^{n} \mathbf{s}(i) \tag{5.38}$$

Spin state is characterized by quantum number of total spin angular momentum (S). For example, anti-parallel-spin coupling (α and β spins) when $S = 0$, and parallel-spin coupling (the same spins) when $S = 1$. The $2S + 1$ value stands for spin multiplicity: 1 (singlet), 2 (doublet), 3 (triplet), 4 (quartet), etc.

The z-component of total spin angular momentum corresponds to the summation of z-component of spin angular momentum.

$$\boldsymbol{S}_z = \sum_{i=1}^{n} s_z(i) \tag{5.39}$$

For example, anti-parallel-spin coupling (α and β spins) appears when $S_z = 0$; parallel-spin coupling of α spins appears when $S_z = +1$; parallel-spin coupling of β spins appears when $S_z = -1$.

The ladder operator of total spin angular momentum corresponds to the summation of ladder operators of spin angular momentum.

$$\boldsymbol{S}_+ - \sum_{i=1}^{n} s_+(i) \tag{5.40}$$

$$\boldsymbol{S}_- = \sum_{i=1}^{n} s_-(i) \tag{5.41}$$

$\boldsymbol{S}^2$ operator is given by

$$\boldsymbol{S}^2 = \sum_{i=1}^{n} s^2(i) \sum_{j=1}^{n} s^2(j) \tag{5.42}$$

By using ladder operators, it is rewritten as

$$\boldsymbol{S}^2 = \boldsymbol{S}_+\boldsymbol{S}_- - \boldsymbol{S}_z + \boldsymbol{S}_z^2 \tag{5.43}$$

The wave-function of total spin angular momentum satisfies the following quantum equations.

$$S^2\Phi_{spin} = S(S+1)\Phi_{spin} \tag{5.44}$$

$$S_z\Phi_{spin} = S_z\Phi_{spin} \tag{5.45}$$

5.3.3 Communication Relation

Let us consider spin-orbital interaction. The Hamiltonian of Schrödinger equation ($\boldsymbol{H}$) is expressed without spin coordinates. There are two commutation relations between $\boldsymbol{H}$ and $\boldsymbol{S}^2$ and between $\boldsymbol{H}$ and $\boldsymbol{S}_z$.

$$\boldsymbol{HS}^2 - \boldsymbol{S}^2\boldsymbol{H} = 0 \tag{5.46}$$

$$\boldsymbol{HS}_z - \mathbf{S}_z\boldsymbol{H} = 0 \tag{5.47}$$

Hence, the exact wave-function of Schrödinger equation (Φ) is expected to be also eigenfunction of $\boldsymbol{S}^2$ and $\boldsymbol{S}_z$ operators.

$$\boldsymbol{S}^2\Phi = S(S+1)\Phi \tag{5.48}$$

$$\boldsymbol{S}_z\Phi = S_z\Phi \tag{5.49}$$

5.3.4 Two-Electron System

Let us consider spin-orbital interaction in closed shell two-electron system, where paired α and β electrons are allocated in the same spatial orbital. By using Slater determinant, the wave-function is expressed as

$$\Phi = |\chi_1(x_1)\chi_2(x_2)\rangle = \chi_1(x_1)\chi_2(x_2) - \chi_2(x_1)\chi_1(x_2) \tag{5.50}$$

By substitutions of both spatial orbitals and spin functions,

$$\Phi = \frac{1}{\sqrt{2}}\{\psi_1(r_1)\psi_2(r_2)\alpha(\omega_1)\beta(\omega_2) - \psi_2(r_1)\psi_1(r_2)\beta(\omega_1)\alpha(\omega_2)\} \tag{5.51}$$

$\boldsymbol{S}^2\boldsymbol{\Phi}$ is divided into the three terms regarding $\boldsymbol{S}_+\boldsymbol{S}_-$, $\boldsymbol{S}_z$ and $\boldsymbol{S}_z^2$ terms. $\boldsymbol{S}_+\boldsymbol{S}_-$ term is rewritten as

$$\boldsymbol{S}_+\boldsymbol{S}_-\Phi = \{s_+(\omega_1)s_-(\omega_1) + s_+(\omega_1)s_-(\omega_2) + s_+(\omega_2)s_-(\omega_1) + s_+(\omega_2)s_-(\omega_2)\}\Phi \tag{5.52}$$

In addition, by substitution of Slater determinant,

$$S_+S_-\Phi = \frac{1}{\sqrt{2}}\{\psi_1(r_1)\psi_2(r_2) - \psi_2(r_1)\psi_1(r_2)\}\{\alpha(\omega_1)\beta(\omega_2) + \beta(\omega_1)\alpha(\omega_2)\} \tag{5.53}$$

By using s_z operator, S_z term is rewritten as

$$S_z\Phi = \{s_z(\omega_1) + s_z(\omega_2)\}\Phi \tag{5.54}$$

In addition, by substitution of Slater determinant,

$$\begin{aligned} S_z\Phi = \frac{1}{\sqrt{2}}\bigg\{&\frac{1}{2}\psi_1(r_1)\psi_2(r_2)\alpha(\omega_1)\beta(\omega_2) - \frac{1}{2}\psi_1(r_1)\psi_2(r_2)\alpha(\omega_1)\beta(\omega_2) \\ &+ \frac{1}{2}\psi_2(r_1)\psi_1(r_2)\beta(\omega_1)\alpha(\omega_2) - \frac{1}{2}\psi_2(r_1)\psi_1(r_2)\beta(\omega_1)\alpha(\omega_2)\bigg\} = 0 \end{aligned} \tag{5.55}$$

By using s_z operators, S_z^2 term is rewritten as

$$S_z^2\Phi = \{s_z(\omega_1)s_z(\omega_1) + s_z(\omega_1)s_z(\omega_2) + s_z(\omega_2)s_z(\omega_1) + s_z(\omega_2)s_z(\omega_2)\}\Phi \tag{5.56}$$

In addition, by substituting of Slater determinant,

$$\begin{aligned} S_z^2\Phi = \frac{1}{\sqrt{2}}\bigg\{&\frac{1}{4}\psi_1(r_1)\psi_2(r_2)\alpha(\omega_1)\beta(\omega_2) - \frac{1}{4}\psi_1(r_1)\psi_2(r_2)\alpha(\omega_1)\beta(\omega_2) \\ &- \frac{1}{4}\psi_1(r_1)\psi_2(r_2)\alpha(\omega_1)\beta(\omega_2) + \frac{1}{4}\psi_1(r_1)\psi_2(r_2)\alpha(\omega_1)\beta(\omega_2) \\ &+ \frac{1}{4}\psi_2(r_1)\psi_1(r_2)\beta(\omega_1)\alpha(\omega_2) - \frac{1}{4}\psi_2(r_1)\psi_1(r_2)\beta(\omega_1)\alpha(\omega_2) \\ &- \frac{1}{4}\psi_2(r_1)\psi_1(r_2)\beta(\omega_1)\alpha(\omega_2) + \frac{1}{4}\psi_2(r_1)\psi_1(r_2)\beta(\omega_1)\alpha(\omega_2)\bigg\} = 0 \end{aligned} \tag{5.57}$$

Finally, we obtain

$$\mathbf{S}^2\Phi = \frac{1}{\sqrt{2}}\{\psi_1(r_1)\psi_2(r_2) - \psi_2(r_1)\psi_1(r_2)\}\{\alpha(\omega_1)\beta(\omega_2) + \beta(\omega_1)\alpha(\omega_2)\} \tag{5.58}$$

In closed shell system, paired α and β electrons are allocated in the same spatial orbital. Namely, $\psi_1 = \psi_2$.

$$\mathbf{S}^2\Phi = \frac{1}{\sqrt{2}}\{\psi_1(r_1)\psi_1(r_2) - \psi_1(r_1)\psi_1(r_2)\}\{\alpha(\omega_1)\beta(\omega_2) + \beta(\omega_1)\alpha(\omega_2)\} = 0 \tag{5.59}$$

It is found that Φ is the eigenfunction of $\boldsymbol{S}^2$, and $S(S + 1)$ eigenvalue is zero, corresponding anti-parallel-spin coupling between α and β spins. On the other hand, when different spatial orbitals are given for α and β electrons, $\boldsymbol{\Phi}$ is not the eigenfunction of $\boldsymbol{S}^2$ any longer. It is because $\boldsymbol{\Phi}$ is not eigenfunction of $\boldsymbol{S}_+\boldsymbol{S}_-$, though it is eigenfunction of $\boldsymbol{S}_z$. It implies that spin symmetry is broken by introduction of different spatial orbitals. In fact, α and β electrons are allocated in the same spatial orbital in ground state of neutral helium, hydrogen molecule and lithium cation.

Let us consider spin-orbital interaction in open shell two-electron system, where two electrons have the same α spin. By substitutions of both spatial orbitals and spin function, the total wave-function is rewritten as

$$\Phi = \frac{1}{\sqrt{2}}\{\psi_1(r_1)\psi_2(r_2) - \psi_2(r_1)\psi_1(r_2)\}\alpha(\omega_1)\alpha(\omega_2) \tag{5.60}$$

$\boldsymbol{S}^2\boldsymbol{\Phi}$ is divided into the three terms regarding $\boldsymbol{S}_+\boldsymbol{S}_-$, $\boldsymbol{S}_z$ and $\boldsymbol{S}_z^2$ terms. $\boldsymbol{S}_+\boldsymbol{S}_-$ term is rewritten in the same manner.

$$\boldsymbol{S}_+\boldsymbol{S}_-\Phi = \frac{2}{\sqrt{2}}\{\psi_1(r_1)\psi_2(r_2) - \psi_2(r_1)\psi_1(r_2)\}\alpha(\omega_1)\alpha(\omega_2) = 2\Phi \tag{5.61}$$

$\boldsymbol{S}_z$ term is rewritten in the same manner:

$$\boldsymbol{S}_z\Phi = \frac{1}{\sqrt{2}}\{\psi_1(r_1)\psi_2(r_2) - \psi_2(r_1)\psi_1(r_2)\}\left(\frac{1}{2} + \frac{1}{2}\right)\alpha(\omega_1)\alpha(\omega_2) = \Phi \tag{5.62}$$

It is found that Φ is eigenfunction of $\boldsymbol{S}_z$ operator, and the eigenvalue is 1. $\boldsymbol{S}_z^2$ term is rewritten in the same manner:

$$\boldsymbol{S}_z^2\Phi = \frac{1}{\sqrt{2}}\{\psi_1(r_1)\psi_2(r_2) - \psi_2(r_1)\psi_1(r_2)\}\left(\frac{1}{4} + \frac{1}{4} + \frac{1}{4} + \frac{1}{4}\right)\alpha(\omega_1)\alpha(\omega_2) = \Phi \tag{5.63}$$

Finally, we obtain

$$\boldsymbol{S}^2\Phi = 2\Phi \tag{5.64}$$

It is found that Φ is the eigenfunction of $\boldsymbol{S}^2$, and $S(S + 1)$ eigenvalue is two, corresponding parallel-spin coupling between α spins.

5.3.5 Three-Electron System

Let us consider spin-orbital interaction in three-electron system, where two α and β electrons are paired, and one electron is unpaired. The total wave-function is given by

$$\begin{aligned}\Phi = &\frac{1}{\sqrt{6}}\psi_2(r_1)\{\psi_3(r_2)\psi_1(r_3) - \psi_1(r_2)\psi_3(r_3)\}\beta(\omega_1)\alpha(\omega_2)\alpha(\omega_3)\\ &+ \frac{1}{\sqrt{6}}\psi_2(r_2)\{\psi_1(r_1)\psi_3(r_3) - \psi_3(r_1)\psi_1(r_3)\}\alpha(\omega_1)\beta(\omega_2)\alpha(\omega_3)\\ &+ \frac{1}{\sqrt{6}}\psi_2(r_3)\{\psi_3(r_1)\psi_1(r_2) - \psi_1(r_1)\psi_3(r_2)\}\alpha(\omega_1)\alpha(\omega_2)\beta(\omega_3)\end{aligned} \tag{5.65}$$

$\boldsymbol{S}^2\boldsymbol{\Phi}$ is divided into the three terms regarding $\boldsymbol{S}_+\boldsymbol{S}_-$, $\boldsymbol{S}_z$ and $\boldsymbol{S}_z^2$ terms. $\boldsymbol{S}_+\boldsymbol{S}_-$ term is rewritten in the same manner.

$$\begin{aligned}\boldsymbol{S}_+\boldsymbol{S}_-\boldsymbol{\Phi} = &\frac{2}{\sqrt{6}}\psi_2(r_1)\{\psi_3(r_2)\psi_1(r_3) - \psi_1(r_2)\psi_3(r_3)\}\{\beta(\omega_1)\alpha(\omega_2)\alpha(\omega_3) + \alpha(\omega_1)\beta(\omega_2)\alpha(\omega_3)\}\\ &+ \frac{2}{\sqrt{6}}\psi_2(r_1)\{\psi_3(r_2)\psi_1(r_3) - \psi_1(r_2)\psi_3(r_3)\}\{\beta(\omega_1)\alpha(\omega_2)\alpha(\omega_3) + \alpha(\omega_1)\alpha(\omega_2)\beta(\omega_3)\}\\ &+ \frac{2}{\sqrt{6}}\psi_2(r_2)\{\psi_1(r_1)\psi_3(r_3) - \psi_3(r_1)\psi_1(r_3)\}\{\alpha(\omega_1)\beta(\omega_2)\alpha(\omega_3) + \beta(\omega_1)\alpha(\omega_2)\alpha(\omega_3)\}\\ &+ \frac{2}{\sqrt{6}}\psi_2(r_2)\{\psi_1(r_1)\psi_3(r_3) - \psi_3(r_1)\psi_1(r_3)\}\{\alpha(\omega_1)\beta(\omega_2)\alpha(\omega_3) + \alpha(\omega_1)\alpha(\omega_2)\beta(\omega_3)\}\\ &+ \frac{2}{\sqrt{6}}\psi_2(r_3)\{\psi_3(r_1)\psi_1(r_2) - \psi_1(r_1)\psi_3(r_2)\}\{\alpha(\omega_1)\alpha(\omega_2)\beta(\omega_3) + \beta(\omega_1)\alpha(\omega_2)\alpha(\omega_3)\}\\ &+ \frac{2}{\sqrt{6}}\psi_2(r_3)\{\psi_3(r_1)\psi_1(r_2) - \psi_1(r_1)\psi_3(r_2)\}\{\alpha(\omega_1)\alpha(\omega_2)\beta(\omega_3) + \alpha(\omega_1)\beta(\omega_2)\alpha(\omega_3)\}\end{aligned} \tag{5.66}$$

In general, $\boldsymbol{\Phi}$ is not eigenfunction of $\boldsymbol{S}_+\boldsymbol{S}_-$ operator. However, as α and β electrons occupy the same spatial orbital (ψ_1 is equivalent to ψ_2), $\boldsymbol{S}_+\boldsymbol{S}_-\boldsymbol{\Phi}$ can be rewritten as

$$\begin{aligned}\boldsymbol{S}_+\boldsymbol{S}_-\boldsymbol{\Phi} = &\frac{2}{\sqrt{6}}\psi_1(r_1)\{\psi_3(r_2)\psi_1(r_3) - \psi_1(r_2)\psi_3(r_3)\}\beta(\omega_1)\alpha(\omega_2)\alpha(\omega_3)\\ &+ \frac{2}{\sqrt{6}}\psi_1(r_2)\{\psi_1(r_1)\psi_3(r_3) - \psi_3(r_1)\psi_1(r_3)\}\alpha(\omega_1)\beta(\omega_2)\alpha(\omega_3)\\ &+ \frac{2}{\sqrt{6}}\psi_1(r_3)\{\psi_3(r_1)\psi_1(r_2) - \psi_1(r_1)\psi_3(r_2)\}\alpha(\omega_1)\alpha(\omega_2)\beta(\omega_3) = 2\boldsymbol{\Phi}\end{aligned} \tag{5.67}$$

Regardless of the spatial orbital, S_z term is rewritten in the same manner.

$$\begin{aligned}S_z\Phi = &\left\{\left(-\frac{1}{2}+\frac{1}{2}+\frac{1}{2}\right)+\left(\frac{1}{2}-\frac{1}{2}+\frac{1}{2}\right)+\left(\frac{1}{2}+\frac{1}{2}-\frac{1}{2}\right)\right\}\\ &\left\{\frac{1}{\sqrt{6}}\psi_2(r_1)\{\psi_3(r_2)\psi_1(r_3)-\psi_1(r_2)\psi_3(r_3)\}\beta(\omega_1)\alpha(\omega_2)\alpha(\omega_3)\right.\\ &+\frac{1}{\sqrt{6}}\psi_2(r_2)\{\psi_1(r_1)\psi_3(r_3)-\psi_3(r_1)\psi_1(r_3)\}\alpha(\omega_1)\beta(\omega_2)\alpha(\omega_3)\\ &\left.+\frac{1}{\sqrt{6}}\psi_2(r_3)\{\psi_3(r_1)\psi_1(r_2)-\psi_1(r_1)\psi_3(r_2)\}\alpha(\omega_1)\alpha(\omega_2)\beta(\omega_3)\right\}=\frac{3}{2}\Phi \end{aligned}\quad (5.68)$$

It is found that Φ is eigenfunction of S_z operator, and the eigenvalue is 3/2. S_z^2 term is rewritten in the same manner.

$$S_z^2\Phi = \frac{1}{4}\Phi \quad (5.69)$$

It is found that Φ is eigenfunction of S_z^2 operator, and the eigenvalue is 3/4. Finally, we obtain

$$\mathbf{S}^2\Phi = \frac{3}{4}\Phi \quad (5.70)$$

It is found that Φ is eigenfunction of S^2, and $S(S + 1)$ eigenvalue is 3/4, corresponding $S = 1/2$. It is concluded that spin symmetry is kept when α and β electrons occupy the same spatial orbital. When paired α and β electrons occupy the same spatial orbital, the Hartree-Fock method is called restricted open shell Hartree-Fock (ROHF) method.

In real three-electron system, the independent spatial orbitals (ψ_1^α and ψ_1^β) for paired α and β electrons are obtained. $\mathbf{\Phi}$ is not the eigenfunction of S^2 any longer, because $\mathbf{\Phi}$ is not eigenfunction of S_+S_- operator (see Eq. 5.56).

5.3.6 Summary

In open shell system, the paired MOs and unpaired MOs (spin source) are obtained. Note that "paired" means the qualitatively same. In paired α and β MOs, molecular orbital coefficients are slightly different. The total wave-function is not eigenfunction of S^2 operator, though it is eigenfunctions of S_z and S_z^2 operators. When spin function is defined as isolated electron, eigen equations of spin function are satisfied. However, in general, they are not satisfied without the restriction of spatial orbital.

5.4 Natural Orbital

Natural orbital is completely different from molecular orbital. It is based on pseudo-quantum mechanics. Natural orbital is derived from the introduction of reduced charge density function. In n-electron system, reduced charge density function is given by

$$\rho\left(x_1^{'}|x_1\right) = n\int \Phi^*\Phi \mathrm{d}x_2\mathrm{d}x_3\cdots\mathrm{d}x_n \tag{5.71}$$

where Φ is the total wave-function of n-electron system.

$$\Phi = |\chi_1(x_1)\chi_2(x_2)\cdots\chi_n(x_n)\rangle \tag{5.72}$$

where χ_i is the i-th spin orbital. Equation (5.71) is rewritten as

$$\rho\left(x_1^{'}|x_1\right) = \sum_{i=1}^{n}\sum_{j=1}^{n}\rho_{ij}\chi_i(x_1)\chi_j(x_1) \tag{5.73}$$

where ρ_{ij} is the coefficient. The matrix expression is

$$\begin{pmatrix} \rho_{11}\chi_1(x_1)\chi_1(x_1) & \rho_{12}\chi_1(x_1)\chi_2(x_1) & & \rho_{1n}\chi_1(x_1)\chi_n(x_1) \\ \rho_{21}\chi_2(x_1)\chi_1(x_1) & \rho_{22}\chi_2(x_1)\chi_2(x_1) & \cdots & \rho_{2n}\chi_2(x_1)\chi_n(x_1) \\ \rho_{21}\chi_3(x_1)\chi_1(x_1) & \rho_{32}\chi_3(x_1)\chi_2(x_1) & & \rho_{3n}\chi_3(x_1)\chi_n(x_1) \\ \vdots & & \ddots & \vdots \\ \rho_{n1}\chi_3(x_1)\chi_1(x_1) & \rho_{n2}\chi_n(x_1)\chi_2(x_1) & \cdots & \rho_{nn}\chi_3(x_1)\chi_n(x_1) \end{pmatrix} \tag{5.74}$$

By diagonalizing the matrix, it is rewritten as

$$\begin{pmatrix} \xi_1\eta_1(x_1)\eta_1(x_1) & 0 & & 0 \\ 0 & \xi_2\eta_2(x_1)\eta_2(x_1) & \cdots & 0 \\ 0 & 0 & & 0 \\ \vdots & & \ddots & \vdots \\ 0 & 0 & \cdots & \xi_n\eta_n(x_1)\eta_n(x_1) \end{pmatrix} \tag{5.75}$$

The reduced charge density function is rewritten as

$$\rho\left(x_1^{'}|x_1\right) = \sum_{i=1}^{n}\xi_i\eta_i(x_1)\eta_i(x_1) \tag{5.76}$$

where η_i is the i-th natural orbital; ζ_i is the i-th occupation number. It implies that the reduced charge density function can be expressed by natural orbitals, instead of spin orbitals. By the integration of Eq. (5.71),

$$\int \rho\left(x_1'|x_1\right)\mathrm{d}x_1 = n\int \Phi^*\Phi\mathrm{d}x_1\mathrm{d}x_2\mathrm{d}x_3\cdots\mathrm{d}x_n = n \tag{5.77}$$

By the integration of Eq. (5.76),

$$\sum_{i=1}^{n}\xi_i\int \eta_i(x_1)\eta_i(x_1)\mathrm{d}x_1 = \xi_1+\xi_2+\cdots+\xi_n \tag{5.78}$$

Finally, we have one equation related to occupation numbers:

$$n = \xi_1+\xi_2+\cdots+\xi_n \tag{5.79}$$

It implies that the total of occupation numbers corresponds to the total number of electrons.

Figure 5.2 depicts the schematic drawing of the comparison between molecular orbital and natural orbital. Molecular orbital is the solution of Hartree-Fock equation. The eigenvalue of Hartree-Fock equation corresponds to orbital energy. As Hartree-Fock equation is based on quantum mechanics, discrete orbital energy is reproduced in molecular orbital. On the other hand, natural orbital is derived from the diagonalization of reduced charge density function. In the process, quantum mechanics is partially neglected. For example, natural orbital is not eigenfunction of Hartree-Fock equation.

$$f_i\eta_i(x_1) \neq \xi_i\eta_i(x_1) \tag{5.80}$$

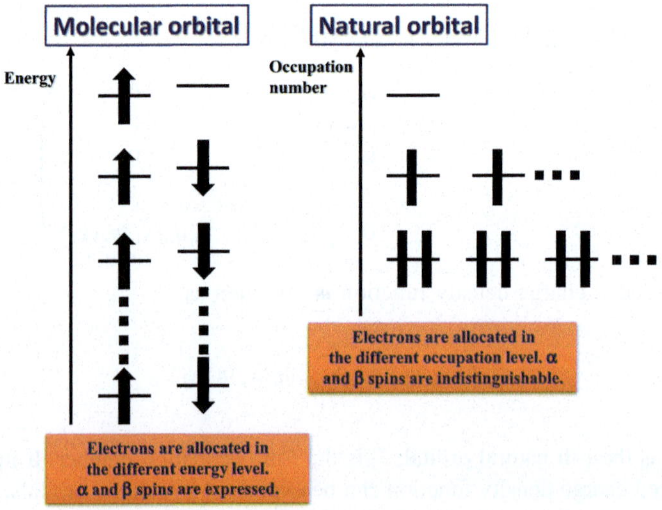

Fig. 5.2 Schematic figure of comparison between molecular orbital and natural orbital

In natural orbital, theoretical concept of the total wave-function are not prepared. In addition, the information of α and β spatial orbitals disappears, and α and β spatial orbitals are mixed. However, natural orbital is sometimes useful, after a deep understanding the serious problems. For example, initial atomic orbitals of spin source are easily characterized, when checking occupation number.

Further Readings

1. Onishi T (2012) Adv Quant Chem 64:43–47
2. Onishi T (2015) Adv Quant Chem 70:35
3. Onishi T (2015) AIP Conf Proc 1702:090002
4. Mulliken RS, Nobel Lecture (1966) Spectroscopy, molecular orbitals, and chemical bonding, 12 December 1966
5. Mulliken RS (1955) J Chem Phys 23(10):1833–1840
6. Mulliken RS (1955) J Chem Phys 23(10):1841–1846
7. Mulliken RS (1955) J Chem Phys 23(12):2338–2342
8. Mulliken RS (1955) J Chem Phys 23(12):2343–2346
9. Szabo A, Ostlund NS (1996) Modern quantum chemistry: introduction to advanced electronic structure theory, Dover Publications Inc., New York, p 97–107, p 149–152, p 252–255
10. de Graaf C, Broer R (2016) Theoretical chemistry and computational modelling magnetic interactions in molecules and solids, Springer, p 5–8
11. Löwdin PO (1955) Phys Rev 97(6):1474–1489
12. Löwdin PO (1955) Phys Rev 97(6):1490–1508
13. Davidson ER (1999) Encyclopedia of computational chemistry. p 1811–1813

Chapter 6
Electron Correlation

Abstract Hartree-Fock method quantitatively reproduces electronic structure. However, electron–electron interaction, which is called electron correlation effect, is theoretically treated in an average manner. For example, Coulomb hole cannot be quantitatively represented, though Fermi hole can be represented. To incorporate electron correlation effect accurately, several calculation methods beyond Hartree-Fock such as configuration interaction (CI), coupled cluster (CC), density functional theory (DFT) have been developed. In CI and CI-based CC methods, it is assumed that the exact wave-function is represented by the combinations of the wave-functions of several excited electron configurations. Though CC method succeeded in reproducing electronic structure of small molecules, CI and CI-based CC methods essentially contain the scientific contradiction that the summation of several Hartree-Fock equations is away from universal quantum concept. DFT has the different concept to incorporate electron correlation effect. The electron correlation effect is directly considered to represent the correct exchange-correlation energy. Though universal exchange-correlation functional has not been developed, DFT predicts correct electronic state in transition metal compounds.

Keywords Fermi hole and Coulomb hole · Electron correlation · Configuration interaction · Coupled cluster · Density functional theory

6.1 Fermi Hole and Coulomb Hole

Electron belongs to Fermi particle. In quantum mechanics, more than two Fermi particles are not allowed to have the same quantum state. Figure 6.1 depicts the schematic drawing of Fermi hole and Coulomb hole. When one α electron exists in the specific spatial orbital, another α electron is not allowed to be allocated in the same spatial orbital. The hole of the spatial orbital is called Fermi hole. On the other hand, two electrons with different spins are allowed to be allocated in the same spatial orbital. However, when Coulomb repulsion between two electrons is much larger, two electrons are not allowed to be allocated in the same spatial orbital.

T. Onishi, *Quantum Computational Chemistry*,
DOI 10.1007/978-981-10-5933-9_6

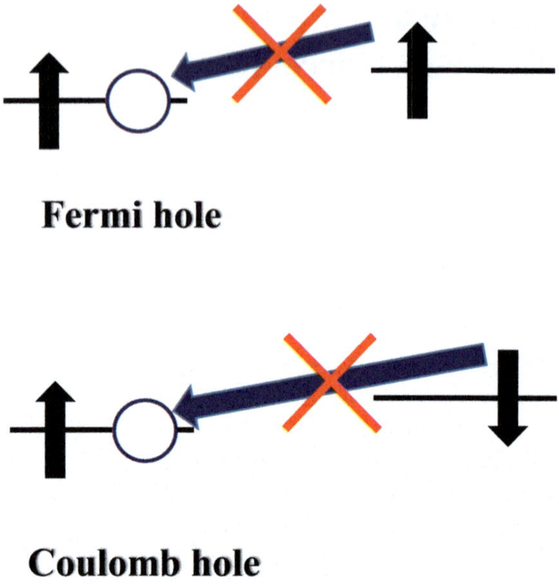

Fig. 6.1 Schematic drawing of Fermi hole and Coulomb hole

The hole is called Coulomb hole. In Hartree-Fock method, though Fermi hole is reproduced, Coulomb hole cannot be quantitatively reproduced. It is because the strength of Coulomb repulsion between two electrons is treated in an average manner.

6.2 Electron Correlation

It is difficult to incorporate accurately electron correlation effect in Hartree-Fock method. In the third term of the Hamiltonian (Eq. 3.2), Coulomb interaction between two electrons is represented in an average manner. In fact, the interaction differs, depending on both shell type and orbital type.

There are two famous theoretical manners to represent electron correlation effect. One is configuration interaction (CI) method, which was proposed from the viewpoint of the correction of Hartree-Fock method. In many-electron system, electron correlation energy (E^{Corr}) is defined as the difference between exact total energy (E^{Exact}) and Hartree-Fock total energy (E^{HF}):

$$E^{\mathrm{Corr}} = E^{\mathrm{Exact}} - E^{\mathrm{HF}} \tag{6.1}$$

In CI method, after finishing Hartree-Fock calculation, correlation energy is estimated as the correction. In general, Hartree-Fock total energy is estimated to be higher than exact total energy.

As is apart from Hartree-Fock method, density functional theory (DFT) was proposed to take electron correlation effect directly into account. Instead of one-electron Hartree-Fock equation, one-electron Kohn–Sham equation is derived by the introduction of electron density to Schrödinger equation. Though electron correlation effect is expressed as exchange-correlation functional, the universal functional has not been developed yet. In fact, they have been determined by several theoretical manners. In present, the best exchange-correlation functional must be selected, depending on considering system.

6.3 Configuration Interaction

In CI method, the theoretical assumption is that the exact wave-function is represented by the combinations of the wave-functions of several excited electron configurations. Let us consider excited electron configurations from Hartree-Fock ground state (see Fig. 6.2). The wave-function of Hartree-Fock ground state is denoted as Ψ(HF).

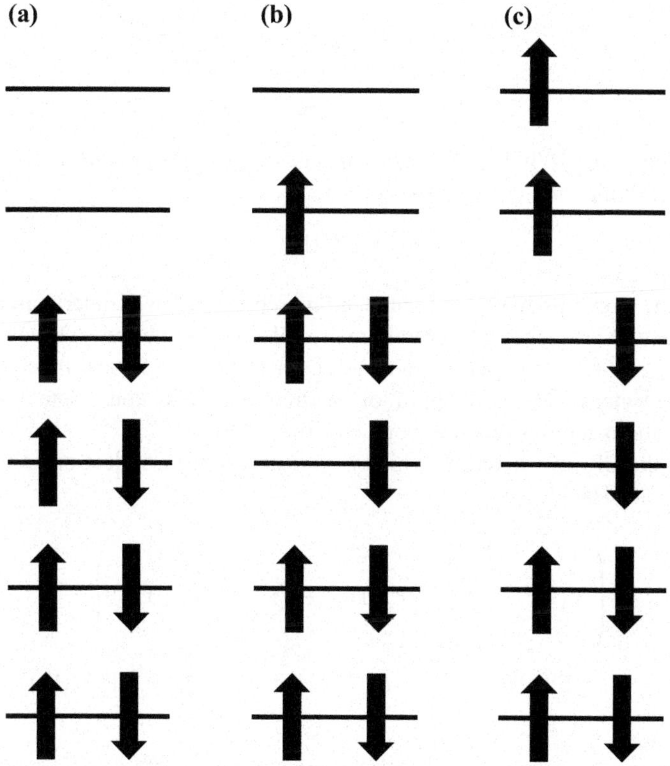

Fig. 6.2 Schematic figure of excited configurations: **a** Hartree-Fock ground state, **b** one-electron excitation, **c** two-electron excitation

When one electron is excited from occupied molecular orbital to unoccupied molecular orbital, one-electron excited configuration appears. For example, the wave-function of one-electron excited configuration is denoted as Ψ_a^p (HF). In the same manner, the wave-function of two-electron excited configuration state is denoted as Ψ_{ab}^{pq} (HF), and the wave-function of multi-electron excited configuration can be defined (see Fig. 6.2). Finally, the full CI wave-function (Ψ(CI)) is given by

$$\Psi(\mathrm{CI}) = \Psi(\mathrm{HF}) + \sum_{a,p} c_a^p \Psi_a^p(\mathrm{HF}) + \sum_{\substack{a<b \\ p<q}} c_{ab}^{pq} \Psi_{ab}^{pq}(\mathrm{HF}) + \cdots \tag{6.2}$$

where c_a^p, c_{ab}^{pq} ... are the coefficients. CI wave-function is assumed to be the exact solution of Hartree-Fock equation.

$$H\Psi(\mathrm{CI}) = E^{\mathrm{CI}}\Psi(\mathrm{CI}) \tag{6.3}$$

where H denotes the Hamiltonian for considering system; E^{CI} denotes CI total energy.

$$\langle\Psi(\mathrm{CI})|\Psi(\mathrm{CI})\rangle = 1 + \sum_{a,p}\left(c_a^p\right)^2 + \sum_{\substack{a<b \\ p<q}}\left(c_{ab}^{pq}\right)^2 + \cdots \tag{6.4}$$

It is found that Ψ(CI) is not normalized, due to the normalization of Ψ(HF). Instead, the following equation is satisfied.

$$\langle\Psi(\mathrm{CI})|\Psi(\mathrm{HF})\rangle = 1 \tag{6.5}$$

In many previous works, CI calculations reproduced well interatomic distance and molecular frequency in several small molecules. However, CI wave-function includes scientific contradiction. In quantum mechanics, one wave-function is given per one electron, as the solution of Hartree-Fock equation. Note that several wave-functions are not given for one electron.

When the CI wave-function, which is truncated until two-electron configuration, operates with Hamiltonian,

$$\begin{aligned} &H\left(\Psi(\mathrm{HF}) + \sum_{a,p} c_a^p \Psi_a^p(\mathrm{HF}) + \sum_{\substack{a<b \\ p<q}} c_{ab}^{pq} \Psi_{ab}^{pq}(\mathrm{HF})\right) \\ &= E\Psi(\mathrm{HF}) + \sum_{a,p} c_a^p E_a^p \Psi_a^p(\mathrm{HF}) + \sum_{a<b} c_{ab}^{pq} E_{ab}^{pq} \Psi_{ab}^{pq}(\mathrm{HF}) \end{aligned} \tag{6.6}$$

The first term of CI wave-function is the solution of Hartree-Fock equation. It satisfies the following equation.

$$H\Psi(\mathrm{HF}) = E\Psi(\mathrm{HF}) \tag{6.7}$$

where E denotes the eigenvalue of Hartree-Fock equation. The second and third terms are also the solution of Hartree-Fock equation.

$$H\Psi_a^p\,(\mathrm{HF}) = E_a^p\,\Psi_a^p\,(\mathrm{HF}) \tag{6.8}$$

$$H\Psi_{ab}^{pq}\,(\mathrm{HF}) = E_{ab}^{pq}\,\Psi_{ab}^{pq}\,(\mathrm{HF}) \tag{6.9}$$

where E_a^p and E_{ab}^{pq} denote total energies of Hartree-Fock equation in one-electron excited and two-electron excited configurations, respectively. It is found that Eq. 6.6 consists of the combination of Eqs. 6.7, 6.8 and 6.9.

In CI method, the further minimization based on variational principle is performed in Eq. 6.6. As the result, CI wave-function pretends to be a solution of one Hartree-Fock equation. However, it cannot be negligible that the scientific contradiction that several Hartree-Fock equations are taken into account at the same time. Quantum mechanics explains that one electron spreads in one spin orbital. The summation of several Hartree-Fock equations is obviously different from universal quantum concept. In CI method, size consistency is not always preserved. The average manner essentially remains in CI method. CI-based calculation predicts wrong electronic structure, especially in transition metal compounds, due to the above problems.

6.4 Coupled Cluster

In coupled cluster (CC) method, the full CI wave function is represented by using cluster operator (T) and Hartree-Fock wave-function (Ψ(HF)). The CC wave-function (Ψ(CC)) is given by

$$\Psi(\mathrm{CC}) = \exp(T)\Psi(\mathrm{HF}) \tag{6.10}$$

Cluster operator is the summation of one-electron excitation operator (T_1), two-electron excitation operator (T_2), …, n-electron excitation operator (T_n).

$$T = T_1 + T_2 + \cdots + T_n \tag{6.11}$$

For example, $T_1\psi$ and $T_2\psi$ satisfy

$$T_1\Psi\,(\mathrm{HF}) = \sum_{a,p} t_a^p\,\Psi_a^p\,(\mathrm{HF}) \tag{6.12}$$

$$T_2\Psi\,(\mathrm{HF}) = \sum_{\substack{a<b\\p<q}} t_{ab}^{pq}\,\Psi_{ab}^{pq}\,(\mathrm{HF}) \tag{6.13}$$

where t_a^p and t_{ab}^{pq} are the coefficients. In coupled cluster singles and doubles (CCSD), as excited electron configuration is truncated until two-electron excitation, T_1 and T_2 are employed. In coupled cluster doubles (CCD), as two-electron excited configuration is only considered, T_2 is employed. By using Taylor expansion, CCD wave-function is written as

$$\Psi(\mathrm{CCD}) = \left(1 + T_2 + \frac{1}{2!}T_2^2 + \frac{1}{3!}T_2^3 + \cdots\right)\Psi(\mathrm{HF}) \tag{6.14}$$

The CCD wave-function is the solution of Hartree-Fock equation.

$$H\Psi(\mathrm{CCD}) = E^{\mathrm{CCD}}\Psi(\mathrm{CCD}) \tag{6.15}$$

By the substitution of Eq. 6.14, it is rewritten as

$$\begin{aligned} &H\left(1 + T_2 + \frac{1}{2!}T_2^2 + \frac{1}{3!}T_2^3 + \ldots\right)\Psi(\mathrm{HF}) \\ &= E^{\mathrm{CCD}}\left(1 + T_2 + \frac{1}{2!}T_2^2 + \frac{1}{3!}T_2^3 + \cdots\right)\Psi(\mathrm{HF}) \end{aligned} \tag{6.16}$$

By multiplying Hartree-Fock wave-function from the left side, the left side is

$$\begin{aligned} &\left\langle \Psi(\mathrm{HF})|H\left(1 + T_2 + \frac{1}{2!}T_2^2 + \frac{1}{3!}T_2^3 + \cdots\right)|\Psi(\mathrm{HF})\right\rangle \\ &= \langle\Psi(\mathrm{HF})|H|\Psi(\mathrm{HF})\rangle + \langle\Psi(\mathrm{HF})|HT_2|\Psi(\mathrm{HF})\rangle \end{aligned} \tag{6.17}$$

It is because there is no coupling between Hartree-Fock ground state and other excited electron configurations, except for two-electron excited configuration. On the other hand, the right side becomes E^{CCD}. It is because the Hartree-Fock wave-function is orthogonal to wave-functions of all excited electron configurations. Finally, we obtain

$$E^{\mathrm{CCD}} = \langle\Psi(\mathrm{HF})|H|\Psi(\mathrm{HF})\rangle + \langle\Psi(\mathrm{HF})|HT_2|\Psi(\mathrm{HF}\rangle) \tag{6.18}$$

The electron correlation energy is represented by the second term. In coupled cluster theory, size consistency is preserved, due to the introduction of exponential. However, it essentially contains the same scientific contradiction as same as configuration interaction. Coupled cluster calculation provides very accurate electronic structure in small molecules. In this book, CCSD method is applied for the species.

6.5 Density Functional Theory

In density functional theory (DFT), Hamiltonian is uniquely represented by using electron density $\rho(\mathrm{r})$.

$$\rho(\mathrm{r}) = \sum_{i=1}^{n} \chi_i^*(x_i)\chi_i(x_i) \tag{6.19}$$

where $\chi_i(x_i)$ is the ith spin orbital in n-electron system. Hartree-Fock equation is rewritten as Kohn–Sham equation.

$$f_i^{\mathrm{KS}}\chi_i^{\mathrm{KS}} = \varepsilon_i^{\mathrm{KS}}\chi_i^{\mathrm{KS}} \tag{6.20}$$

where f_i^{KS} denotes Kohn–Sham operator; $\varepsilon_i^{\mathrm{KS}}$ denotes Kohn–Sham orbital energy; χ_i^{KS} denotes the wave-function of Kohn–Sham molecular orbital. In this book, the Kohn–Sham molecular orbital is called just "molecular orbital".

In Kohn–Sham equation, the kinetic energy is calculated under the assumption of non-interacting electrons, as same as Hartree-Fock equation. The DFT total energy is generally expressed as

$$E^{\mathrm{DFT}}(\rho) = T_{\mathrm{exact}}(\rho) + E_{\mathrm{ne}}(\rho) + J(\rho) + E_{\mathrm{XC}}(\rho) \tag{6.21}$$

where the first and second terms denote the exact kinetic energy and the Coulomb interaction energy between atomic-nucleus and electrons, respectively; the third and fourth terms denote Coulomb interaction energy between electrons, and exchange interaction energy between electrons, respectively. Note that the exact kinetic energy is obtained in non-interacting n-electron system.

It is known that Hartree-Fock method provides about 99% kinetic energy ($T(\rho)$). The energy difference with Hartree-Fock method is incorporated into $E_{\mathrm{XC}}(\rho)$.

$$E_{\mathrm{XC}}(\rho) = \{T(\rho) - T_{\mathrm{exact}}(\rho)\} + \{E_{\mathrm{ee}}(\rho) - J(\rho)\} \tag{6.22}$$

where $E_{\mathrm{ee}}(\rho)$ denotes all-electron interaction energy. If the universal exchange-correlation energy is given, DFT provides the exact solution. However, no universal exchange-correlation energy is defined at present. When basis sets are introduced in Kohn–Sham equation, the problem is converted to obtain the expansion coefficients and orbital energies numerically by SCF calculation.

Previously, many useful functionals of exchange-correlation energy have been developed, by fitting functional to experimental results, physical conditions and so on. In local density approximation (LDA), density is locally treated as uniform electron gas. Local spin density approximation (LSDA) is applied for open shell system.

Table 6.1 Several exchange and correlation functionals of density functional theory

Type	Functional	
Pure exchange	Slater	
	Becke	
Pure correlation	VWN	Vosko–Wilk–Nusair correlation
	LYP	Lee–Yang–Parr correlation
Combination	SVWN	Slater exchange + VWN exchange
	BLYP	Becke exchange + LYP correlation
	PBE[a]	PBE exchange + PBE correlation
Hybrid	BHHLYP	Hartree-Fock exchange + Becke exchange + LYP correlation
	B3LYP	Becke exchange + Slater exchange + Hartree-Fock exchange + LYP correlation + VWN correlation

[a]PBE = Perdew–Burke–Ernzerhof

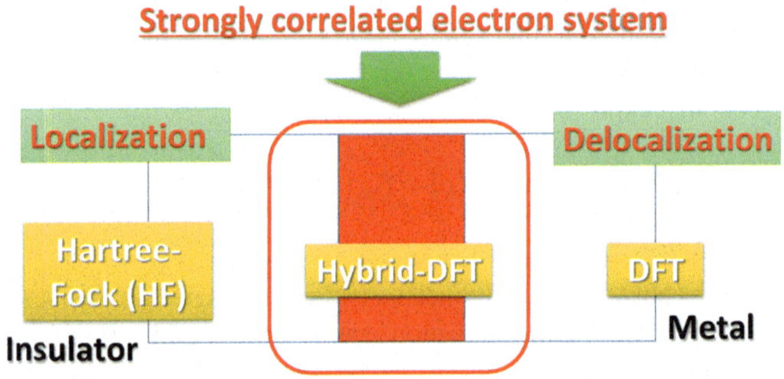

Fig. 6.3 Schematic drawing of hybrid DFT

Vosko, Wilk and Nusair (VWN) is also based on a uniform electron gas. In order to treat as non-uniform electron gas, generalized gradient approximation (GGA) was developed. Becke 1988 exchange functional and Lee-Yang-Parr (LYP) correlation functional were developed based on GGA. Table 6.1 summarizes the several exchange and correlation functionals.

When there is no correlation energy, Kohn–Sham molecular orbitals are identical to Hartree-Fock molecular orbitals. The exact exchange energy is represented as the Hartree-Fock exchange energy. The exchange-correlation functional including the exact (Hartree-Fock) exchange functional is called hybrid functional. For example, let us consider transition metal compounds, which belong to strongly correlated electron system. They contain both localization and delocalized properties. In hybrid DFT, Hartree-Fock exchange functional represents localization property, and exchange and correlation functionals represents delocalized property, as shown in Fig. 6.3. It is well known that hybrid DFT method reproduces well electronic structure in transition metal compounds.

Further Readings

1. Buijse MA (1991) Electron correlation-Fermi and coulomb holes dynamical and nondynamical correlation, VRIJE UNIVERSITEIT
2. Szabo A, Ostlund NS (1996) Modern quantum chemistry: introduction to advanced electronic structure theory. Dover Publications Inc., New York, pp 231–319
3. Löwdin PO (1955) Phys Rev 97(6):1509–1520
4. Sherrill CD, Schaefer HF III (1999) Adv Quant Chem 34:143–269
5. Gauss J (1999) Encyclopedia of computational chemistry, pp 615–636
6. Bartlett RJ, Musiał M (2007) Rev Mod Phys 79:291–352
7. Pople J (1998) Nobel lecture: quantum chemical models
8. Parr RG, Weitao Y (1994) Density-functional theory of atoms and molecules. Oxford University Press
9. Cohen AJ, Mori-Sánchez P, Weitao Y (2012) Chem Rev 112(1):289–320
10. Kohn W (1999) Nobel lecture: electronic structure of matter-wave functions and density functionals
11. Kohn W (1999) Rev Mod Phys 71(5):1253–1266
12. Hohenberg P, Kohn W (1964) Phys Rev 136(3B):B864–B871
13. Kohn W, Sham LJ (1965) Phys Rev 140(4A):A1133–A1138
14. Jensen F (1999) Introduction to computational chemistry. Wiley, pp 177–194
15. Onishi T (2012) Adv Quant Chem 64:43–47

Part II
Atomic Orbital, and Molecular Orbital of Diatomic Molecule

Chapter 7
Atomic Orbital Calculation

Abstract Atomic orbital analysis is based on initial atomic orbital, which is designated by basis set. Orbital hybridization is observed in atomic orbital. Regarding 3d orbital, 6D expression is often utilized. The real $3d_{3z^2-r^2}$ orbital is different from 6D $3d_{z^2}$ orbital. It is represented by the hybridization between 6D $3d_{z^2}$, $3d_{x^2}$ and $3d_{y^2}$ orbitals. Electron configuration rule empirically predicts an atomic electron configuration. Electrons are allocated to realize maximum spin multiplicity in a subshell. In this chapter, coupled cluster calculations are performed for typical atoms. In one-electron system such as neutral hydrogen, helium cation and divalent lithium, the exact solution of Schrödinger equation can be obtained. The calculation results are compared with the exact solution. In many-electron system, there is a flexibility of electronic structure. Different formal charges and different electron configurations are considered. The calculation results of hydrogen, helium, lithium, boron, carbon, nitrogen, oxygen and fluorine are introduced.

Keywords Hybridization · Electron configuration rule · Atomic orbital · Exact solution · Coupled cluster

7.1 Hybridization of Initial Atomic Orbital

When coefficients appear in different initial atomic orbitals (IAOs), orbital is hybridized. Note that IAO is just called "orbital" in this book. Though 2s orbital tends to be hybridized with 1s orbital, the signs of coefficients are normally opposite. It is called inversion hybridization.

In principal, the wave-functions of three 2p orbitals such as $2p_x$, $2p_y$, and $2p_z$ orbitals have the same radial wave-function, but they have the different angular wave-function. In atom, when different 2p orbitals are hybridized, orbital rotation is caused, due to the hybridization of different angular wave-functions. The hybridization between 3p IAOs also causes orbital rotation, due to the same reason.

The 3d orbitals have the five different 3d orbitals such as $3d_{x^2-y^2}$, $3d_{3z^2-r^2}$, $3d_{xy}$, $3d_{yz}$ and $3d_{xz}$ orbitals. The orbital rotation occurs as same as 2p orbital. Though five

T. Onishi, *Quantum Computational Chemistry*,
DOI 10.1007/978-981-10-5933-9_7

different 3d orbitals actually exist, 6D expression is often utilized in molecular orbital calculation. In 6D expression, 3d orbital is represented by six different 3d orbitals such as $3d_{x^2}$, $3d_{y^2}$, $3d_{z^2}$, $3d_{xy}$, $3d_{yz}$, $3d_{xz}$ orbitals. For example, $3d_{x^2-y^2}$ orbital is represented by the hybridization between $3d_{x^2}$ and $3d_{y^2}$ orbitals, and $3d_{3z^2-r^2}$ orbital is represented by the hybridization between $3d_{z^2}$, $3d_{x^2}$ and $3d_{y^2}$ orbitals. Note that $3d_{3z^2-r^2}$ orbital is often denoted as $3d_{z^2}$ orbital in chemistry.

7.2 Electron Configuration Rule

A quantum number of orbital angular momentum (l) designates a subshell type. Electron configurable rule is empirical, but it is useful to predict electron configuration in orbitals with the same l value. In electron configuration rule, electrons are allocated to realize maximum spin multiplicity in the same subshell. It is because electron–electron repulsion energy may be minimized.

In neutral carbon, two electrons exist in 2p orbitals. From electron configuration rule, two electrons are allocated in triplet electron configuration. The notation of electron configuration of neutral carbon is

$$\mathrm{C{:}1s^2 2s^2 2p_x^1 2p_y^1} \tag{7.1}$$

In neutral nitrogen, three electrons exist in 2p orbitals. From electron configuration rule, three electrons are allocated in quartet electron configuration. The notation of electron configuration of neutral nitrogen is

$$\mathrm{N{:}1s^2 2s^2 2p_x^1 2p_y^1 2p_z^1} \tag{7.2}$$

In neutral oxygen, four electrons exist in 2p orbitals. From electron configuration rule, four electrons are allocated in triplet electron configuration. Spin paired 2p orbital is arbitrary,

$$\mathrm{O{:}1s^2 2s^2 2p_x^2 2p_y^1 2p_z^1} \tag{7.3}$$

By using helium electron configuration [He], the brief notation is possible. Equations 7.1, 7.2 and 7.3 are rewritten as

$$\mathrm{C{:}[He]2s^2 2p_x^1 2p_y^1} \tag{7.4}$$

$$\mathrm{N{:}[He]2s^2 2p_x^1 2p_y^1 2p_z^1} \tag{7.5}$$

$$\mathrm{O{:}[He]2s^2 2p_x^2 2p_y^1 2p_z^1} \tag{7.6}$$

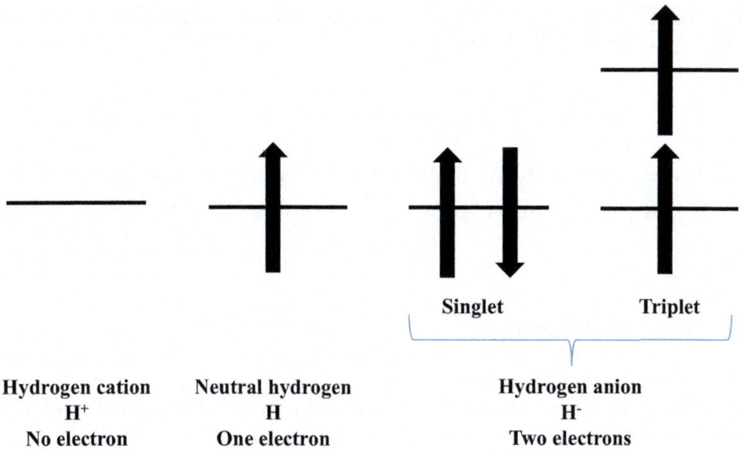

Fig. 7.1 Three electronic structures of hydrogen atom

7.3 Hydrogen Atom

CCSD method, which is based on Hartree-Fock method, has succeeded in accurate calculation in small molecules. Here, CCSD/aug-cc-pVTZ calculation is performed for hydrogen atom. Figure 7.1 depicts three electronic structures of hydrogen atom: (1) hydrogen cation (proton): H^+, (2) neutral hydrogen with doublet spin state: H and (3) hydrogen anion with singlet or triplet spin state: H^-. In H^+ and H, there is no electron–electron interaction. As the special case, in H^+ and H, the calculation results can be compared with the exact solution of Schrödinger equation.

7.3.1 Proton

Table 7.1 summarizes the calculated total energy and orbital energy of hydrogen atom. The total energy becomes zero in proton (H^+). It is because unoccupied

Table 7.1 Calculated total energy and orbital energy of hydrogen atom

	Spin State	Total energy		AO1	AO2
H^+		0.00000		−0.49982	−0.12399
H	Doublet	−0.49982	α	−0.49982	0.05775
			β	0.01560	0.12374
H^-	Singlet	−0.52656		−0.04571	0.24205
	Triplet	−0.44283	α	−0.32716	0.05772
			β	0.17363	0.28772

*Energy is shown in au

atomic orbital (AO) is only given, due to no existence of electron. The obtained wave-function of unoccupied AO1 is

$$\psi_{\mathrm{AO1}}(\mathrm{H}^+) = 0.24\phi_{\mathrm{H(1s')}} + 0.51\phi_{\mathrm{H(1s'')}} + 0.38\phi_{\mathrm{H(1s''')}} \tag{7.7}$$

1s orbital is represented by three Gaussian basis functions: $\phi_{\mathrm{H(1s')}}$, $\phi_{\mathrm{H(1s'')}}$ and $\phi_{\mathrm{H(1s''')}}$. It is found that unoccupied AO1 consists of only 1s orbital.

7.3.2 *Neutral Hydrogen*

Neutral hydrogen (H) is open shell system with doublet electron configuration. One electron is occupied in AO1α. The obtained wave-function of AO1α is

$$\psi_{\mathrm{AO1\alpha}}(\mathrm{H}) = 0.24\phi_{\mathrm{H(1s')}} + 0.51\phi_{\mathrm{H(1s'')}} + 0.38\phi_{\mathrm{H(1s''')}} \tag{7.8}$$

As the wave-function of AO1α corresponds to the total wave-function of H, the orbital energy of AO1α corresponds to the total energy. It is found that the calculated total energy (−0.49982 au) reproduces well the exact total energy (−0.5 au).

In neutral hydrogen, which belongs to one-electron system, Coulomb and kinetic integrals are not defined. The orbital energy is given by

$$\varepsilon_{\mathrm{AO1\alpha}}(\mathrm{H}) = \langle \Psi_{\mathrm{AO1\alpha}} | h_1 | \Psi_{\mathrm{AO1\alpha}} \rangle \tag{7.9}$$

where $\Psi_{\mathrm{AO1\alpha}}$ denotes the wave-function of atomic orbital; h_1 is one-electron operator (see Eq. 3.8). Note that one electron occupies AO1α with doublet electron configuration. On the other hand, in proton, the orbital energy of unoccupied AO1 is defined by the allocation of one electron in unoccupied AO1 virtually. It is given by

$$\varepsilon_{\mathrm{AO1}}(\mathrm{H}^+) = \langle \Psi_{\mathrm{AO1}} | h_1 | \Psi_{\mathrm{AO1}} \rangle \tag{7.10}$$

As $\psi_{\mathrm{AO1\alpha}}(H)$ is equivalent to $\psi_{\mathrm{AO1}}(\mathrm{H}^+)$, it is found that $\varepsilon_{\mathrm{AO1\alpha}}(\mathrm{H})$ corresponds to $\varepsilon_{\mathrm{AO1}}(\mathrm{H}^+)$. The exact orbital energy of AO2 is estimated to be −0.25 from Eq. 2.17. On the other hand, the calculated orbital energies of unoccupied AO2α and AO2β are 0.05775 and 0.12374 au, respectively. It is found that unoccupied AOs do not correspond to the exact AO.

7.3.3 *Hydrogen Anion*

Two electron configurations are considered in hydrogen anion. One is singlet electron configuration where two electrons are allocated in the same AO1.

The other is triplet electron configuration, where two electrons are allocated in two different AO1α and AO2α.

In singlet electron configuration, two electrons occupy AO1. The obtained wave-function of occupied AO1 is

$$\psi_{\mathrm{AO1}}(\mathrm{H}^-) = 0.16\phi_{\mathrm{H(1s')}} + 0.27\phi_{\mathrm{H(1s'')}} + 0.41\phi_{\mathrm{H(1s''')}} + 0.37\phi_{\mathrm{H(2s)}} \tag{7.11}$$

Hybridization occurs between 1s and 2s orbitals. It is found that AO1(H^-) is different from AO1α(H). The calculated orbital energy ($\varepsilon_{\mathrm{AO1}}(\mathrm{H}^-)$) is −0.04571 au. From orbital energy rule, it is found that AO1 electrons can be removed with much smaller energy, in comparison with neutral hydrogen.

In triplet electron configuration, two electrons are allocated in different alpha AO1α and AO2α, though AO1β unoccupied. The obtained wave-functions of occupied AO1α and AO2α are

$$\psi_{\mathrm{AO1}\alpha}(\mathrm{H}^-) = 0.24\phi_{\mathrm{H(1s')}} + 0.51\phi_{\mathrm{H(1s'')}} + 0.37\phi_{\mathrm{H(1s''')}} \tag{7.12}$$

$$\psi_{\mathrm{AO2}\alpha}(\mathrm{H}^-) = -0.71\phi_{\mathrm{H(1s''')}} + 1.36\phi_{\mathrm{H(2s)}} \tag{7.13}$$

Though AO1α consists of 1s orbital, inversion hybridization occurs between 1s and 2s orbitals in AO2α. The calculated orbital energies of AO1α and AO2α are −0.32716 and 0.05772 au, respectively. From orbital energy rule, it is considered that the electron of AO2α is easily removed, due to positive value.

The total energy of singlet electron configuration (−0.52656 au) is smaller than triplet electron configuration (−0.44283 au). The energy difference is 0.084478 au (2.30 eV). Provided an energy to singlet electron configuration by external field, triplet electron configuration could be realized. However, electron of AO2α is not stabilized.

One may think that electron is coercively moved from AO1 to AO2, keeping the electronic structure of singlet electron configuration, as shown in Fig. 7.2. We call it "virtual excitation". This idea is similar to frontier orbital theory. It is explained that the excitation reaction occurs through electron transfer from highest occupied AO (AO1) to lowest unoccupied AO (AO2). In fact, the wave-functions and orbital energies in singlet electron configuration are different from triplet electron configuration. We must pay attention to adapt the concept of virtual excitation.

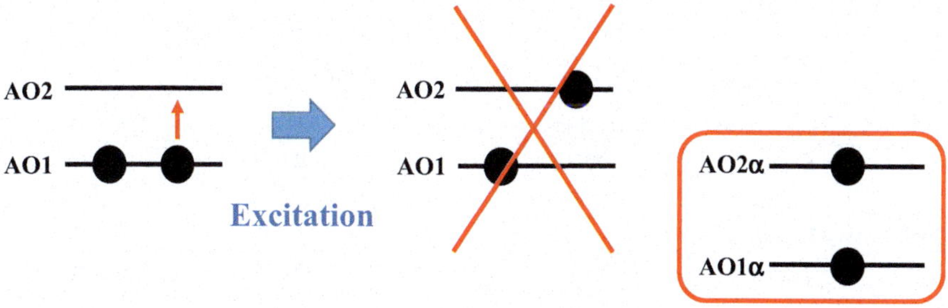

Fig. 7.2 Schematic drawing of electron excitation in hydrogen anion

7.4 Helium Atom

Helium exists as colourless, odourless and inert gas. Helium is the second lightest and abundant element in the universe. Figure 7.3 depicts four electronic structures that is considered for helium atom: (1) singlet neutral helium, (2) triplet neutral helium, (3) doublet helium cation (He^+) and (4) doublet helium anion (He^-). CCSD/aug-cc-pVTZ calculation is performed for them. As the special case, the Schrödinger equation of He^+ can be analytically solved. The calculation results are compared with the exact solution.

7.4.1 Neutral Helium

Table 7.2 summarizes the calculated total energy and orbital energy for helium atom. Neutral helium has singlet electron configuration, or triplet electron configuration, respectively. The total energy of singlet electron configuration is 0.73 au

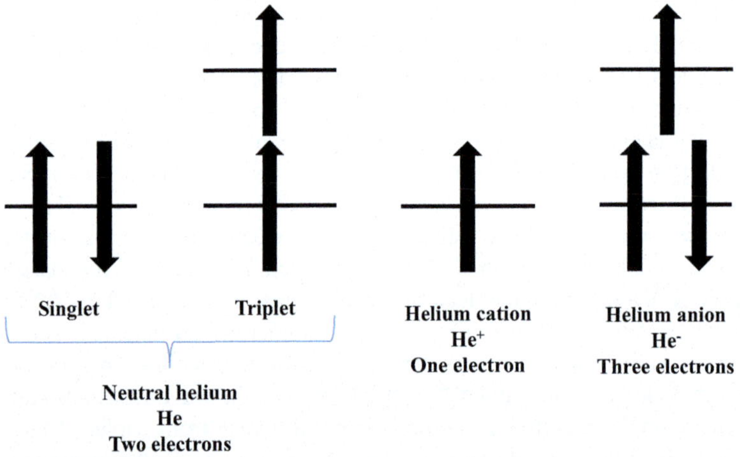

Fig. 7.3 Four electronic structures of helium atom

Table 7.2 Calculated total energy and orbital energy of helium atom

	Spin State	Total energy		AO1	AO2
He	Singlet	−2.90060		−0.91787	
	Triplet	−2.16989	α	−1.70884	−0.16997
He^+	Doublet	−1.99892	α	−1.99892	
He^-	Triplet	−2.79199	α	−0.67342	0.11030
			β	−0.64837	

*Energy is shown in au

lower than triplet electron configuration. It is found that the ground state of neutral helium is singlet, and singlet-triplet excitation is caused. In singlet electron configuration, two electrons occupy AO1. The obtained wave-function of AO1 is

$$\psi_{\mathrm{AO1}}(\mathrm{He}) = 0.35\phi_{\mathrm{He}(1s')} + 0.48\phi_{\mathrm{He}(1s'')} + 0.30\phi_{\mathrm{He}(1s''')} \tag{7.14}$$

AO1 consists of only 1s orbital. On the other hand, in triplet electron configuration, two electrons occupy AO1α and AO2α. The obtained wave-functions of AO1α and AO2α are

$$\psi_{\mathrm{AO1}\alpha}(\mathrm{He}) = 0.46\phi_{\mathrm{He}(1s')} + 0.56\phi_{\mathrm{He}(1s'')} \tag{7.15}$$

$$\psi_{\mathrm{AO2}\alpha}(\mathrm{He}) = -0.10\phi_{\mathrm{He}(1s')} - 0.19\phi_{\mathrm{He}(1s'')} - 0.16\phi_{\mathrm{He}(1s''')} + 1.14\phi_{\mathrm{He}(2s)} \tag{7.16}$$

AO1α consists of only 1s orbital. In AO2α, there is inversion hybridization between 1s and 2s orbitals. It implies that the electron is delocalized over 1s and 2s IAOs. The orbital energy of AO1α is 0.79097 au lower than occupied AO1 of singlet electron configuration. It is due to the difference of electron repulsion. In fact, the total electron–electron repulsion energies of singlet and triplet electron configurations are 1.02545 and 0.29008 au, respectively. Instead, occupied AO2α is destabilized.

7.4.2 Helium Cation

Helium cation is open shell system with doublet electron configuration (see Fig. 7.3). One electron occupies AO1α. The obtained wave-function of AO1α is

$$\psi_{\mathrm{AO1}\alpha}(\mathrm{He}^{+}) = 0.46\phi_{\mathrm{He}(1s')} + 0.56\phi_{\mathrm{He}(1s'')} \tag{7.17}$$

AO1α consists of only 1s orbital. The calculated total energy (−1.9989 au) reproduces well the exact total energy (−2.0 au).

7.4.3 Helium Anion

Helium anion is open shell system with doublet electron configuration (see Fig. 7.3). Though it has more electron than neutral helium, the total energy is higher than singlet neutral helium. It implies that helium anion is destabilized. The obtained wave-functions of AO1α and AO1β are

$$\psi_{AO1\alpha}(He^-) = 0.36\phi_{He(1s')} + 0.48\phi_{He(1s'')} + 0.29\phi_{He(1s''')} \quad (7.18)$$

$$\psi_{AO1\beta}(He^-) = 0.35\phi_{He(1s')} + 0.47\phi_{He(1s'')} + 0.31\phi_{He(1s''')} \quad (7.19)$$

They consist of only 1s orbital. Though two wave-functions are qualitatively the same, the different orbital energies are given. The orbital energies of AO1α and AO1β are −0.67342 and −0.64837 au, respectively. They are larger than occupied AO1 of singlet neutral helium. It implies that AO1α and AO1β are destabilized. The obtained wave-functions of AO2α is

$$\psi_{AO2\alpha}(He^-) = -0.64\phi_{He(1s''')} + 1.34\phi_{He(2s)} \quad (7.20)$$

There is inversion hybridization between 1s and 2s orbitals. It implies that electron is delocalized over 1s and 2s IAOs. The orbital energy of AO2α is positive. From orbital energy rule, it is considered that the electron of AO2α is easily removed.

7.5 Lithium Atom

Lithium is categorized as alkali metal. As it is the lightest metal under normal condition, it has been widely used for lithium ion battery, where lithium cation (Li^+) migrates as conductive ion. Figure 7.4 depicts four electronic structures of lithium atom: (1) divalent lithium cation (Li^{+2}), (2) monovalent lithium cation (Li^+), (3) neutral lithium (Li), (4) lithium anion (Li^-). CCSD/aug-cc-pVTZ calculation is performed for them. In aug-cc-pVTZ basis sets, 1s orbital is represented by one contracted basis function, and valence 2s orbital is represented by one contracted basis function and two basis functions. 3s orbital is represented by one basis function.

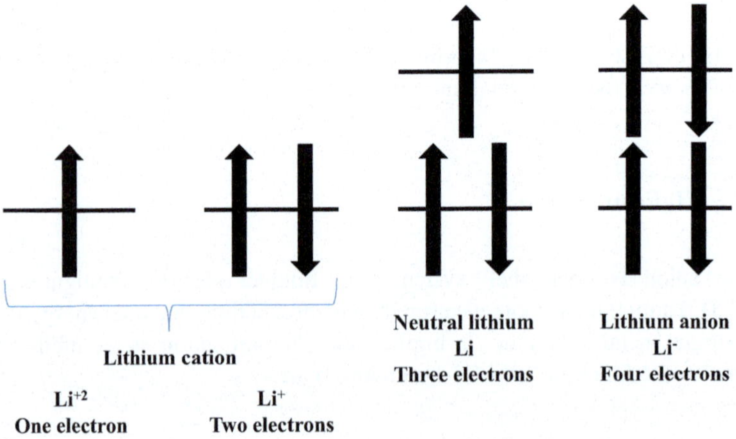

Fig. 7.4 Electronic structures of lithium atom

Table 7.3 Calculated total energy and orbital energy of lithium atom

	Spin State	Total energy		AO1	AO2
Li^{+2}	Doublet	−4.49888	α	−4.49888	
Li^{+}	Singlet	−7.23638		−2.79236	
Li	Doublet	−7.43271	α	−2.48668	−0.19636
			β	−2.46883	
Li^{-}	Singlet	−7.45528		−2.32252	−0.01432

7.5.1 Divalent Lithium Cation

Table 7.3 summarizes the calculated total energy and orbital energy of lithium atom. In divalent lithium cation, the exact solution of Schrödinger equation is given, as same as neutral hydrogen. From the equation of the exact total energy (Eq. 2.17), the exact total energy of lithium cation is nine times ($Z^2 = 3^2$) larger than neutral hydrogen (−0.5 au). The calculated total energy of Li^{+2} (−4.49888 au) reproduces well the exact total energy (−4.5 au). The obtained wave-function of AO1α is

$$\psi_{\mathrm{AO1}\alpha}(\mathrm{Li}^{+2}) = 0.92\phi_{\mathrm{Li(1s)}} - 0.10\phi_{\mathrm{Li(2s')}} \tag{7.21}$$

There is inversion hybridization between 1s and 2s orbitals. The main coefficient is for 1s orbital. It implies that electron is delocalized over 1s and 2s orbitals.

7.5.2 Monovalent Lithium Cation

Monovalent lithium is closed shell system. Two electrons occupy AO1. The obtained wave-function of AO1 is expressed as

$$\psi_{\mathrm{AO1}}(\mathrm{Li}^{+}) = 0.76\phi_{\mathrm{Li(1s)}} - 0.28\phi_{\mathrm{Li(2s')}} \tag{7.22}$$

There is inversion hybridization between 1s and 2s orbitals. The main coefficient is for 1s orbital. It implies that electron is delocalized over 1s and 2s orbitals. In comparison with divalent lithium cation, though electron–electron repulsion between two electrons exists, the total energy is smaller. It is found that monovalent lithium cation is more stabilized.

7.5.3 Neutral Lithium

Neutral lithium with three electrons is open shell system with doublet electron configuration (see Fig. 7.4). The obtained wave-functions of AO1α and AO1β are

$$\psi_{\mathrm{AO1\alpha}}(\mathrm{Li}) = 0.76\phi_{\mathrm{Li(1s)}} - 0.28\phi_{\mathrm{Li(2s')}} \quad (7.23)$$

$$\psi_{\mathrm{AO1\beta}}(\mathrm{Li}) = 0.76\phi_{\mathrm{Li(1s)}} - 0.29\phi_{\mathrm{Li(2s')}} \quad (7.24)$$

The wave-functions of AO1α and AO1β are qualitatively the same. AO1α and AO1β are paired. Though inversion hybridization occurs between 1s and 2s orbitals, the main component is for 1s orbital. The orbital energies of AO1α and AO1β are much smaller than AO2α. It implies that 1s orbital exists in inner shell. The obtained wave-functions of AO2α is

$$\psi_{\mathrm{AO2\alpha}}(\mathrm{Li}) = -0.12\phi_{\mathrm{Li(1s)}} + 0.17\phi_{\mathrm{Li(2s')}} + 0.57\phi_{\mathrm{Li(2s'')}} + 0.52\phi_{\mathrm{Li(2s''')}} \quad (7.25)$$

In AO2α, there is also inversion hybridization between 1s and 2s orbitals. The main components are for 2s orbital. As the orbital energy of AO2α is larger than AO1α and AO1β, it is considered that 2s electron is more reactive.

7.5.4 Lithium Anion

Lithium anion with four electrons is closed shell system (see Fig. 7.4). Four electrons occupy AO1 and AO2. The obtained wave-functions of AO1 and AO2 are

$$\psi_{\mathrm{AO1}}(\mathrm{Li}^-) = 0.76\phi_{\mathrm{Li(1s)}} - 0.28\phi_{\mathrm{Li(2s')}} \quad (7.26)$$

$$\psi_{\mathrm{AO2}}(\mathrm{Li}^-) = 0.11\phi_{\mathrm{Li(2s')}} + 0.25\phi_{\mathrm{Li(2s'')}} + 0.45\phi_{\mathrm{Li(2s''')}} + 0.45\phi_{\mathrm{Li(3s)}} \quad (7.27)$$

In AO1, inversion hybridization occurs between 1s and 2s orbitals. In addition, in AO2, 2s and 3s orbitals are hybridized. It is found that electrons spread from 1s, 2s and 3s orbitals. As the orbital energy of AO2 is close to zero, it is considered that 2s electron is more reactive.

7.6 Boron Atom

It is known that boron atom forms covalent bonding with other atoms. In neutral boron, two electrons occupy K shell, and three electrons occupy L shell. Figure 7.5 depicts the electronic structures of neutral boron. Possible two electron configurations are considered: (1) doublet electron configuration, (2) quartet electron configuration. Note that electronic structure of L shell is only shown. CCSD/aug-cc-pVTZ calculation is performed for them.

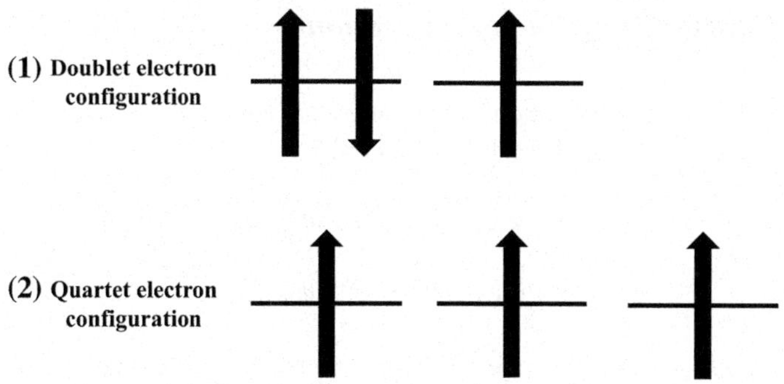

Fig. 7.5 Electronic structures of neutral boron: *1* doublet electron configuration, *2* quartet electron configuration. Electrons of L shell are only shown

7.6.1 Doublet Electron Configuration

In doublet electron configuration, three alpha and two beta AOs are occupied. The obtained wave-functions of AO1α and AO1β are

$$\psi_{\mathrm{AO1\alpha}}(\mathrm{B}) = 0.98\phi_{\mathrm{B(1s)}} \tag{7.28}$$

$$\psi_{\mathrm{AO1\beta}}(\mathrm{B}) = 0.98\phi_{\mathrm{B(1s)}} \tag{7.29}$$

AO1α and AO1β are paired. They represent 1s orbital. The obtained wave-functions of AO2α and AO2β are

$$\psi_{\mathrm{AO2\alpha}}(\mathrm{B}) = -0.20\phi_{\mathrm{B(1s)}} + 0.57\phi_{\mathrm{B(2s')}} + 0.12\phi_{\mathrm{B(2s'')}} + 0.39\phi_{\mathrm{B(2s''')}} \tag{7.30}$$

$$\psi_{\mathrm{AO2\beta}}(\mathrm{B}) = -0.19\phi_{\mathrm{B(1s)}} + 0.54\phi_{\mathrm{B(2s')}} + 0.11\phi_{\mathrm{B(2s'')}} + 0.42\phi_{\mathrm{B(2s''')}} \tag{7.31}$$

They are qualitatively the same, though coefficients are slightly different. There is inversion hybridization between 1s and 2s orbitals. The main component is for 2s orbital. The obtained wave-function of AO3α is

$$\psi_{\mathrm{AO3\alpha}}(\mathrm{B}) = 0.34\phi_{\mathrm{B(2p'_z)}} + 0.51\phi_{\mathrm{B(2p''_z)}} + 0.34\phi_{\mathrm{B(2p'''_z)}} \tag{7.32}$$

AO3α has no paired AO and is responsible for spin density. It consists of only $2p_z$ orbital.

7.6.2 Quartet Electron Configuration

In quartet electron configuration, four alpha and one beta AOs are occupied. The obtained wave-functions of AO1α and AO1β are

$$\psi_{AO1\alpha}(B) = 0.98\phi_{B(1s)} \tag{7.33}$$

$$\psi_{AO1\beta}(B) = 0.98\phi_{B(1s)} \tag{7.34}$$

AO1α and AO1β are paired. They represent 1s orbital. The obtained wave-functions of AO2α, AO3α and AO4α are

$$\psi_{AO2\alpha}(B) = -0.20\phi_{B(1s)} + 0.61\phi_{B(2s')} + 0.13\phi_{B(2s'')} + 0.35\phi_{B(2s''')} \tag{7.35}$$

$$\psi_{AO3\alpha}(B) = 0.35\phi_{B(2p_z')} + 0.54\phi_{B(2p_z'')} + 0.30\phi_{B(2p_z''')} \tag{7.36}$$

$$\psi_{AO4\alpha}(B) = 0.35\phi_{B(2p_y')} + 0.54\phi_{B(2p_y'')} + 0.30\phi_{B(2p_y''')} \tag{7.37}$$

In AO2α, inversion hybridization occurs between 1s and 2s orbitals. The main components are for 2s orbital. The figures of AO3α and AO4α are the same, though the directions are different. The wave-functions of AO3α and AO4α are along z direction and y direction, respectively. As they have the same orbital energy (−0.35638 au), it is found that they are degenerated. In neutral boron, the total energies for doublet and quartet electron configurations are −24.53217 and −24.45136 au, respectively. The doublet electron configuration is more stable, corresponding to building-up principle.

7.7 Carbon Atom

Carbon exhibits strong covalency. Two electrons occupy K shell, and four electrons occupy L shell. Figure 7.6 depicts the electronic structures of neutral carbon. Possible three electron configurations are considered: (1) singlet electron configuration, (2) triplet electron configuration, (3) quintet electron configuration. CCSD/aug-cc-pVTZ calculation is performed for them.

7.7.1 Singlet Electron Configuration

Neutral carbon with singlet electron configuration is closed shell system. Three alpha and three beta AOs are occupied. The obtained wave-function of AO1 is

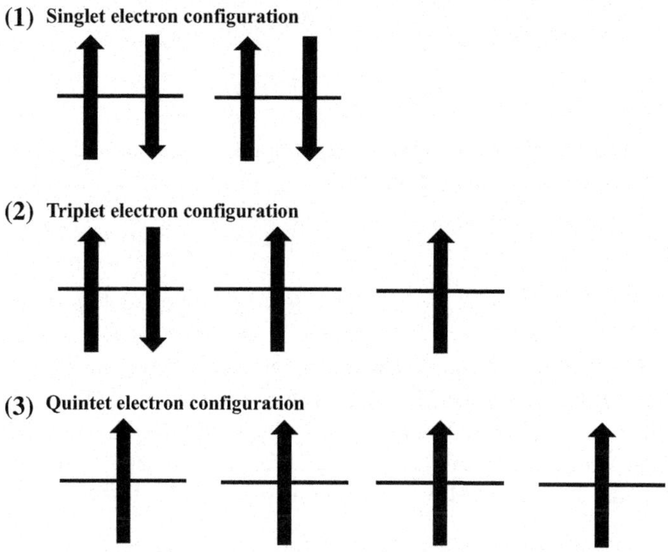

Fig. 7.6 Electronic structures of neutral carbon: *1* singlet electron configuration, *2* triplet electron configuration, *3* quintet electron configuration. Electrons of L shell are only shown

$$\psi_{\mathrm{AO1}}(\mathrm{C}) = 0.98\phi_{\mathrm{C(1s)}} \tag{7.38}$$

AO1 consists of only 1s orbital. The obtained wave-functions of AO2 and AO3 are

$$\psi_{\mathrm{AO2}}(\mathrm{C}) = -0.21\phi_{\mathrm{C(1s)}} + 0.57\phi_{\mathrm{C(2s')}} + 0.15\phi_{\mathrm{C(2s'')}} + 0.38\phi_{\mathrm{C(2s''')}} \tag{7.39}$$

$$\psi_{\mathrm{AO3}}(\mathrm{C}) = 0.35\phi_{\mathrm{C(2p_z')}} + 0.47\phi_{\mathrm{C(2p_z'')}} + 0.37\phi_{\mathrm{C(2p_z''')}} \tag{7.40}$$

AO2 represents 2s orbital. In AO2, inversion hybridization occurs between 1s and 2s orbitals. The main components are for 2s orbital. On the other hand, AO3 consists of only $2p_z$ orbital. The orbital energies of AO2 and AO3 are −0.72600 and −0.35825 au, respectively. $2p_z$ orbital is more reactive.

7.7.2 Triplet Electron Configuration

In triplet electron configuration, four alpha and two beta AOs are occupied. The obtained wave-functions of AO1α and AO1β are

$$\psi_{AO1\alpha}(C) = 0.98\phi_{C(1s)} \tag{7.41}$$

$$\psi_{AO1\beta}(C) = 0.98\phi_{C(1s)} \tag{7.42}$$

AO1α and AO1β are paired. They represent 1s orbital. The obtained wave-functions of AO2α and AO2β are

$$\psi_{AO2\alpha}(C) = -0.21\phi_{C(1s)} + 0.59\phi_{C(2s')} + 0.15\phi_{C(2s'')} + 0.36\phi_{C(2s''')} \tag{7.43}$$

$$\psi_{AO2\beta}(C) = -0.20\phi_{C(1s)} + 0.52\phi_{C(2s')} + 0.13\phi_{C(2s'')} + 0.42\phi_{C(2s''')} \tag{7.44}$$

In AO2α and AO2β, inversion hybridization occurs between 1s and 2s orbitals. The main components are for 2s orbital. The coefficients of AO2α and AO2β are slightly different, and orbital energies of AO2α and AO2β are −0.82958 and −0.58414 au, respectively. AO2α and AO2β are paired, due to qualitative same wave-functions. The obtained wave-functions of AO3α and AO4α are

$$\psi_{AO3\alpha}(C) = 0.36\phi_{C(2p'_y)} + 0.51\phi_{C(2p''_y)} + 0.33\phi_{C(2p'''_y)} \tag{7.45}$$

$$\psi_{AO4\alpha}(C) = 0.36\phi_{C(2p'_x)} + 0.51\phi_{C(2p''_x)} + 0.33\phi_{C(2p'''_x)} \tag{7.46}$$

The figures of AO3α and AO4α are the same, though the directions are different. The wave-functions of AO3α and AO4α are along y direction and x direction, respectively. As they have the same orbital energy (−0.43882 au), it is found that they are degenerated.

7.7.3 *Quintet Electron Configuration*

In quintet electron configuration, five alpha and one beta AOs are occupied. The obtained wave-functions of AO1α and AO1β are

$$\psi_{AO1\alpha}(C) = 0.98\phi_{C(1s)} \tag{7.47}$$

$$\psi_{AO1\beta}(C) = 0.98\phi_{C(1s)} \tag{7.48}$$

AO1α and AO1β are paired. They represent 1s orbital. The obtained wave-function of AO2α is

$$\psi_{AO2\alpha}(C) = -0.22\phi_{C(1s)} + 0.61\phi_{C(2s')} + 0.15\phi_{C(2s'')} + 0.34\phi_{C(2s''')} \tag{7.49}$$

In AO2α, inversion hybridization occurs between 1s and 2s orbitals. The main components are for 2s orbitals. The obtained wave-functions of AO3α, AO4α and AO5α are

$$\psi_{AO3\alpha}(C) = 0.36\phi_{C(2p'_z)} + 0.53\phi_{C(2p''_z)} + 0.30\phi_{C(2p'''_z)} \quad (7.50)$$

$$\psi_{AO4\alpha}(C) = 0.36\phi_{C(2p'_x)} + 0.53\phi_{C(2p''_x)} + 0.30\phi_{C(2p'''_x)} \quad (7.51)$$

$$\psi_{AO5\alpha}(C) = 0.36\phi_{C(2p'_y)} + 0.53\phi_{C(2p''_y)} + 0.30\phi_{C(2p'''_y)} \quad (7.52)$$

The figures of AO3α, AO4α and AO5α are the same, though the directions are different. The wave-functions of AO3α, AO4α and AO5α are along z, x and y directions, respectively. As they have the same orbital energy (−0.47897 au), it is found that they are degenerated. In neutral carbon, the total energies for singlet, triplet and quintet electron configurations are −37.60305, −37.69181 and −37.59680 au, respectively. It is found that quintet electron configuration is destabilized by the formation of unpaired 2s AO, corresponding to building-up principle. In comparison with singlet electron configuration, the stabilization of triplet electron configuration follows electron configuration rule.

7.8 Nitrogen Atom

Figure 7.7 depicts the electronic structures of nitrogen atom. Neutral nitrogen has five electrons in L shell. Two electron configurations are considered: (1) doublet electron configuration; (2) quintet electron configuration. In solids, the formal charge of nitrogen atom is often −3. Trivalent nitrogen anion is closed shell system. CCSD/aug-cc-pVTZ calculation is performed for them.

7.8.1 *Doublet Neutral Nitrogen*

In doublet neural nitrogen, four alpha three beta AOs are occupied. The obtained wave-functions of AO1α and AO1β are

$$\psi_{AO1\alpha}(N) = 0.98\phi_{N(1s)} \quad (7.53)$$

$$\psi_{AO1\beta}(N) = 0.98\phi_{N(1s)} \quad (7.54)$$

AO1α and AO1β are paired. They represent 1s orbital. The obtained wave-functions of AO2α and AO2β are

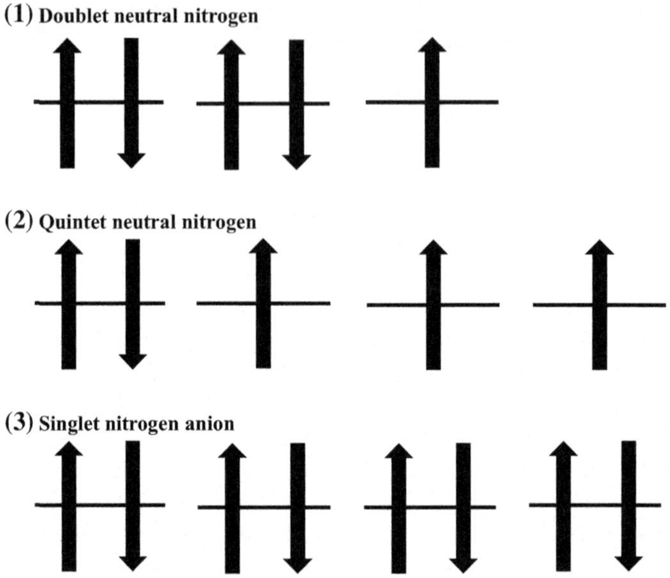

Fig. 7.7 Electronic structures of nitrogen atom: *1* doublet electron configuration of neutral nitrogen; *2* quintet electron configuration of neutral nitrogen; *3* singlet electron configuration of nitrogen anion. Electrons of L shell are only shown

$$\psi_{AO2\alpha}(N) = -0.22\phi_{N(1s)} + 0.58\phi_{N(2s')} + 0.16\phi_{N(2s'')} + 0.36\phi_{N(2s''')} \quad (7.55)$$

$$\psi_{AO2\beta}(N) = -0.21\phi_{N(1s)} + 0.55\phi_{N(2s')} + 0.15\phi_{N(2s'')} + 0.39\phi_{N(2s''')} \quad (7.56)$$

In AO2α and AO2β, hybridization occurs between 1s and 2s orbitals. Though the coefficients of AO2α and AO2β are slightly different, AO2α and AO2β are paired, due to the qualitative same wave-functions. The main coefficients are for 2s orbital. The orbital energies of AO2α and AO2β are −1.04737 and −0.89817 au, respectively. AO2α is more stabilized than AO2β. The obtained wave-functions of AO3α, AO4α and AO3β are

$$\psi_{AO3\alpha}(N) = 0.37\phi_{N(2p_z')} + 0.50\phi_{N(2p_z'')} + 0.32\phi_{N(2p_z''')} \quad (7.57)$$

$$\psi_{AO3\beta}(N) = 0.35\phi_{N(2p_x')} + 0.46\phi_{N(2p_x'')} + 0.38\phi_{N(2p_x''')} \quad (7.58)$$

$$\psi_{AO4\alpha}(N) = 0.36\phi_{N(2p_x')} + 0.48\phi_{N(2p_x'')} + 0.35\phi_{N(2p_x''')} \quad (7.59)$$

AO3α and AO4α consist of $2p_z$ and $2p_x$ orbitals, respectively. On the other hand, AO3β consists of $2p_x$ orbital. Though the coefficients of AO4α and AO3β are slightly different, AO4α and AO3β are paired, due to qualitative same wave-functions. The orbital energies of AO3α and AO4α are −0.56702 and −0.47856 au, respectively. It is because the electron correlation is different in AO3α and AO4α. For example, AO3α has no paired AO, and AO4α has paired AO.

7.8.2 *Quintet Neutral Nitrogen*

In quintet neutral nitrogen, five alpha AOs and two beta AOs are occupied. The obtained wave-functions of AO1α and AO1β are

$$\psi_{\mathrm{AO1}\alpha}(\mathrm{N}) = 0.98\phi_{\mathrm{N(1s)}} \tag{7.60}$$

$$\psi_{\mathrm{AO1}\beta}(\mathrm{N}) = 0.98\phi_{\mathrm{N(1s)}} \tag{7.61}$$

AO1α and AO1β are paired. They represent 1s orbital. The obtained wave-functions of AO2α and AO2β are

$$\psi_{\mathrm{AO2}\alpha}(\mathrm{N}) = -0.22\phi_{\mathrm{N(1s)}} + 0.60\phi_{\mathrm{N(2s')}} + 0.16\phi_{\mathrm{N(2s'')}} + 0.34\phi_{\mathrm{N(2s''')}} \tag{7.62}$$

$$\psi_{\mathrm{AO2}\beta}(\mathrm{N}) = -0.21\phi_{\mathrm{N(1s)}} + 0.51\phi_{\mathrm{N(2s')}} + 0.14\phi_{\mathrm{N(2s'')}} + 0.43\phi_{\mathrm{N(2s''')}} \tag{7.63}$$

In AO2α and AO2β, inversion hybridization occurs between 1s and 2s orbitals. Though the coefficients of AO2α and AO2β are slightly different, AO2α and AO2β paired, due to the qualitative same wave-functions. The main coefficients are for 2s orbital. The orbital energies of AO2α and AO2β are −1.16360 and −0.72695 au, respectively. AO2α is more stabilized than AO2β. The obtained wave-functions of AO2α, AO3α and AO3α are

$$\psi_{\mathrm{AO3}\alpha}(\mathrm{N}) = 0.37\phi_{\mathrm{N}(2\mathrm{p}'_x)} + 0.50\phi_{\mathrm{N}(2\mathrm{p}''_x)} + 0.32\phi_{\mathrm{N}(2\mathrm{p}'''_x)} \tag{7.64}$$

$$\psi_{\mathrm{AO4}\alpha}(\mathrm{N}) = 0.37\phi_{\mathrm{N}(2\mathrm{p}'_y)} + 0.50\phi_{\mathrm{N}(2\mathrm{p}''_y)} + 0.32\phi_{\mathrm{N}(2\mathrm{p}'''_y)} \tag{7.65}$$

$$\psi_{\mathrm{AO5}\alpha}(\mathrm{N}) = 0.37\phi_{\mathrm{N}(2\mathrm{p}'_z)} + 0.50\phi_{\mathrm{N}(2\mathrm{p}''_z)} + 0.32\phi_{\mathrm{N}(2\mathrm{p}'''_z)} \tag{7.66}$$

The figures of AO3α, AO4α and AO5α are the same, though the directions are different. The wave-functions of AO3α, AO4α and AO5α are along x, y and z directions, respectively. As they have the same orbital energy (−0.57074 au), they are degenerated. The total energies of doublet and quintet neutral nitrogen are −54.26529 and −54.40116 au, respectively. Quintet electron configuration is more stable than doublet electron configuration, following electron configuration rule.

7.8.3 *Singlet Nitrogen Anion*

In nitrogen anion, eight electrons occupy all AOs of L shell. The obtained wave-functions of AO1 and AO2 are

$$\psi_{AO1}(N) = 0.98\phi_{N(1s)} \tag{7.67}$$

$$\psi_{AO2}(N) = -0.21\phi_{N(1s)} + 0.54\phi_{N(2s')} + 0.15\phi_{N(2s'')} + 0.33\phi_{N(2s''')} + 0.11\phi_{N(3s)} \tag{7.68}$$

Though AO1 consists of 1s orbital, AO2 consists of 1s, 2s and 3s orbitals. In AO2, inversion hybridization occurs between 1s and other orbitals. The main components are for 2s orbital. The obtained wave-functions of AO3, AO4 and AO5 are

$$\psi_{AO3}(N) = 0.24\phi_{N(2p'_y)} + 0.35\phi_{N(2p''_y)} + 0.17\phi_{N(2p'''_y)} + 0.60\phi_{N(3p_y)} \tag{7.69}$$

$$\psi_{AO4}(N) = 0.24\phi_{N(2p'_x)} + 0.35\phi_{N(2p''_x)} + 0.17\phi_{N(2p'''_x)} + 0.60\phi_{N(3p_x)} \tag{7.70}$$

$$\psi_{AO5}(N) = 0.24\phi_{N(2p'_z)} + 0.35\phi_{N(2p''_z)} + 0.17\phi_{N(2p'''_z)} + 0.60\phi_{N(3p_z)} \tag{7.71}$$

The figures of AO3, AO4 and AO5 are the same, though the directions are different. The wave-functions of AO3, AO4 and AO5 are along y, x and z directions, respectively. As they have the same positive orbital energy (0.38587 au), they are degenerated. As they consist of not only 2p and but also 3p orbitals, electrons are delocalized over both 2p and 3p orbitals. From the delocalization and orbital energy rule, charge transfer easily occurs from nitrogen to other atoms in molecule or solid.

7.9 Oxygen Atom

Figure 7.8 depicts the electronic structures of oxygen atom. Neutral oxygen has six electrons in L shell. Two electron configurations are considered: (1) singlet electron configuration, (2) triplet electron configuration. In solids, the formal charge of oxygen is often −2. Divalent oxygen anion is closed shell system. CCSD/aug-cc-pVTZ calculation is performed for them.

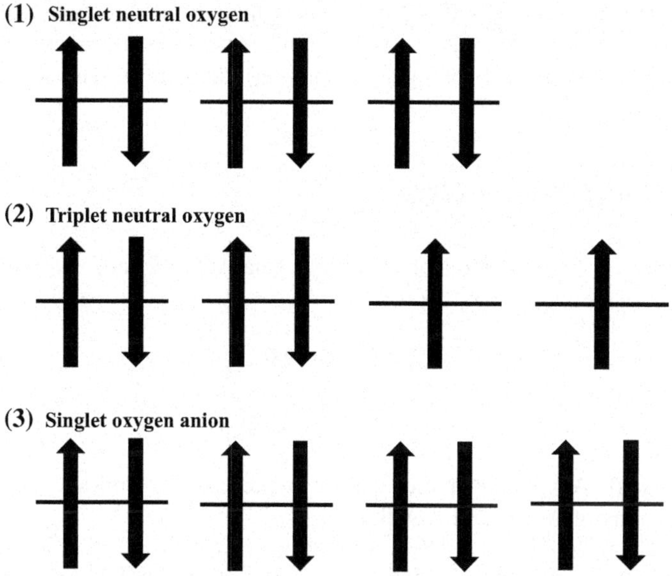

Fig. 7.8 Electronic structures of oxygen atom: *1* singlet electron configuration of neutral oxygen, *2* triplet electron configuration of neutral oxygen, *3* singlet electron configuration of oxygen anion. Electrons of L shell are only shown

7.9.1 Singlet Neutral Oxygen

In singlet neural oxygen, eight electrons occupy four AOs. The obtained wave-functions of AO1 and AO2 are

$$\psi_{\mathrm{AO1}}(\mathrm{O}) = 0.98\phi_{\mathrm{O(1s)}} \tag{7.72}$$

$$\psi_{\mathrm{AO2}}(\mathrm{O}) = -0.22\phi_{\mathrm{O(1s)}} + 0.57\phi_{\mathrm{O(2s')}} + 0.16\phi_{\mathrm{O(2s'')}} + 0.37\phi_{\mathrm{O(2s''')}} \tag{7.73}$$

AO1 consists of only 1s orbital. In AO2, inversion hybridization occurs between 1s and 2s orbitals. The main components are for 2s orbital. The obtained wave-functions of AO3 and AO4 are

$$\psi_{\mathrm{AO3}}(\mathrm{O}) = 0.38\phi_{\mathrm{O}(2\mathrm{p}'_x)} + 0.48\phi_{\mathrm{O}(2\mathrm{p}''_x)} + 0.34\phi_{\mathrm{O}(2\mathrm{p}'''_x)} \tag{7.74}$$

$$\psi_{\mathrm{AO4}}(\mathrm{O}) = 0.38\phi_{\mathrm{O}(2\mathrm{p}''_z)} + 0.48\phi_{\mathrm{O}(2\mathrm{p}''_z)} + 0.34\phi_{\mathrm{O}(2\mathrm{p}'''_z)} \tag{7.75}$$

The figures of AO3 and AO4 are the same, though the directions are different. The wave-functions of AO3 and AO4 are along x and z directions, respectively. As they have the same orbital energy (−0.58664 au), they are degenerated.

7.9.2 Triplet Neutral Oxygen

In triplet neutral oxygen, five alpha and three beta AOs are occupied. The obtained wave-functions of AO1α and AO1β are

$$\psi_{AO1\alpha}(O) = 0.98\phi_{O(1s)} \tag{7.76}$$

$$\psi_{AO1\beta}(O) = 0.98\phi_{O(1s)} \tag{7.77}$$

AO1α and AO1β are paired. They represent 1s orbital. The obtained wave-functions of AO2α and AO2β are

$$\psi_{AO2\alpha}(O) = -0.22\phi_{O(1s)} + 0.59\phi_{O(2s')} + 0.17\phi_{O(2s'')} + 0.34\phi_{O(2s''')} \tag{7.78}$$

$$\psi_{AO2\beta}(O) = -0.22\phi_{O(1s)} + 0.54\phi_{O(2s')} + 0.15\phi_{O(2s'')} + 0.40\phi_{O(2s''')} \tag{7.79}$$

In AO2α and AO2β, inversion hybridization occurs between 1s and 2s orbitals. Though the coefficients of AO2α and AO2β are slightly different, AO2α and AO2β are paired, due to the qualitative same wave-functions. The main coefficients are for 2s orbital. The orbital energies of AO2α and AO2β are −1.41956 and −1.07729 au, respectively. AO2α is more stabilized than AO2β. The obtained wave-functions of AO3α, AO3β, AO4α and AO5α are

$$\psi_{AO3\alpha}(O) = 0.40\phi_{O(2p'_x)} + 0.51\phi_{O(2p''_x)} + 0.30\phi_{O(2p'''_x)} \tag{7.80}$$

$$\psi_{AO3\beta}(O) = 0.37\phi_{O(2p'_z)} + 0.46\phi_{O(2p''_z)} + 0.37\phi_{O(2p'''_z)} \tag{7.81}$$

$$\psi_{AO4\alpha}(O) = 0.40\phi_{O(2p'_y)} + 0.51\phi_{O(2p''_y)} + 0.30\phi_{O(2p'''_y)} \tag{7.82}$$

$$\psi_{AO5\alpha}(O) = 0.39\phi_{O(2p'_z)} + 0.49\phi_{O(2p''_z)} + 0.33\phi_{O(2p'''_z)} \tag{7.83}$$

The figures of AO3α and AO4α are the same, though the directions are different. The wave-functions of AO3α and AO4α are along x and y directions, respectively. As the orbital energies of AO3α and AO4α are the same (−0.71100 au), they are degenerated. On the other hand, the orbital energy of AO5α (−0.61182 au) is larger than AO3α and AO4α. AO5α and AO3β consist of $2p_z$ orbital. Though the coefficients of AO5α and AO3β are slightly different, AO5α and AO3β are paired, due to the qualitative same wave-functions. In neutral oxygen, the total energies of singlet and triplet electron configurations are −74.88724 and −74.97552 au, respectively. The triplet electron configuration is more stable, following electron configuration rule.

7.9.3 Singlet Oxygen Anion

In oxygen anion, eight electrons occupy all AOs of L shell. The obtained wave-function of AO1 and AO2 are

$$\psi_{\mathrm{AO1}}(\mathrm{O}) = 0.98\phi_{\mathrm{O(1s)}} \tag{7.84}$$

$$\psi_{\mathrm{AO2}}(\mathrm{O}) = -0.22\phi_{\mathrm{O(1s)}} + 0.53\phi_{\mathrm{O(2s')}} + 0.15\phi_{\mathrm{O(2s'')}} + 0.37\phi_{\mathrm{O(2s''')}} \tag{7.85}$$

Though AO1 consists of 1s orbital, AO2 consists of 1s and 2s orbitals. In AO2, inversion hybridization occurs between 1s and 2s orbitals. The main components are for 2s orbital. The obtained wave-function of AO3, AO4 and AO5 are

$$\psi_{\mathrm{AO3}}(\mathrm{O}) = 0.31\phi_{\mathrm{O}(2p'_z)} + 0.39\phi_{\mathrm{O}(2p''_z)} + 0.31\phi_{\mathrm{O}(2p'''_z)} + 0.34\phi_{\mathrm{O}(3p_z)} \tag{7.86}$$

$$\psi_{\mathrm{AO4}}(\mathrm{O}) = 0.31\phi_{\mathrm{O}(2p'_y)} + 0.39\phi_{\mathrm{O}(2p''_y)} + 0.31\phi_{\mathrm{O}(2p'''_y)} + 0.34\phi_{\mathrm{O}(3p_y)} \tag{7.87}$$

$$\psi_{\mathrm{AO5}}(\mathrm{O}) = 0.31\phi_{\mathrm{O}(2p'_x)} + 0.39\phi_{\mathrm{O}(2p''_x)} + 0.31\phi_{\mathrm{O}(2p'''_x)} + 0.34\phi_{\mathrm{O}(3p_x)} \tag{7.88}$$

The figures of AO3, AO4 and AO5 are the same, though the directions are different. The wave-functions of AO3, AO4 and AO5 are along z, y and x directions, respectively. As they have the same positive orbital energy (0.19348 au), they are degenerated. As they consist of not only 2p and but also 3p orbitals, electrons are delocalized over both 2p and 3p orbitals. From the delocalization and orbital energy rule, charge transfer easily occurs from nitrogen to other atoms in molecule or solid.

7.10 Fluorine Atom

Figure 7.9 depicts the electronic structures of fluorine atom. As neutral fluorine has seven electrons in L shell, the spin multiplicity is doublet. In solids, the formal charge of fluorine is −1. Monovalent fluorine anion is closed shell system. CCSD/aug-cc-pVTZ calculation is performed for them.

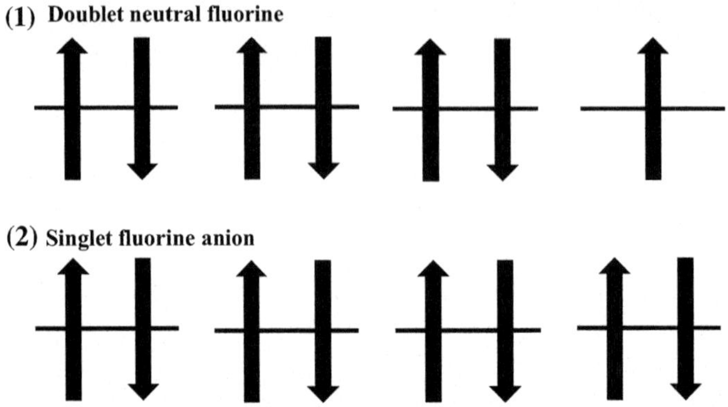

Fig. 7.9 Electronic structures for fluorine atom: *1* doublet electron configuration of neutral fluorine, *2* singlet electron configuration of fluorine anion. Electrons of L shell are only shown

7.10.1 Neutral Fluorine

The obtained wave-functions of AO1α and AO1β are

$$\psi_{AO1\alpha}(F) = 0.98\phi_{F(1s)} \tag{7.89}$$

$$\psi_{AO1\beta}(F) = 0.98\phi_{F(1s)} \tag{7.90}$$

AO1α and AO1β are paired. They represent 1s orbital. The obtained wave-functions of AO2α and AO2β are

$$\psi_{AO2\alpha}(F) = -0.23\phi_{F(1s)} + 0.58\phi_{F(2s')} + 0.18\phi_{F(2s'')} + 0.35\phi_{F(2s''')} \tag{7.91}$$

$$\psi_{AO2\beta}(F) = -0.22\phi_{F(1s)} + 0.55\phi_{F(2s')} + 0.17\phi_{F(2s'')} + 0.38\phi_{F(2s''')} \tag{7.92}$$

In AO2α and AO2β, inversion hybridization occurs between 1s and 2s orbitals. Though the coefficients of AO2α and AO2β are slightly different, AO2α and AO2β paired, due to the qualitative same wave-functions. The main coefficients are for 2s orbital. The orbital energies of AO2α and AO2β are −1.67449 and −1.47926 au, respectively. AO2α is more stabilized than AO2β. The obtained wave-functions of AO3α, AO3β, AO4α, AO4β and AO5α

$$\psi_{AO3\alpha}(F) = 0.41\phi_{F(2p'_x)} + 0.51\phi_{F(2p''_x)} + 0.29\phi_{F(2p'''_x)} \tag{7.93}$$

$$\psi_{AO3\beta}(F) = 0.39\phi_{O(2p'_z)} + 0.48\phi_{O(2p''_z)} + 0.34\phi_{O(2p'''_z)} \tag{7.94}$$

$$\psi_{AO4\alpha}(F) = 0.40\phi_{O(2p'_z)} + 0.49\phi_{O(2p''_z)} + 0.32\phi_{O(2p'''_z)} \quad (7.95)$$

$$\psi_{AO4\beta}(F) = 0.39\phi_{O(2p'_y)} + 0.48\phi_{O(2p''_y)} + 0.34\phi_{O(2p'''_y)} \quad (7.96)$$

$$\psi_{AO5\alpha}(F) = 0.40\phi_{O(2p'_y)} + 0.49\phi_{O(2p''_y)} + 0.32\phi_{O(2p'''_y)} \quad (7.97)$$

The figures of AO4α and AO5α are the same, though the directions are different. The wave-functions of AO4α and AO5α are along z and y directions, respectively. As the orbital energies of AO4α and AO5α are the same (−0.73207 au), they are degenerated. The figures of AO3β and AO4β are the same, though the directions are different. The wave-functions of AO3β and AO4β are along z and y directions, respectively. As the orbital energies of AO3β and AO4β are the same value (−0.68029 au), they are degenerated. AO4α and AO3β are paired, and AO5α and AO4β are paired, due to the qualitative same wave-functions. AO3α, which consists of $2p_x$ orbital, has no paired beta AO. As the orbital energy (−1.67449 au) is smaller than AO4α and AO5α, it is stabilized.

7.10.2 Fluorine Anion

In fluorine anion, eight electrons occupy all AOs of L shell. The obtained wave-function of AO1 and AO2 are

$$\psi_{AO1}(F) = 0.98\phi_{F(1s)} \quad (7.98)$$

$$\psi_{AO2}(F) = -0.22\phi_{F(1s)} + 0.54\phi_{F(2s')} + 0.16\phi_{F(2s'')} + 0.38\phi_{F(2s''')} \quad (7.99)$$

Though AO1 consists of 1s orbital, AO2 consists of 1s and 2s orbitals. In AO2, inversion hybridization occurs between 1s and 2s orbitals. The main components are for 2s orbital. The obtained wave-functions of AO3, AO4 and AO5 are

$$\psi_{AO3}(F) = 0.36\phi_{O(2p'_x)} + 0.44\phi_{O(2p''_x)} + 0.36\phi_{O(2p'''_x)} + 0.12\phi_{O(3p_x)} \quad (7.100)$$

$$\psi_{AO4}(F) = 0.36\phi_{O(2p'_y)} + 0.44\phi_{O(2p''_y)} + 0.36\phi_{O(2p'''_y)} + 0.12\phi_{O(3p_y)} \quad (7.101)$$

$$\psi_{AO5}(F) = 0.36\phi_{O(2p'_z)} + 0.44\phi_{O(2p''_z)} + 0.36\phi_{O(2p'''_z)} + 0.12\phi_{O(3p_z)} \quad (7.102)$$

The figures of AO3, AO4 and AO5 are the same, though the directions are different. The wave-functions of AO3, AO4 and AO5 are along x, y and z directions, respectively. As they have the same orbital energy (−0.18095 au), they are degenerated. As they consist of not only 2p but also 3p orbitals, electrons are delocalized over both 2p and 3p orbitals.

Further Readings

1. Gaussian 09, Frisch MJ, Trucks GW, Schlegel HB, Scuseria GE, Robb MA, Cheeseman JR, Scalmani G, Barone V, Mennucci B, Petersson GA, Nakatsuji H, Caricato M, Li X, Hratchian HP, Izmaylov AF, Bloino J, Zheng G, Sonnenberg JL, Hada M, Ehara M, Toyota K, Fukuda R, Hasegawa J, Ishida M, Nakajima T, Honda Y, Kitao O, Nakai H, Vreven T, Montgomery JA Jr, Peralta JE, Ogliaro F, Bearpark M, Heyd JJ, Brothers E, Kudin KN, Staroverov VN, Kobayashi R, Normand J, Raghavachari K, Rendell A, Burant JC, Iyengar SS, Tomasi J, Cossi M, Rega N, Millam JM, Klene M, Knox JE, Cross JB, Bakken V, Adamo C, Jaramillo J, Gomperts R, Stratmann RE, Yazyev O, Austin AJ, Cammi R, Pomelli C, Ochterski JW, Martin RL, Morokuma K, Zakrzewski VG, Voth GA, Salvador P, Dannenberg JJ, Dapprich S, Daniels AD, Farkas Ö, Foresman JB, Ortiz JV, Cioslowski J, Fox DJ, Gaussian, Inc., Wallingford CT, 2009
2. Dunning TH Jr. (1989) J Chem Phys 90(2): 1007.1023
3. Woon DE, Dunning TH Jr (1994) J Chem Phys 100(4):2975–2988
4. Woon DE, Dunning TH Jr (1994) J Chem Phys 98(2):1358–1371
5. Dunning TH Jr., Peterson KA, Woon DE (1999) Encyclopedia of Computational Chemistry: pp 88–115

Chapter 8
Molecular Orbital Calculation of Diatomic Molecule

Abstract Covalent bonding is formed through orbital overlap between orbitals of different atoms. It is classified into σ-type and π-type by the difference of the interaction between lobes. In this chapter, chemical bonding formation of diatomic molecule is clearly explained from the viewpoints of molecular orbital analysis and energetics. In homonuclear diatomic molecule, chemical bonding formations of hydrogen molecule, lithium dimer, nitrogen molecule and oxygen molecule are explained through concrete calculation results. Triplet and singlet spin states are compared in oxygen molecule. The stability of triplet oxygen molecule is clearly explained. The high reactivity of superoxide is also discussed. On the other hand, in heteronuclear diatomic molecule, chemical bonding formations of hydrogen fluoride, hydrogen chloride, hydroxide and carbon oxide are explained through concrete calculation results. The difference of acidity is discussed in comparison with hydrogen fluoride and hydrogen chloride. In comparison with hydroxide, the reactivity of hydroxide radical is also discussed. Point charge notation has been used for atom and molecule. However, the limit of point charge denotation is pointed out.

Keywords Orbital overlap · Covalent bonding · Inversion covalent bonding · Homonuclear diatomic molecule · Heteronuclear diatomic molecule

8.1 Orbital Overlap

In many-electron atom, initial atomic orbitals (IAOs) are hybridized in the same atom. It is called orbital hybridization. Note that IAO is designated by basis set. In this book, IAO is just called "orbital".

In many-electron molecule, molecular orbital (MO) is often represented by the combination of IAOs of different atoms. It is called orbital overlap. There are two orbital overlap patterns. One is conventional orbital overlap, when MO coefficients of different atoms have the same sign. The other is inversion orbital overlap, when the coefficients of different atoms have the different signs. In inversion orbital

T. Onishi, *Quantum Computational Chemistry*,
DOI 10.1007/978-981-10-5933-9_8

overlap, the wave-function is annihilated, due to node between different atoms. Node is where the wave-function is zero. Hence, the orbital energy of inversion covalent bonding is higher than conventional covalent bonding.

Lobe is orbital figure with the same sign. s and p orbitals have one lobe and two lobes, respectively. d orbital has three or four lobes. The use of lobe makes it possible to explain the difference of σ-type and π-type covalent bonds.

Figure 8.1 depicts σ-type covalent bonding patterns between orbitals of different atoms. In σ-type covalent bonding, one lobe interacts with one lobe of different atom. When the sign of another lobe is opposite, inversion σ-type covalent bonding is formed. The orbital energy of inversion σ-type covalent bonding is higher than corresponding σ-type covalent bonding, due to the existence of node.

Figure 8.2 depicts π-type covalent bonding patterns between orbitals of different atoms. In π-type covalent bonding, two lobes interact with two lobes of different atom. When the sign of another lobes are opposite, inversion π-type covalent bonding is formed. The orbital energy of inversion π-type covalent bonding is higher than corresponding π-type covalent bonding, due to the existence of node.

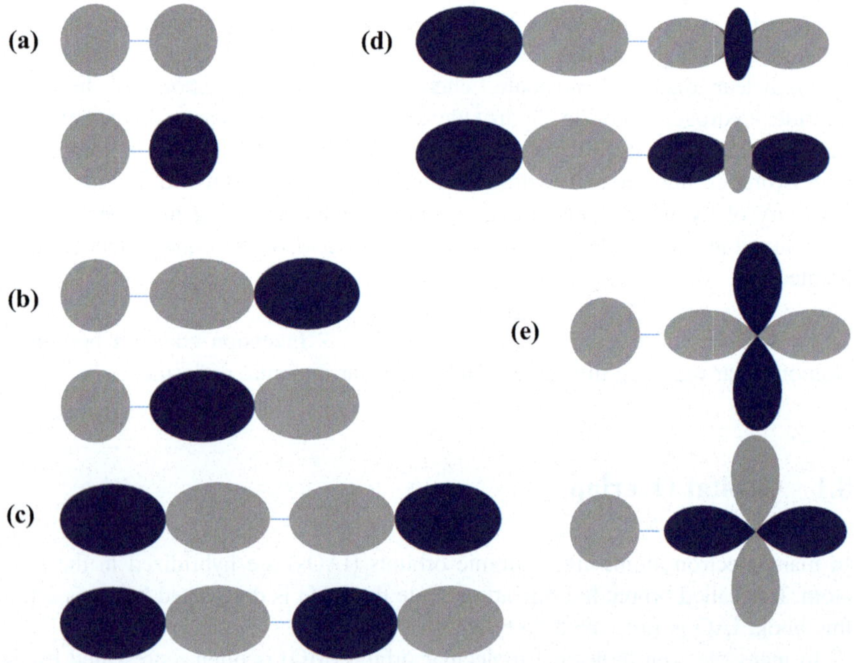

Fig. 8.1 The σ-type covalent bonding patterns between orbitals of different atoms: **a** s and s orbitals, **b** s and p orbitals, **c** p and p orbitals, **d** p and d orbitals, **e** s and d orbitals. The corresponding inversion σ-type covalent bonding patterns are also shown. The *grey and blue lobes* has the positive and negative coefficients

Fig. 8.2 The π-type covalent bonding patterns between orbitals of different atoms: **a** p and p orbitals, **b** p and d orbitals, **c** d and d orbitals. The corresponding inversion π-type covalent bonding patterns are also shown. The *grey and blue lobes* has the positive and negative coefficients

8.2 Hydrogen Molecule

8.2.1 Hydrogen Molecule

Hydrogen molecule, which is denoted as H_2, is a homonuclear diatomic molecule. The lowest and second lowest orbital energies are given in MO1 and MO2, respectively. Though two electrons occupy MO1, MO2 is unoccupied. MO1 and MO2 correspond to the highest occupied molecular orbital (HOMO) and lowest unoccupied molecular orbital (LUMO), respectively.

B3LYP/6-31G* calculation is performed for H_2 (H1–H2) with closed shell electron configuration. In closed shell system, the number of MOs corresponds to the total number of basis functions. As 6-31G* basis sets of hydrogen has two basis functions, four MO are produced. The obtained wave-function of occupied MO1 is

$$\psi_{MO1}(H_2) = 0.33\phi_{H1(1s')} + 0.27\phi_{H1(1s'')} + 0.33\phi_{H2(1s')} + 0.27\phi_{H2(1s'')} \quad (8.1)$$

One H1 lobe interacts with one H2 lobe. The signs of H1 and H2 coefficients are positive. From chemical bonding rule, it is found that the σ-type covalent bonding is formed between two hydrogen 1s orbitals. MO1 is symmetric to middle point between two hydrogen atoms. The obtained wave-function of unoccupied MO2 is

$$\psi_{MO2}(H_2) = -0.18\phi_{H1(1s')} - 1.64\phi_{H1(1s'')} + 0.18\phi_{H2(1s')} + 1.64\phi_{H2(1s'')} \quad (8.2)$$

One H1 lobe interacts with one H2 lobe. Though the absolute values of H1 and H2 coefficients are the same, the signs are different. From chemical bonding rule, it is found that inversion σ-type covalent bonding is formed between two hydrogen 1s orbitals.

Figure 8.3 depicts the orbital energy level diagram and molecular orbitals of H_2 molecule, and atomic orbital of H atom. It can be also understood that covalent bonding is formed through the combinations of two hydrogen 1s atomic orbitals (AOs), and inversion covalent bonding is formed through hydrogen 1s and inversion 1s AOs. One may think that the explanation based on independent AOs is natural. The explanation is not always applicable.

When changing the interatomic H1–H2 distance (r), the change of total energy is investigated (see Fig. 8.4). The local minimum is given at 0.734 Å, corresponding to H_2 bond length. When r is smaller than local minimum, higher total energy is given, due to electron–electron repulsion. On the other hand, when r is larger than local minimum, higher energy is also given. Bond dissociation energy is a useful indication of bond dissociation. In general, it can be estimated from the total energy difference between the local minimum and completely dissociated point.

$$E_{\text{dissociation}}(H_2) = E(H\cdot) + E(H\cdot) - E(H_2) \quad (8.3)$$

Note that it is assumed that two hydrogen radicals exist at completely dissociated point. The bond dissociation energy is estimated to be 108.6 kcal/mol. The zero-point vibration energy is 6.292 kcal/mol. It is much smaller than the bond dissociation energy.

8.2.2 Hydrogen Molecule Cation

In hydrogen molecule cation (H_2^+), though there is only one electron, Schrödinger equation cannot be analytically solved, due to three-body problem. One electron occupies MO1α with the lowest orbital energy.

B3LYP/6-31G* calculation is performed for H_2^+ (H1–H2) with open shell electron configuration. The spin state is doublet, due to no paired beta MO. The number of alpha or beta MOs corresponds to the total number of basis functions.

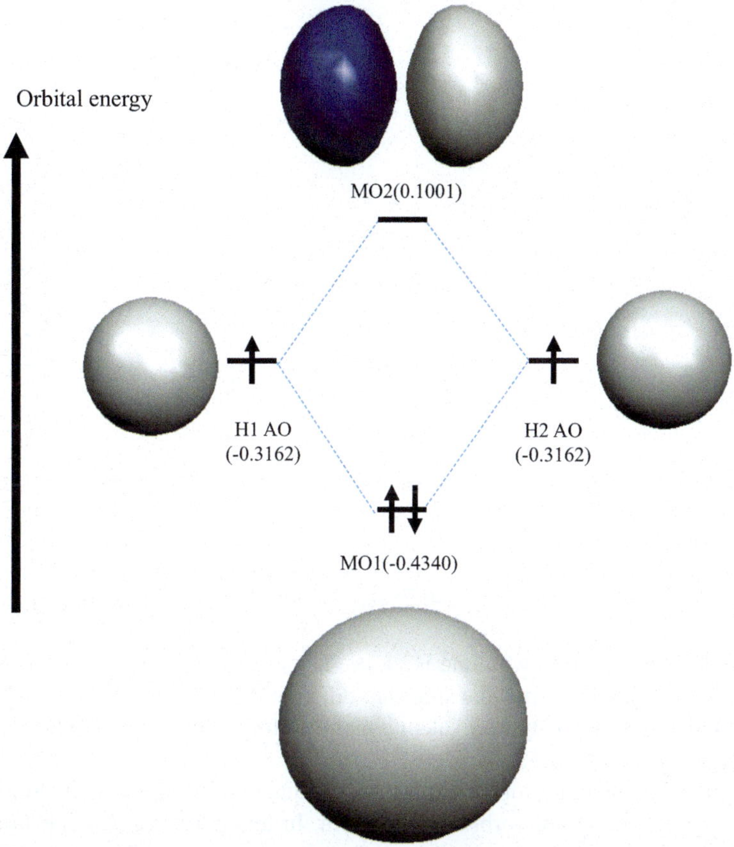

Fig. 8.3 Orbital energy level diagram and molecular orbitals of hydrogen molecule (H_2), and atomic orbital of hydrogen atom (H). The calculated orbital energy is shown in *parentheses* (B3LYP/6-31G*)

Four alpha and four beta MOs are produced, because 6-31G* basis set of hydrogen atom has two basis functions. MO1β, MO2α, MO2β, MO3α, MO3β, MO4α and MO4β are unoccupied. The obtained wave-function of MO1α is

$$\psi_{MO1\alpha}(H_2{}^+) = 0.39\phi_{H1(1s')} + 0.26\phi_{H1(1s'')} + 0.39\phi_{H2(1s')} + 0.26\phi_{H2(1s'')} \quad (8.4)$$

One H1 lobe interacts with one H2 lobe. The signs of H1 and H2 coefficients are positive. From chemical bonding rule, it is found that σ-type covalent bonding is formed between two hydrogen 1s orbitals. MO1α is symmetric to the middle point between two hydrogen atoms. The obtained wave-function of unoccupied MO2α is

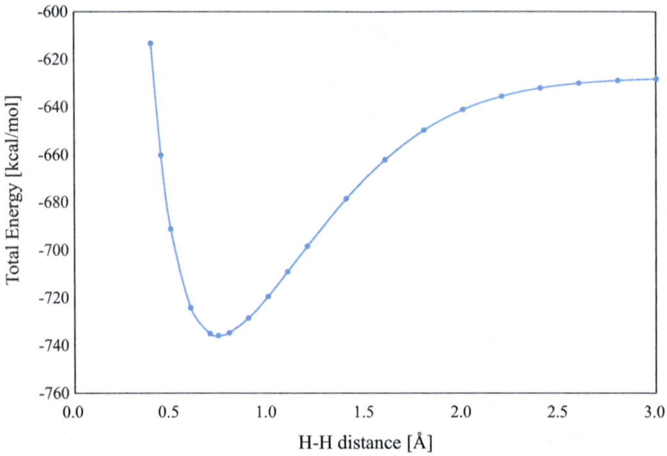

Fig. 8.4 Potential energy curve of hydrogen molecule, changing the interatomic H1–H2 distance (CCSD/aug-cc-pVTZ)

$$\psi_{\mathrm{MO2}\alpha}(\mathrm{H_2}^+) = -0.40\phi_{\mathrm{H1(1s')}} - 0.78\phi_{\mathrm{H1(1s'')}} + 0.40\phi_{\mathrm{H2(1s')}} + 0.78\phi_{\mathrm{H2(1s'')}} \quad (8.5)$$

One H1 lobe interacts with one H2 lobe. Though the absolute values of H1 and H2 coefficients are the same, the signs are different. From chemical bonding rule, it is found that inversion σ-type covalent bonding is formed between two hydrogen 1s orbitals.

Figure 8.5 depicts the orbital energy level diagram and molecular orbitals for H_2^+ molecule, and atomic orbitals of H atom. In this case, it is difficult to understand chemical bonding formation through the combinations of AOs. It is because no electron exists in AO of H2. In MO1α, one electron is shared between H1 and H2.

When changing the interatomic H1–H2 distance (r), the local minimum is given at 1.058 Å (see Fig. 8.6). It is found that H_2^+ bond length is larger than H_2 bond length. Bond dissociation energy can be estimated from the total energy difference between the local minimum and completely dissociated point.

$$E_{\mathrm{dissociation}}(\mathrm{H_2}^+) = E(\mathrm{H}\cdot) + E(\mathrm{H}^+) - E(\mathrm{H_2}^+) \quad (8.6)$$

Note that it is assumed that one hydrogen radical and proton exist at the completely dissociated point. In proton, where there exists no electron, the total energy is zero, from the definition of the total energy in molecular orbital calculation. The bond dissociation energy is estimated to be 64.31 kcal/mol. It is smaller than H_2. The zero-point vibration energy is 3.319 kcal/mol. It is much smaller than the bond dissociation energy.

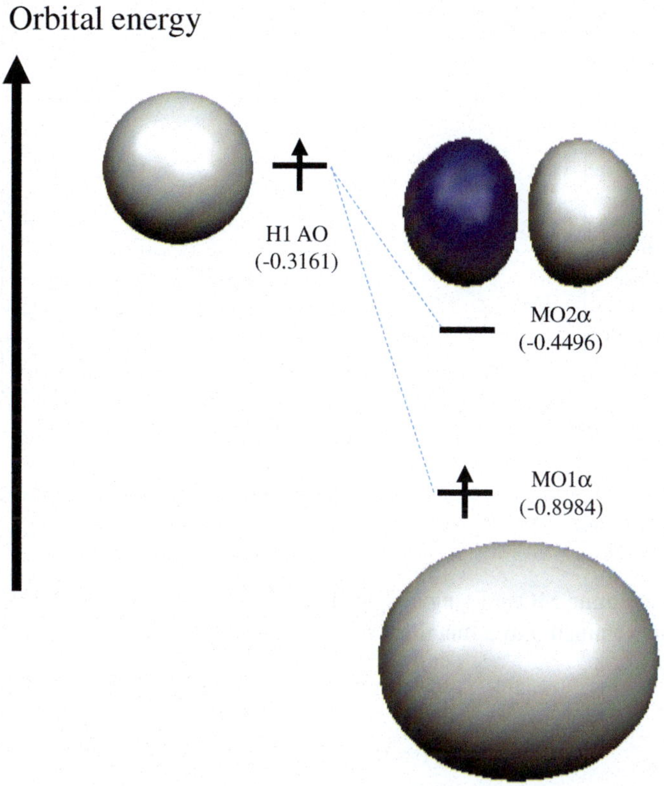

Fig. 8.5 Orbital energy level diagram and molecular orbitals of hydrogen molecule cation (H_2^+), and atomic orbital of neutral hydrogen atom (H). The calculated orbital energy is shown in *parentheses* (B3LYP/6-31G*)

8.3 Lithium Dimer

8.3.1 Lithium Dimer

In Chap. 7, AOs for lithium atom were explained. Two electrons occupy paired AO1α and AO1β, and one electron occupies AO2α. The electron configuration of lithium atom is written as Li: $1s^2 2s^1$. When two lithium atoms are bound, lithium dimer (Li_2) is formed.

B3LYP/6-31G* calculations is performed for lithium dimer (Li1–Li2). The total energies of singlet and triplet spin states are −15.01475 au and −14.98267 au, respectively. It is found that singlet spin state is more stable than triplet spin state. Thirty MOs are produced, because 6-31G* basis set of lithium atom has fifteen basis functions.

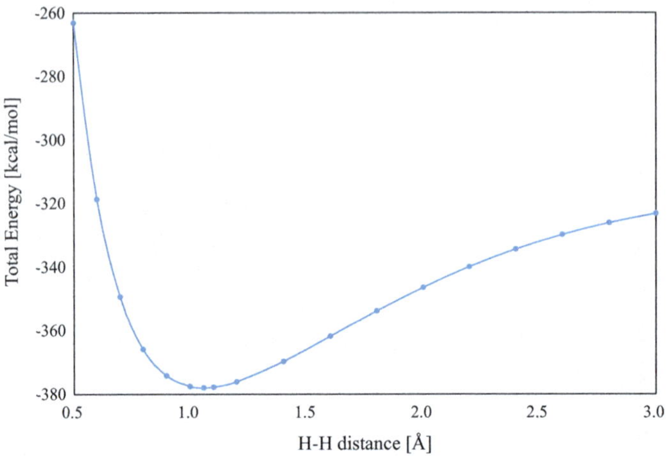

Fig. 8.6 Potential energy curve of hydrogen molecule cation, changing the interatomic H1–H2 distance (CCSD/aug-cc-pVTZ)

Let us examine MOs of singlet spin state. Six electrons occupy MO1, MO2 and MO3. The obtained wave-functions of MO1 and MO2 are

$$\psi_{\mathrm{MO1}}(\mathrm{Li_2}) = 0.70\phi_{\mathrm{Li1(1s)}} + 0.70\phi_{\mathrm{Li2(1s)}} \quad (8.7)$$

$$\psi_{\mathrm{MO2}}(\mathrm{Li_2}) = 0.70\phi_{\mathrm{Li1(1s)}} - 0.70\phi_{\mathrm{Li2(1s)}} \quad (8.8)$$

Li1 1s orbital overlaps with Li2 1s orbital. One Li1 lobe interacts with one Li2 lobe. From chemical bonding rule, it is found that the σ-type covalent bonding is formed between two lithium 1s orbitals. Though Li1 and Li2 coefficients are positive in MO1, they are different in MO2. The inversion σ-type covalent bonding is formed in MO2. The obtained wave-functions of MO3 is

$$\begin{aligned}\psi_{\mathrm{MO3}}(\mathrm{Li_2}) = {} & 0.15\phi_{\mathrm{Li1(1s)}} - 0.24\phi_{\mathrm{Li1(2s')}} - 0.33\phi_{\mathrm{Li1(2s'')}} - 0.11\phi_{\mathrm{Li1}(2p'_x)} \\ & + 0.15\phi_{\mathrm{Li2(1s)}} - 0.24\phi_{\mathrm{Li2(2s)}} - 0.33\phi_{\mathrm{Li2(2s'')}} - 0.11\phi_{\mathrm{Li2}(2p'_x)}\end{aligned} \quad (8.9)$$

In Li1 and Li2, it is found that inversion hybridization occurs between 1s and 2s orbitals, and $2p_x$ orbital is also hybridized. The main coefficients are for Li1 and Li2 2s orbitals. One Li1 lobe interacts with one Li2 lobe. From chemical bonding rule, it is found that the σ-type covalent bonding is formed between Li1 and Li2 2s orbitals.

Figure 8.7 shows the orbital energy level diagram and molecular orbitals of lithium dimer, and atomic orbitals for lithium atom. Regarding MO1 and MO2, it can be understood that covalent bonding is formed through the combination of two 1s AOs, and inversion covalent bonding is formed through the combination of two

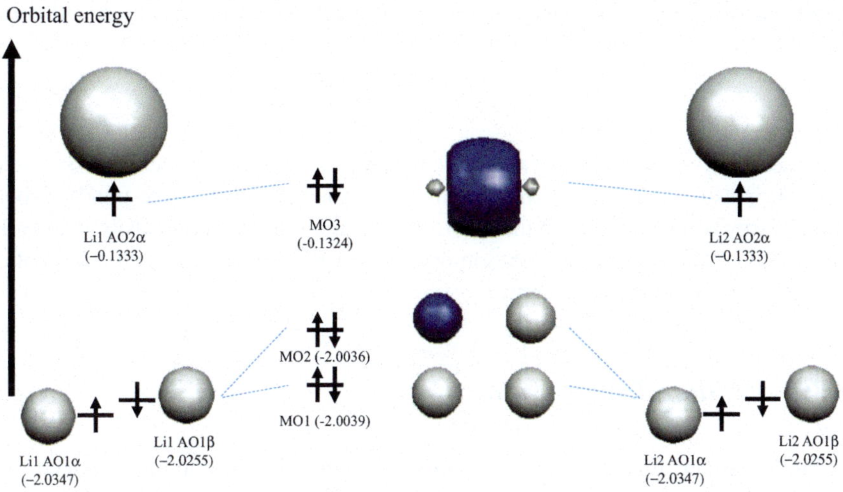

Fig. 8.7 Orbital energy level diagram and molecular orbitals of lithium dimer (Li_2), and atomic orbital of lithium atom (Li). The calculated orbital energy is shown in *parentheses* (B3LYP/6-31G*)

1s and inversion 1s AOs. However, in MO3, it is not easy to explain chemical bonding formation through the combination of AOs. It is because $2p_x$ orbital is also hybridized in MO3.

Figure 8.8 depicts the potential energy curve of lithium dimer, changing the interatomic Li1–Li2 distance (r). The local minimum is given at 2.727 Å, corresponding to the bond length of lithium dimer. Bond dissociation energy can be

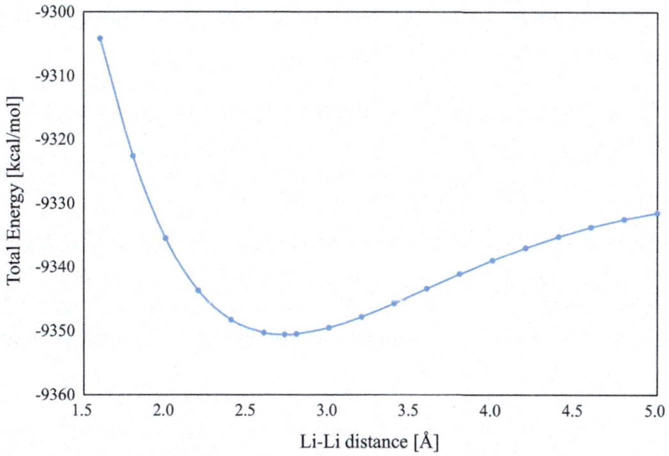

Fig. 8.8 Potential energy curve of lithium dimer, changing the interatomic Li1–Li2 distance (CCSD/aug-cc-pVTZ)

estimated from the total energy difference between the local minimum and completely dissociated point.

$$E_{\text{dissociation}}(\text{Li}_2) = E(\text{Li}\cdot) + E(\text{Li}\cdot) - E(\text{Li}_2) \tag{8.10}$$

Note that it is assumed that two lithium radicals exist at completely dissociated point. It is estimated to be 22.67 kcal/mol. As it is smaller than the bond dissociation energy of H_2, it is found that Li–Li dissociation occurs more easily than H–H dissociation. The zero-point vibration energy is 0.484 kcal/mol. It is much smaller than the bond dissociation energy.

8.3.2 *Lithium Dimer Cation*

In lithium dimer, lithium atoms are bound through covalent bonding between two outer shell electrons. In lithium dimer cation (Li_2^+), there is one outer shell electron, and spin state is doublet.

B3LYP/6-31G* calculation is performed for Li_2^+ (Li1–Li2) with open shell electron configuration. Thirty alpha and beta MOs are produced, because 6-31G* basis set of lithium atom has fifteen basis functions. Four electrons occupy MO1α, MO1β, MO2α and MO2β. The obtained wave-functions of MO1α and MO1β are

$$\psi_{\text{MO1}\alpha}(\text{Li}_2{}^+) = -0.70\phi_{\text{Li1(1s)}} - 0.70\phi_{\text{Li2(1s)}} \tag{8.11}$$

$$\psi_{\text{MO1}\beta}(\text{Li}_2{}^+) = -0.70\phi_{\text{Li1(1s)}} - 0.70\phi_{\text{Li2(1s)}} \tag{8.12}$$

MO1α and MO1β are paired. Li 1s orbital overlaps with Li2 orbital. One Li1 lobe interacts with one Li2 lobe. From chemical bonding rule, it is found that the σ-type covalent bonding is formed between two lithium 1s orbitals. The obtained wave-functions of MO2α and MO2β are

$$\psi_{\text{MO2}\alpha}(\text{Li}_2{}^+) = 0.70\phi_{\text{Li1(1s)}} - 0.70\phi_{\text{Li2(1s)}} \tag{8.13}$$

$$\psi_{\text{MO2}\beta}(\text{Li}_2{}^+) = 0.70\phi_{\text{Li1(1s)}} - 0.70\phi_{\text{Li2(1s)}} \tag{8.14}$$

MO2α and MO2β are paired. Li 1s orbital overlaps with Li2 orbital. One Li1 lobe interacts with one Li2 lobe. The signs of Li1 and Li2 coefficients are different. From chemical bonding rule, it is found that the inversion σ-type covalent bonding is formed between two lithium 1s orbitals. The obtained wave-functions of MO3α is

$$\begin{aligned}\psi_{\text{MO3}\alpha}(\text{Li}_2{}^+) = &-0.13\phi_{\text{Li1(1s)}} + 0.37\phi_{\text{Li1(2s')}} + 0.14\phi_{\text{Li1(2s'')}} + 0.27\phi_{\text{Li1}(2\text{p}'_x)} \\ &- 0.13\phi_{\text{Li2(1s)}} + 0.37\phi_{\text{Li2(2s')}} + 0.14\phi_{\text{Li2(2s'')}} + 0.27\phi_{\text{Li2}(2\text{p}'_x)}\end{aligned} \tag{8.15}$$

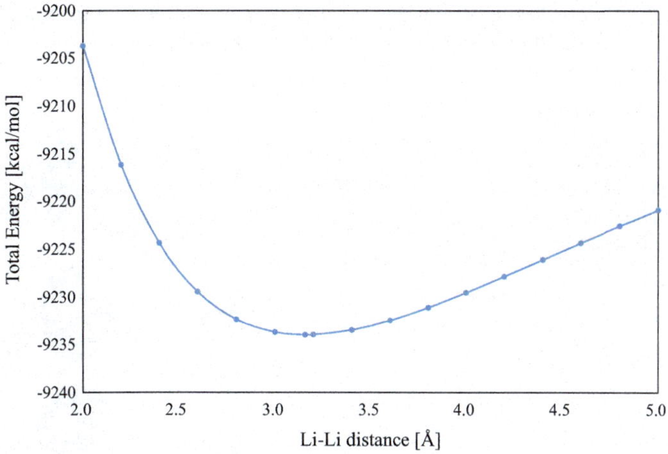

Fig. 8.9 Potential energy curve of lithium dimer cation, changing the interatomic Li1–Li2 distance (CCSD/aug-cc-pVTZ)

In Li1 and Li2, inversion hybridization occurs between 1s and 2s orbitals, and $2p_x$ orbital is also hybridized. The main coefficients are for Li1 2s and Li2 2s orbitals. One Li1 lobe interacts with one Li2 lobe. From chemical bonding rule, it is found that the σ-type covalent bonding is formed between Li1 and Li2 2s orbitals.

Figure 8.9 depicts the potential energy curve of lithium dimer cation, changing the interatomic Li1–Li2 distance (r). The local minimum is given at 3.158 Å, which is larger than lithium dimer (2.727 Å). It is because covalency of Li_2^+ is more subsided than Li_2, due to one outer shell electron. Bond dissociation energy can be estimated from the energy difference between the local minimum and completely dissociated point.

$$E_{\text{dissociation}}(\mathrm{Li_2}^+) = E(\mathrm{Li}^+) + E(\mathrm{Li}\cdot) - E(\mathrm{Li_2}^+) \tag{8.16}$$

Note that it is assumed that lithium cation and lithium radical exist at the completely dissociated point. It is 29.24 kcal/mol. The zero point vibration energy is 0.372 kcal/mol. It is much smaller than the bond dissociation energy.

8.4 Nitrogen Molecule

Nitrogen atom has two electrons in K shell and five electrons in L shell. When two nitrogen atoms are bound, nitrogen molecule is formed. B3LYP/6-31G* calculation is performed for N_2 (N1–N2) with closed shell electron configuration. Thirty MOs are produced, because 6-31G* basis set of nitrogen atom has fifteen basis functions. Fourteen electrons occupy MO1, MO2, MO3, MO4, MO5, MO6 and MO7.

Figure 8.10 depicts the orbital energy diagram and MOs of nitrogen molecule at optimized structure. The obtained wave-function of MO1 is

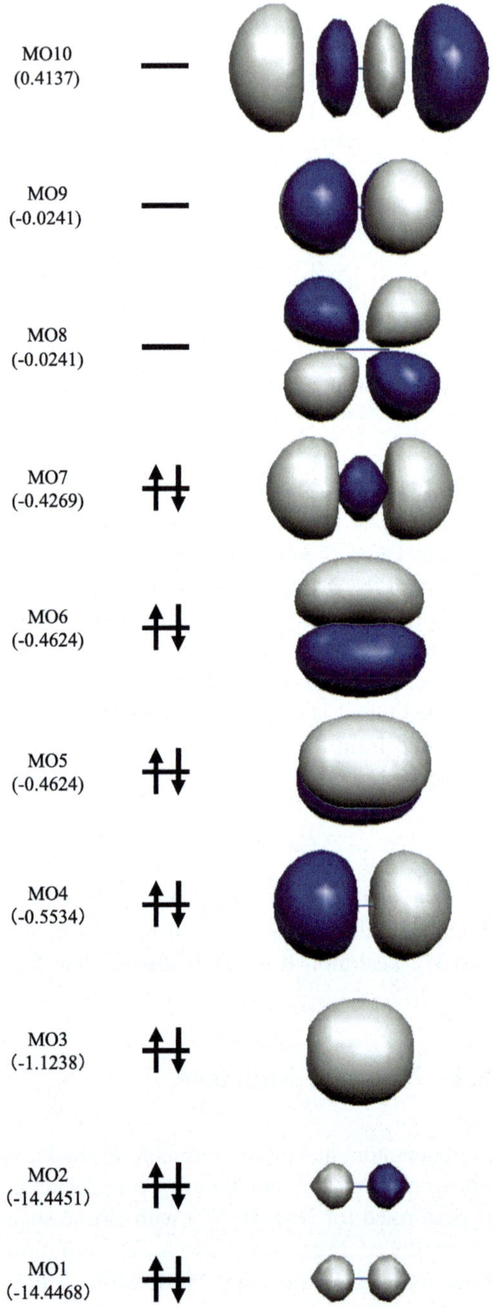

Fig. 8.10 Orbital energy diagram and molecular orbitals of nitrogen molecule at optimized structure (B3LYP/6-31G*). The calculated orbital energy is shown in *parentheses*

$$\psi_{\text{MO1}}(\text{N}_2) = 0.70\phi_{\text{N1(1s)}} + 0.70\phi_{\text{N2(1s)}} \tag{8.17}$$

N1 1s orbital overlaps with N2 1s orbital. One N1 lobe interacts with one N2 lobe. From chemical bonding rule, it is found that σ-type covalent bonding is formed between nitrogen 1s orbitals. The obtained wave-functions of MO2 is

$$\psi_{\text{MO2}}(\text{N}_2) = 0.70\phi_{\text{N1(1s)}} - 0.70\phi_{\text{N2(1s)}} \tag{8.18}$$

The signs of N1 and N2 coefficients are different in MO2. The inversion σ-type covalent bonding is formed between nitrogen 1s orbitals. The obtained wave-functions of MO3 and MO4 are

$$\begin{aligned}\psi_{\text{MO3}}(\text{N}_2) = &-0.16\phi_{\text{N1(1s)}} + 0.34\phi_{\text{N1(2s')}} + 0.19\phi_{\text{N1(2s'')}} + 0.23\phi_{\text{N1}(2\text{p}'_x)} \\ &- 0.16\phi_{\text{N2(1s)}} + 0.34\phi_{\text{N2(2s')}} + 0.19\phi_{\text{N2(2s'')}} - 0.23\phi_{\text{N2}(2\text{p}'_x)}\end{aligned} \tag{8.19}$$

$$\begin{aligned}\psi_{\text{MO4}}(\text{N}_2) = &\ 0.15\phi_{\text{N1(1s)}} - 0.33\phi_{\text{N1(2s')}} - 0.53\phi_{\text{N1(2s'')}} + 0.21\phi_{\text{N1}(2\text{p}'_x)} \\ &- 0.15\phi_{\text{N2(1s)}} + 0.33\phi_{\text{N2(2s')}} + 0.53\phi_{\text{N2(2s'')}} + 0.21\phi_{\text{N2}(2\text{p}'_x)}\end{aligned} \tag{8.20}$$

In MO3 and MO4, there is inversion hybridization between 1s and 2s orbitals, and 2p_x orbital is also hybridized. The main coefficients are for 2s orbital. One N1 lobe interacts with one N2 lobe. In MO4, the signs of the coefficients of N1 and N2 2s orbitals are different. From chemical bonding rule, it is found that the σ-type covalent bonding is formed between 2s orbitals in MO3, and inversion σ-type covalent bonding is formed between 2s orbitals in MO4. Thc orbital energies of MO3 and MO4 are much larger than MO1 and MO2. It is because four electrons of MO1 and MO2 are in inner K shell.

The coefficients of MO5, MO6 and MO7 are for outer shell electrons (2s and 2p electrons). The obtained wave-function of MO5 and MO6 are

$$\begin{aligned}\psi_{\text{MO5}}(\text{N}_2) = &\ 0.13\phi_{\text{N1}(2\text{p}'_y)} + 0.44\phi_{\text{N1}(2\text{p}'_z)} + 0.22\phi_{\text{N1}(2\text{p}''_z)} \\ &+ 0.13\phi_{\text{N2}(2\text{p}'_y)} + 0.44\phi_{\text{N2}(2\text{p}'_z)} + 0.22\phi_{\text{N2}(2\text{p}''_z)}\end{aligned} \tag{8.21}$$

$$\begin{aligned}\psi_{\text{MO6}}(\text{N}_2) = &\ 0.44\phi_{\text{N1}(2\text{p}'_y)} + 0.22\phi_{\text{N1}(2\text{p}''_y)} - 0.13\phi_{\text{N1}(2\text{p}'_z)} \\ &+ 0.44\phi_{\text{N2}(2\text{p}'_y)} + 0.22\phi_{\text{N2}(2\text{p}''_y)} - 0.13\phi_{\text{N2}(2\text{p}'_z)}\end{aligned} \tag{8.22}$$

In MO5 and MO6, as there is hybridization between 2p_y and 2p_z orbitals, rotated 2p orbital is given. Two N1 lobes interact with two N2 lobes. From chemical bonding rule, it is found that π-type covalent bonding is formed between N1 and N2 2p orbitals. As the orbital energies of MO5 and MO6 are the same, they are degenerated. Note that the wave-functions of MO5 and MO6 are different as quantum mechanics, due to direction difference. The obtained wave-function of MO7 is

$$\begin{aligned}\psi_{\mathrm{MO7}}(\mathrm{N_2}) = 0.11\phi_{\mathrm{N1(2s')}} + 0.34\phi_{\mathrm{N1(2s'')}} - 0.46\phi_{\mathrm{N1}(2\mathrm{p}'_x)} - 0.19\phi_{\mathrm{N1}(2\mathrm{p}''_x)} \\ + 0.11\phi_{\mathrm{N2(2s')}} + 0.34\phi_{\mathrm{N2(2s'')}} + 0.46\phi_{\mathrm{N2}(2\mathrm{p}'_x)} + 0.19\phi_{\mathrm{N2}(2\mathrm{p}''_x)}\end{aligned} \tag{8.23}$$

There is hybridization between $2\mathrm{p}_x$ and 2s orbitals. The main coefficients are for $2\mathrm{p}_x$ orbital. The coefficients of N1 and N2 2s orbitals are positive, and the signs of the coefficients of N1 and N2 $2\mathrm{p}_x$ orbitals are different. One N1 lobe interacts with one N2 lobe. From chemical bonding rule, σ-type covalent bonding is formed between N1 and N2 $2\mathrm{p}_x$ orbitals.

MO8, MO9 and MO10 are unoccupied. The obtained wave-function of MO8, MO9 and MO10 are

$$\psi_{\mathrm{MO8}}(\mathrm{N_2}) = -0.50\phi_{\mathrm{N1}(2\mathrm{p}'_y)} - 0.56\phi_{\mathrm{N1}(2\mathrm{p}''_y)} + 0.50\phi_{\mathrm{N2}(2\mathrm{p}'_y)} + 0.56\phi_{\mathrm{N2}(2\mathrm{p}''_y)} \tag{8.24}$$

$$\psi_{\mathrm{MO9}}(\mathrm{N_2}) = -0.50\phi_{\mathrm{N1}(2\mathrm{p}'_z)} - 0.56\phi_{\mathrm{N1}(2\mathrm{p}''_z)} + 0.50\phi_{\mathrm{N2}(2\mathrm{p}'_z)} + 0.56\phi_{\mathrm{N2}(2\mathrm{p}''_z)} \tag{8.25}$$

$$\begin{aligned}\psi_{\mathrm{MO10}}(\mathrm{N_2}) = -0.24\phi_{\mathrm{N1(2s')}} - 3.85\phi_{\mathrm{N1(2s'')}} - 0.12\phi_{\mathrm{N1}(2\mathrm{p}'_x)} - 2.58\phi_{\mathrm{N1}(2\mathrm{p}''_x)} \\ + 0.24\phi_{\mathrm{N2(2s')}} + 3.85\phi_{\mathrm{N2(2s'')}} - 0.12\phi_{\mathrm{N2}(2\mathrm{p}'_x)} - 2.58\phi_{\mathrm{N2}(2\mathrm{p}''_x)}\end{aligned} \tag{8.26}$$

It is found that inversion π-type covalent bonding is formed in MO8 and MO9, and inversion σ-type covalent bonding is formed in MO10. The orbital energies of MO8 and MO9 are the same, due to the degeneracy.

Figure 8.11 shows the potential energy curve of nitrogen molecule, changing interatomic N1–N2 distance. The local minimum is given at 1.097 Å, corresponding to the N_2 bond length. Bond dissociation energy can be estimated from the total energy difference between the local minimum and completely dissociated point.

$$E_{\mathrm{dissociation}}(\mathrm{N_2}) = E(\mathrm{N_{quintet}}) + E(\mathrm{N_{quintet}}) - E(\mathrm{N_2}) \tag{8.27}$$

Note that it is assumed that two neutral nitrogen atoms with quintet spin state exist at the completely dissociated point. The bond dissociated energy is estimated to be 208.9 kcal/mol. It is much larger than hydrogen molecule. The zero-point vibration energy is 3.458 kcal/mol. It is much smaller than the dissociation energy.

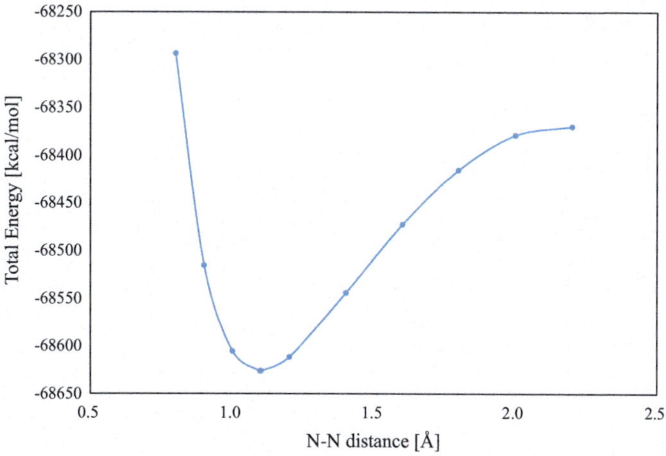

Fig. 8.11 Potential energy curve of nitrogen molecule, changing the interatomic N1–N2 distance (CCSD/aug-cc-pVTZ)

8.5 Oxygen Molecule

8.5.1 Triplet and Singlet Oxygen Molecules

It is well known that oxygen molecule (O1–O2) exhibits triplet spin state. Let us confirm the fact, from the viewpoint of energetics. B3LYP/6-31G* calculation is performed for singlet and triplet oxygen molecules.

Figures 8.12 and 8.13 depict the potential energy curves of triplet and singlet oxygen molecules, changing the interatomic O1–O2 distance. The local minima are given at 1.202 Å in triplet O_2 and 1.209 Å in singlet O_2. The bond lengths are almost the same. The total energies of triplet and singlet spin states are −94,203.08 and −94,170.22 kcal/mol, respectively. The total energy of triplet spin state is 32.87 kcal/mol lower.

Bond dissociation energy can be estimated from the total energy difference between the local minimum and completely dissociated point.

$$E_{\text{dissociation}}(O_2) = E(O_{\text{triplet}}) + E(O_{\text{triplet}}) - E(O_2) \tag{8.28}$$

Note that it is assumed that two triplet oxygen atoms appear at the completely dissociated point. The dissociation energies of triplet and singlet spin states are 107.30 and 74.44 kcal/mol, respectively. In this point, it is found that oxygen atoms are strongly bound in triplet state. In comparison with nitrogen molecule, both values are about half. It implies that oxygen molecule is more reactive than nitrogen molecule.

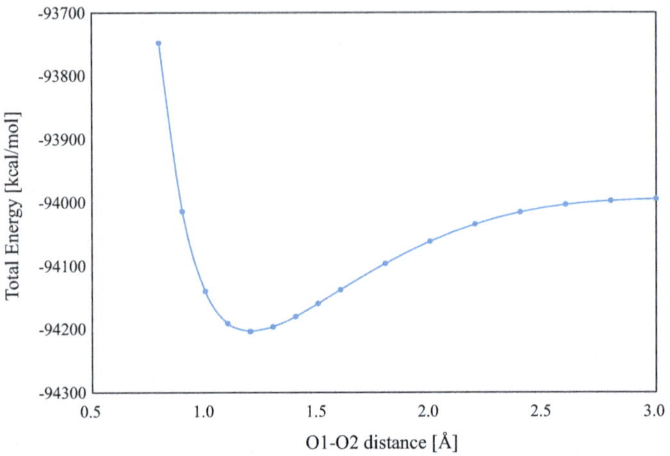

Fig. 8.12 Potential energy curve of triplet oxygen molecule, changing the interatomic O1–O2 distance (CCSD/aug-cc-pVTZ)

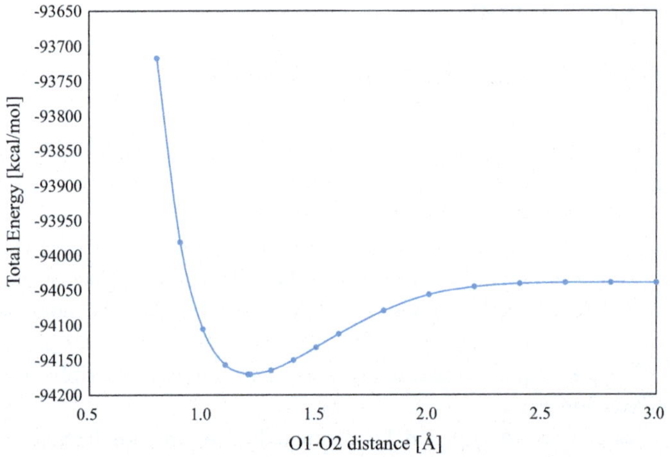

Fig. 8.13 Potential energy curve for singlet oxygen molecule, changing the interatomic O1–O2 distance (CCSD/aug-cc-pVTZ)

8.5.2 Molecular Orbital of Triplet Oxygen Molecule

In triplet oxygen molecule, thirty alpha and beta MOs are produced, because 6-31G* basis set of oxygen atom has fifteen basis functions. Nine electrons occupy nine alpha MOs such as MO1α, MO2α, MO3α, MO4α, MO5α, MO6α, MO7α, MO8α and MO9α. Seven electrons occupy seven beta MOs such as MO1β, MO2β, MO3β, MO4β, MO5β, MO6β and MO7β.

Figure 8.14 depicts the orbital energy diagram and molecular orbitals of triplet oxygen molecule. The obtained wave-functions of MO1α, MO1β MO2α and MO2β are

$$\psi_{MO1\alpha}(O_2) = -0.70\phi_{O1(1s)} - 0.70\phi_{O2(1s)} \tag{8.29}$$

$$\psi_{MO1\beta}(O_2) = -0.70\phi_{O1(1s)} - 0.70\phi_{O2(1s)} \tag{8.30}$$

$$\psi_{MO2\alpha}(O_2) = 0.70\phi_{O1(1s)} - 0.70\phi_{O2(1s)} \tag{8.31}$$

$$\psi_{MO2\beta}(O_2) = -0.70\phi_{O1(1s)} + 0.70\phi_{O2(1s)} \tag{8.32}$$

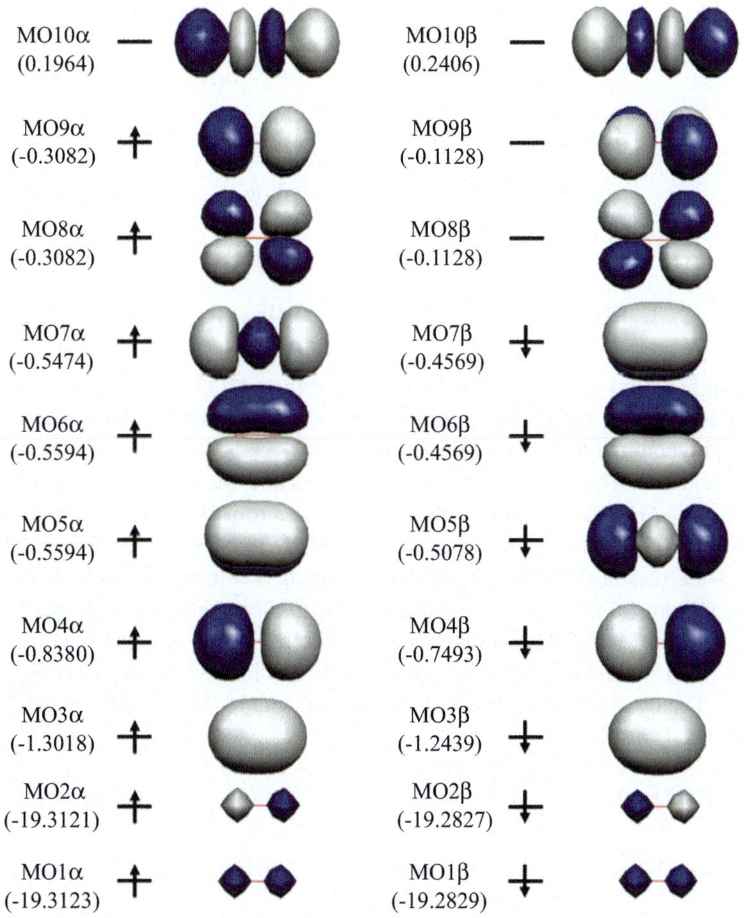

Fig. 8.14 Orbital energy diagram and molecular orbitals of triplet oxygen molecule at optimized structure (B3LYP/6-31G*). The calculated orbital energy is shown in *parentheses*

In MO1α and MO1β, though orbital energies are different, wave-functions are the same. MO1α and MO1β are paired. Due to the same reason, MO2α and MO2β are paired. MO1α, MO1β, MO2α and MO2β consist of O1 and O2 1s orbitals. O1 1s orbital overlaps O2 1s orbital. One O1 lobe interacts with one O2 lobe. From chemical bonding rule, it is found that σ-type covalent bonding is formed between O1 and O2 1s orbitals. In MO2α and MO2β, as the signs of O1 and O2 coefficients are different, inversion σ-type covalent bonding is formed. The lower orbital energies are given in MO1α, MO1β, MO2α and MO2β because electrons are in inner K shell. The obtained wave-functions of MO3α, MO3β MO4α and MO4β are

$$\begin{aligned}\psi_{\mathrm{MO3\alpha}}(\mathrm{O_2}) = & -0.15\phi_{\mathrm{O1(1s)}} + 0.36\phi_{\mathrm{O1(2s')}} + 0.22\phi_{\mathrm{O1(2s'')}} + 0.18\phi_{\mathrm{O1(2p'_x)}} \\ & - 0.15\phi_{\mathrm{O2(1s)}} + 0.36\phi_{\mathrm{O2(2s')}} + 0.22\phi_{\mathrm{O2(2s'')}} - 0.18\phi_{\mathrm{O2(2p'_x)}}\end{aligned} \tag{8.33}$$

$$\begin{aligned}\psi_{\mathrm{MO3\beta}}(\mathrm{O_2}) = & -0.15\phi_{\mathrm{O1(1s)}} + 0.34\phi_{\mathrm{O1(2s')}} + 0.23\phi_{\mathrm{O1(2s'')}} + 0.19\phi_{\mathrm{O1(2p'_x)}} \\ & - 0.15\phi_{\mathrm{O2(1s)}} + 0.34\phi_{\mathrm{O2(2s')}} + 0.22\phi_{\mathrm{O2(2s'')}} - 0.19\phi_{\mathrm{O2(2p'_x)}}\end{aligned} \tag{8.34}$$

$$\begin{aligned}\psi_{\mathrm{MO4\alpha}}(\mathrm{O_2}) = & \ 0.17\phi_{\mathrm{O1(1s)}} - 0.40\phi_{\mathrm{O1(2s')}} - 0.47\phi_{\mathrm{O1(2s'')}} + 0.13\phi_{\mathrm{O1(2p'_x)}} \\ & - 0.17\phi_{\mathrm{O2(1s)}} + 0.40\phi_{\mathrm{O2(2s')}} + 0.47\phi_{\mathrm{O2(2s'')}} + 0.13\phi_{\mathrm{O2(2p'_x)}}\end{aligned} \tag{8.35}$$

$$\begin{aligned}\psi_{\mathrm{MO4\beta}}(\mathrm{O_2}) = & -0.16\phi_{\mathrm{O1(1s)}} + 0.38\phi_{\mathrm{O1(2s')}} + 0.48\phi_{\mathrm{O1(2s'')}} - 0.14\phi_{\mathrm{O1(2p'_x)}} \\ & + 0.16\phi_{\mathrm{O2(1s)}} - 0.38\phi_{\mathrm{O2(2s')}} - 0.48\phi_{\mathrm{O2(2s'')}} - 0.14\phi_{\mathrm{O2(2p'_x)}}\end{aligned} \tag{8.36}$$

Though orbital energy of MO3α is smaller than MO3β, the wave-functions of MO3α and MO3β are qualitatively the same. MO3α and MO3β are paired. In O1 and O2, inversion hybridization occurs between 1s and 2s orbitals, and $2p_x$ orbital is also hybridized. The main coefficients of MO3α and MO3β are for O1 and O2 2s orbitals. One O1 lobe interacts with one O2 lobe. From chemical bonding rule, it is found that σ-type covalent bonding is formed between O1 and O2 2s orbitals. Due to the same reason, MO4α and MO4β are paired. The main coefficients of MO4α and MO4β are for O1 and O2 2s orbitals. One O1 lobe interacts with one O2 lobe. The signs of O1 and O2 2s coefficient are different. From chemical bonding rule, it is found that inversion σ-type covalent bonding is formed between O1 and O2 2s orbitals. The obtained wave-functions of MO5α, MO5β, MO6α, MO6β, MO7α and MO7β are

$$\psi_{\mathrm{MO5\alpha}}(\mathrm{O_2}) = 0.48\phi_{\mathrm{O1(2p'_z)}} + 0.25\phi_{\mathrm{O1(2p''_z)}} + 0.48\phi_{\mathrm{O2(2p'_z)}} + 0.25\phi_{\mathrm{O2(2p''_z)}} \tag{8.37}$$

$$\begin{aligned}\psi_{\mathrm{MO5\beta}}(\mathrm{O_2}) = & -0.11\phi_{\mathrm{O1(2s')}} - 0.30\phi_{\mathrm{O1(2s'')}} + 0.46\phi_{\mathrm{O1(2p'_x)}} + 0.22\phi_{\mathrm{O1(2p''_x)}} \\ & - 0.11\phi_{\mathrm{O2(2s')}} - 0.22\phi_{\mathrm{O2(2s')}} - 0.46\phi_{\mathrm{O2(2p'_x)}} - 0.22\phi_{\mathrm{O2(2p''_x)}}\end{aligned} \tag{8.38}$$

$$\psi_{MO6\alpha}(O_2) = -0.48\phi_{O1(2p'_y)} - 0.25\phi_{O1(2p''_y)} - 0.48\phi_{O2(2p'_y)} - 0.25\phi_{O2(2p''_y)} \tag{8.39}$$

$$\begin{aligned}\psi_{MO6\beta}(O_2) = &-0.42\phi_{O1(2p'_y)} - 0.27\phi_{O1(2p''_y)} + 0.14\phi_{O1(2p'_z)} \\ &- 0.42\phi_{O2(2p'_y)} - 0.27\phi_{O2(2p''_y)} + 0.14\phi_{O2(2p'_z)}\end{aligned} \tag{8.40}$$

$$\begin{aligned}\psi_{MO7\alpha}(O_2) = &\,0.12\phi_{O1(2s')} + 0.28\phi_{O1(2s'')} - 0.47\phi_{O1(2p'_x)} - 0.22\phi_{O1(2p''_x)} \\ &+ 0.12\phi_{O2(2s')} + 0.28\phi_{O2(2s'')} + 0.47\phi_{O2(2p'_x)} + 0.22\phi_{O2(2p''_x)}\end{aligned} \tag{8.41}$$

$$\begin{aligned}\psi_{MO7\beta}(O_2) = &\,0.14\phi_{O1(2p'_y)} + 0.42\phi_{O1(2p'_z)} + 0.27\phi_{O1(2p''_z)} \\ &+ 0.14\phi_{O2(2p'_y)} + 0.42\phi_{O2(2p'_z)} + 0.27\phi_{O2(2p''_z)}\end{aligned} \tag{8.42}$$

MO5α consists of O1 and O2 $2p_z$ orbitals. In MO7β, though O1 and O2 $2p_y$ orbitals are hybridized, the main coefficients are for O1 and O2 $2p_z$ orbitals. MO5α and MO7β are paired. Two O1 lobes interact with two O2 lobes. From chemical bonding rule, it is found that π-type covalent bonding is formed between O1 and O2 2p orbitals in MO5α and MO7β. MO5α and MO6α are degenerated. MO6α consists of O1 and O2 $2p_y$ orbitals. In MO6β, though O1 and O2 $2p_z$ orbitals are hybridized, the main coefficients are for O1 and O2 $2p_y$ orbitals. MO6α and MO6β are paired. Two O1 lobes interact with two O2 lobes. From chemical bonding rule, it is found that the π-type covalent bonding is formed between O1 2p and O2 2p orbitals in MO6α and MO6β. In MO7α and MO5β, though O1 and O2 2s orbitals are hybridized, the main coefficients are for $2p_x$ orbital. MO5β and MO7α are paired. One O1 lobe interacts with one O2 lobe. From chemical bonding rule, it is found that σ-type covalent bonding is formed between O1 and O2 2p orbitals in MO7α and MO5β. The obtained wave-functions of MO8α, MO8β, MO9α, MO9β, MO10α and MO10β are

$$\begin{aligned}\psi_{MO8\alpha}(O_2) = &-0.53\phi_{O1(2p'_y)} - 0.37\phi_{O1(2p''_y)} + 0.13\phi_{O1(2p'_z)} \\ &+ 0.53\phi_{O2(2p'_y)} + 0.37\phi_{O2(2p''_y)} - 0.13\phi_{O2(2p'_z)}\end{aligned} \tag{8.43}$$

$$\psi_{MO8\beta}(O_2) = 0.51\phi_{O1(2p'_y)} + 0.43\phi_{O1(2p''_y)} - 0.51\phi_{O2(2p'_y)} - 0.43\phi_{O2(2p''_y)} \tag{8.44}$$

$$\begin{aligned}\psi_{MO9\alpha}(O_2) = &-0.13\phi_{O1(2p'_y)} - 0.53\phi_{O1(2p'_z)} - 0.37\phi_{O1(2p''_z)} \\ &+ 0.13\phi_{O2(2p'_y)} + 0.53\phi_{O2(2p'_z)} + 0.37\phi_{O2(2p''_z)}\end{aligned} \tag{8.45}$$

$$\psi_{\mathrm{MO9\beta}}(\mathrm{O_2}) = 0.51\phi_{\mathrm{O1}(2\mathrm{p}'_z)} + 0.43\phi_{\mathrm{O1}(2\mathrm{p}''_z)} - 0.51\phi_{\mathrm{O2}(2\mathrm{p}'_z)} - 0.43\phi_{\mathrm{O2}(2\mathrm{p}''_z)} \tag{8.46}$$

$$\begin{aligned}\psi_{\mathrm{MO10\alpha}}(\mathrm{O_2}) = {} & 0.24\phi_{\mathrm{O1}(2\mathrm{s}')} + 1.02\phi_{\mathrm{O1}(2\mathrm{s}'')} + 0.52\phi_{\mathrm{O1}(2\mathrm{p}'_x)} + 0.99\phi_{\mathrm{O1}(2\mathrm{p}''_x)} \\ & - 0.24\phi_{\mathrm{O2}(2\mathrm{s}')} - 1.02\phi_{\mathrm{O2}(2\mathrm{s}'')} + 0.52\phi_{\mathrm{O2}(2\mathrm{p}'_x)} + 0.99\phi_{\mathrm{O2}(2\mathrm{p}''_x)}\end{aligned} \tag{8.47}$$

$$\begin{aligned}\psi_{\mathrm{MO10\beta}}(\mathrm{O_2}) = {} & -0.24\phi_{\mathrm{O1}(2\mathrm{s}')} - 1.12\phi_{\mathrm{O1}(2\mathrm{s}'')} - 0.51\phi_{\mathrm{O1}(2\mathrm{p}'_x)} - 1.05\phi_{\mathrm{O1}(2\mathrm{p}''_x)} \\ & + 0.24\phi_{\mathrm{O2}(2\mathrm{s}')} + 1.12\phi_{\mathrm{O2}(2\mathrm{s}'')} - 0.51\phi_{\mathrm{O2}(2\mathrm{p}'_x)} - 1.05\phi_{\mathrm{O2}(2\mathrm{p}''_x)}\end{aligned} \tag{8.48}$$

Degenerated MO8α and MO9α are occupied and are responsible for spin density. In MO8α and MO9α, though $2p_z$ and $2p_y$ orbitals are also hybridized, the main coefficients are for $2p_y$ and $2p_z$ orbitals, respectively. There are orbital overlaps between O1 and O2 $2p_y$ orbitals in MO8α, and between O1 and O2 $2p_z$ orbitals in MO9α. The sign of O1 coefficient is opposite to the sign of O2 coefficient in MO8α and MO9α. From chemical bonding rule, it is found that inversion π-type covalent bonding is formed between O1 and O2 2p orbitals. MO8β, MO9β, MO10α and MO10β are unoccupied. In MO8β and MO9β, inversion π-type covalent bonding is formed. In MO10α and MO10β, inversion σ-type covalent bonding is formed.

8.5.3 Molecular Orbital of Singlet Oxygen Molecule

In singlet oxygen molecule, thirty MOs are produced, because 6-31G* basis set of oxygen atom has fifteen basis functions. Sixteen electrons occupy eight MOs with spin pairs.

Figure 8.15 depicts the orbital energy diagram and molecular orbitals of singlet oxygen molecule at optimized structure (B3LYP/6-31G*). The obtained wave-functions of MO1, MO2, MO3 and MO4 are

$$\psi_{\mathrm{MO1}}(\mathrm{O_2}) = -0.70\phi_{\mathrm{O1}(1\mathrm{s})} - 0.70\phi_{\mathrm{O2}(1\mathrm{s})} \tag{8.49}$$

$$\psi_{\mathrm{MO2}}(\mathrm{O_2}) = -0.70\phi_{\mathrm{O1}(1\mathrm{s})} + 0.70\phi_{\mathrm{O2}(1\mathrm{s})} \tag{8.50}$$

$$\begin{aligned}\psi_{\mathrm{MO3}}(\mathrm{O_2}) = {} & -0.15\phi_{\mathrm{O1}(1\mathrm{s})} + 0.35\phi_{\mathrm{O1}(2\mathrm{s}')} + 0.22\phi_{\mathrm{O1}(2\mathrm{s}'')} + 0.18\phi_{\mathrm{O1}(2\mathrm{p}'_x)} \\ & - 0.15\phi_{\mathrm{O2}(1\mathrm{s})} + 0.35\phi_{\mathrm{O2}(2\mathrm{s}')} + 0.22\phi_{\mathrm{O2}(2\mathrm{s}'')} - 0.18\phi_{\mathrm{O2}(2\mathrm{p}'_x)}\end{aligned} \tag{8.51}$$

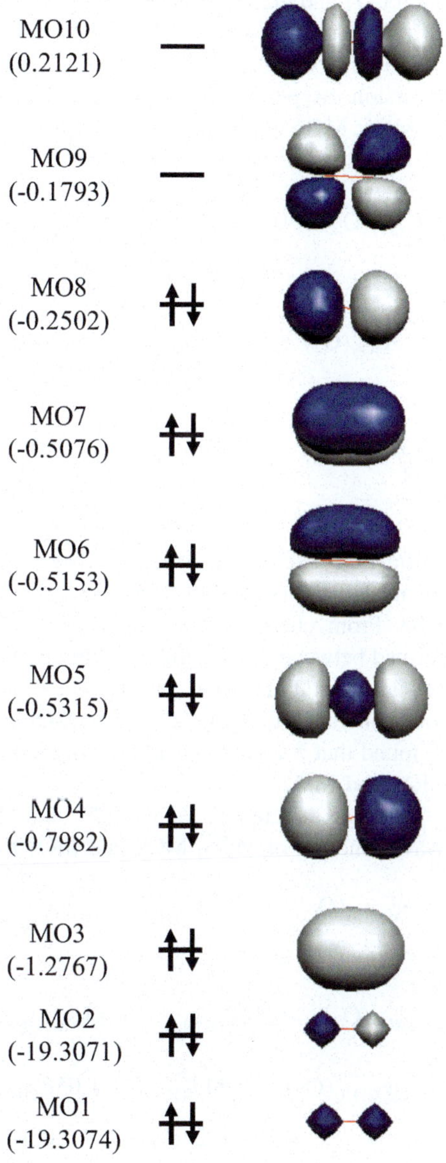

Fig. 8.15 Orbital energy diagram and molecular orbitals of singlet oxygen molecule at optimized structure (B3LYP/6-31G*). The calculated orbital energy is shown in *parentheses*

$$\begin{aligned}\psi_{\mathrm{MO4}}(\mathrm{O_2}) = &-0.17\phi_{\mathrm{O1(1s)}} + 0.39\phi_{\mathrm{O1(2s')}} + 0.48\phi_{\mathrm{O1(2s'')}} - 0.13\phi_{\mathrm{O1}(2\mathrm{p}'_x)} \\ &+ 0.17\phi_{\mathrm{O2(1s)}} - 0.39\phi_{\mathrm{O2(2s')}} - 0.48\phi_{\mathrm{O2(2s'')}} - 0.13\phi_{\mathrm{O2}(2\mathrm{p}'_x)}\end{aligned} \quad (8.52)$$

As same as triplet oxygen molecule, σ-type covalent bonding is formed between O1 and O2 1s orbitals in MO1, and inversion σ-type covalent bonding is formed between O1 and O2 1s orbitals in MO2. In MO3 and MO4, the main coefficients are for O1 and O2 2s orbitals, though there are inversion hybridizations between O1 1s and 2s orbitals, and between O2 1s and 2s orbitals, combined with hybridizations of O1 and O2 $2p_x$ orbitals. σ-type covalent bonding is formed in MO3, and inversion σ-type covalent bonding is formed in MO4.

MO5, MO6 and MO7 are for 2p orbital. The obtained wave-functions of MO5, MO6 and MO7 are

$$\begin{aligned}\psi_{MO5}(O_2) = {} & 0.12\phi_{O1(2s')} + 0.29\phi_{O1(2s'')} - 0.47\phi_{O1(2p'_x)} - 0.22\phi_{O1(2p''_x)} \\ & + 0.12\phi_{O2(2s')} + 0.29\phi_{O2(2s'')} + 0.47\phi_{O2(2p'_x)} + 0.22\phi_{O2(2p''_x)}\end{aligned} \tag{8.53}$$

$$\psi_{MO6}(O_2) = -0.46\phi_{O1(2p'_y)} - 0.27\phi_{O1(2p''_y)} - 0.46\phi_{O2(2p'_y)} - 0.27\phi_{O2(2p''_y)} \tag{8.54}$$

$$\psi_{MO7}(O_2) = -0.47\phi_{O1(2p'_z)} - 0.26\phi_{O1(2p''_z)} - 0.47\phi_{O2(2p'_z)} - 0.26\phi_{O2(2p''_z)} \tag{8.55}$$

In MO5, there is hybridization between $2p_x$ and 2s orbitals in O1 and O2. The main coefficients are for O1 and O2 $2p_x$ orbitals. One O1 lobe interacts with one O2 lobe. From chemical bonding rule, it is found that σ-type covalent bonding is formed between O1 and O2 $2p_x$ orbitals. There are orbital overlaps between O1 and O2 $2p_y$ orbitals in MO6, and between O1 and O2 $2p_z$ orbitals in MO7. Two O1 lobes interact with two O2 lobes in MO6 and MO7. From chemical bonding rule, it is found that π-type covalent bonding is formed between O1 and O2 2p orbitals in MO6 and MO7.

MO8 is occupied, and MO9 and MO10 are unoccupied. The obtained wave-function of MO8, MO9 and MO10 are

$$\psi_{MO8}(O_2) = -0.53\phi_{O1(2p'_z)} - 0.39\phi_{O1(2p''_z)} + 0.53\phi_{O2(2p'_z)} + 0.39\phi_{O2(2p''_z)} \tag{8.56}$$

$$\psi_{MO9}(O_2) = 0.52\phi_{O1(2p'_y)} + 0.41\phi_{O1(2p''_y)} - 0.52\phi_{O2(2p'_y)} - 0.41\phi_{O2(2p''_y)} \tag{8.57}$$

$$\begin{aligned}\psi_{MO10}(O_2) = {} & 0.24\phi_{O1(2s')} + 1.05\phi_{O1(2s'')} + 0.52\phi_{O1(2p'_x)} + 1.01\phi_{O1(2p''_x)} \\ & - 0.24\phi_{O2(2s')} - 1.05\phi_{O2(2s'')} + 0.52\phi_{O2(2p'_x)} + 1.01\phi_{O2(2p''_x)}\end{aligned} \tag{8.58}$$

In MO8, there is orbital overlap between O1 and O2 $2p_z$ orbitals. Two O1 lobes interact with two O2 lobes. The sign of O1 coefficients are opposite to the sign of O2 coefficients. From chemical bonding rule, it is found that inversion π-type covalent bonding is formed between O1 and O2 2p orbitals. MO9 and MO10 are

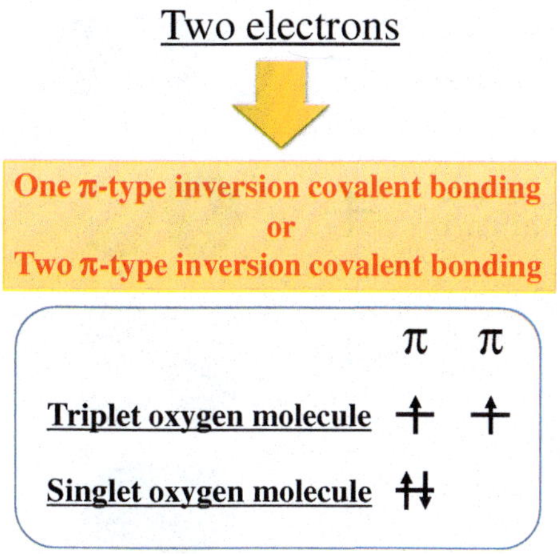

Fig. 8.16 Difference between triplet and singlet oxygen molecules

unoccupied. In MO9, inversion π-type covalent bonding is formed between O1 and O2 2p orbitals. In MO10, inversion σ-type covalent bonding is formed between O1 and O2 2p orbitals.

Figure 8.16 summarizes the difference between triplet and singlet oxygen molecules. In triplet oxygen molecule, two electrons are allocated in two MOs of inversion π-type covalent bonding. On the other hand, in singlet oxygen molecule, two electrons are allocated in one MO of inversion π-type covalent bonding.

8.5.4 Superoxide

It is well known that superoxide, hydroxyl radical and singlet oxygen molecule have high chemical reactivity. They are called reactive oxygen species. Superoxide denotes monovalent oxygen molecule anion (O_2^-) with doublet spin state. Let us examine MOs of superoxide. B3LYP/6-31G* calculation is performed for superoxide. In superoxide, thirty alpha and beta MOs are produced, because 6-31G* basis set of oxygen atom has fifteen basis functions.

Figure 8.17 depicts the orbital energy diagram and molecular orbitals of superoxide. Alpha electrons are occupied up to MO9α, and beta electrons are occupied up to MO8β. The obtained wave-functions of MO1α, MO1β, MO2α and MO2β are

$$\psi_{MO1\alpha}(O_2^-) = -0.70\phi_{O1(1s)} - 0.70\phi_{O2(1s)} \tag{8.59}$$

$$\psi_{MO1\beta}(O_2^-) = -0.70\phi_{O1(1s)} - 0.70\phi_{O2(1s)} \tag{8.60}$$

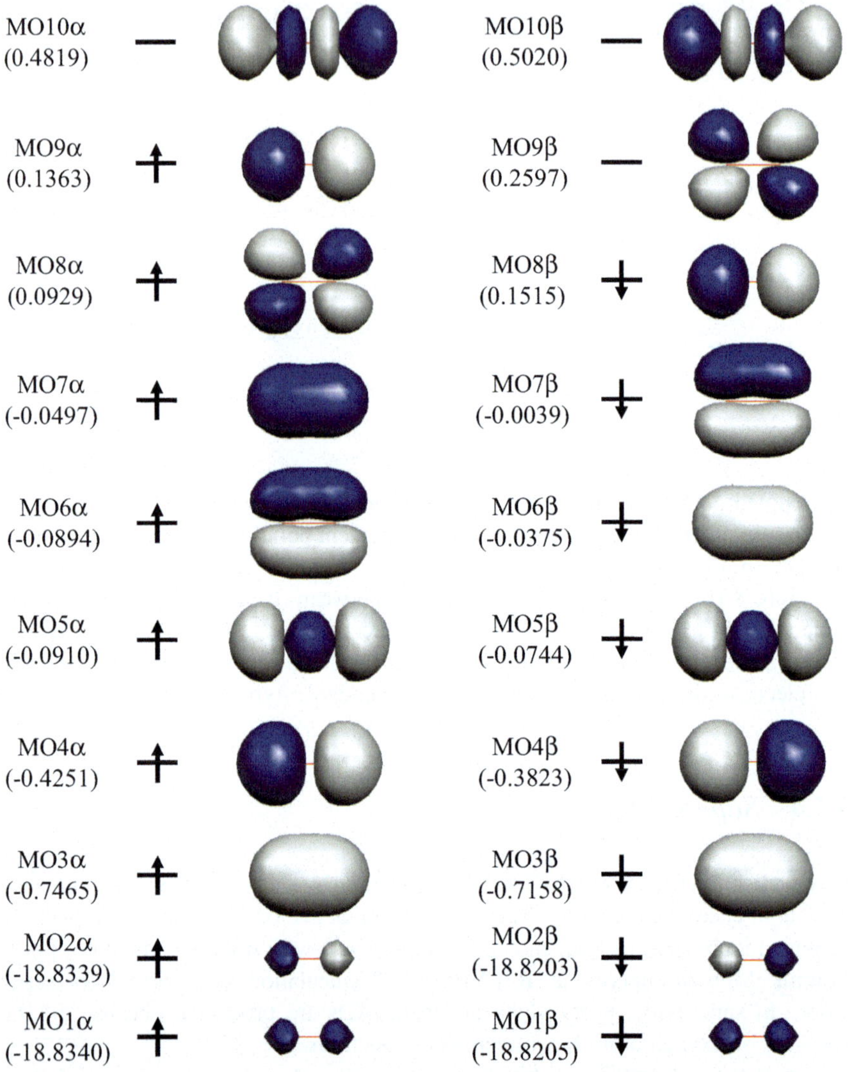

Fig. 8.17 Orbital energy diagram and molecular orbitals of superoxide at optimized structure (B3LYP/6-31G*). The calculated orbital energy is shown in *parentheses*

$$\psi_{MO2\alpha}(O_2{}^-) = -0.70\phi_{O1(1s)} + 0.70\phi_{O2(1s)} \tag{8.61}$$

$$\psi_{MO2\beta}(O_2{}^-) = 0.70\phi_{O1(1s)} - 0.70\phi_{O2(1s)} \tag{8.62}$$

As same as oxygen molecule, σ-type covalent bonding is formed between O1 and O2 1s orbitals in MO1α and MO1β. MO1α and MO1β are paired. Inversion σ-type covalent bonding is formed between O1 and O2 1s orbitals in MO2α and

MO2β. MO2α and MO2β are paired. The obtained wave-functions of MO3α, MO3β, MO4α and MO4β are

$$\psi_{\mathrm{MO3\alpha}}(\mathrm{O_2}^-) = -0.15\phi_{\mathrm{O1(1s)}} + 0.35\phi_{\mathrm{O1(2s')}} + 0.29\phi_{\mathrm{O1(2s'')}} + 0.13\phi_{\mathrm{O1}(2p'_x)} - 0.15\phi_{\mathrm{O2(1s)}} + 0.35\phi_{\mathrm{O2(2s')}} + 0.29\phi_{\mathrm{O2(2s'')}} - 0.13\phi_{\mathrm{O2}(2p'_x)} \quad (8.63)$$

$$\psi_{\mathrm{MO3\beta}}(\mathrm{O_2}^-) = -0.15\phi_{\mathrm{O1(1s)}} + 0.33\phi_{\mathrm{O1(2s')}} + 0.29\phi_{\mathrm{O1(2s'')}} + 0.14\phi_{\mathrm{O1}(2p'_x)} - 0.15\phi_{\mathrm{O2(1s)}} + 0.33\phi_{\mathrm{O2(2s')}} + 0.29\phi_{\mathrm{O2(2s'')}} - 0.14\phi_{\mathrm{O2}(2p'_x)} \quad (8.64)$$

$$\psi_{\mathrm{MO4\alpha}}(\mathrm{O_2}^-) = 0.17\phi_{\mathrm{O1(1s)}} - 0.37\phi_{\mathrm{O1(2s')}} - 0.48\phi_{\mathrm{O1(2s'')}} - 0.17\phi_{\mathrm{O2(1s)}} + 0.37\phi_{\mathrm{O2(2s')}} + 0.48\phi_{\mathrm{O2(2s'')}} \quad (8.65)$$

$$\psi_{\mathrm{MO4\beta}}(\mathrm{O_2}^-) = -0.17\phi_{\mathrm{O1(1s)}} + 0.36\phi_{\mathrm{O1(2s')}} + 0.49\phi_{\mathrm{O1(2s'')}} + 0.17\phi_{\mathrm{O2(1s)}} - 0.36\phi_{\mathrm{O2(2s')}} - 0.49\phi_{\mathrm{O2(2s'')}} \quad (8.66)$$

In MO3α and MO3β, the main coefficients are for O1 and O2 2s orbitals, though there are inversion hybridizations between O1 1s and 2s orbitals, and between O2 1s and 2s orbitals, combined with hybridizations of O1 and O2 $2p_x$ orbitals. MO3α and MO3β are paired. One O1 lobe interacts with one O2 lobe. From chemical bonding rule, it is found that σ-type covalent bonding is formed between O1 and O2 2s orbitals. In MO4α and MO4β, the main coefficients are for O1 and O2 2s orbitals, though there are inversion hybridizations between O1 1s and 2s orbitals, and between O2 1s and 2s orbitals. MO4α and MO4β are paired. One O1 lobe interacts with one O2 lobe. The sign of O1 coefficients is opposite to the sign of O2 coefficients. From chemical bonding rule, it is found that inversion σ-type covalent bonding is formed between O1 and O2 1s 2s orbitals.

MO5α, MO5β, MO6α, MO6β, MO7α and MO7β are for 2p orbital. The obtained wave-functions of MO5α, MO5β, MO6α, MO6β, MO7α and MO7β are

$$\psi_{\mathrm{MO5\alpha}}(\mathrm{O_2}^-) = 0.28\phi_{\mathrm{O1(2s'')}} - 0.45\phi_{\mathrm{O1}(2p'_x)} - 0.27\phi_{\mathrm{O1}(2p''_x)} + 0.28\phi_{\mathrm{O2(2s'')}} + 0.45\phi_{\mathrm{O2}(2p'_x)} + 0.27\phi_{\mathrm{O2}(2p''_x)} \quad (8.67)$$

$$\psi_{\mathrm{MO5\beta}}(\mathrm{O_2}^-) = 0.29\phi_{\mathrm{O1(2s'')}} - 0.45\phi_{\mathrm{O1}(2p'_x)} - 0.27\phi_{\mathrm{O1}(2p''_x)} + 0.29\phi_{\mathrm{O2(2s'')}} + 0.45\phi_{\mathrm{O2}(2p'_x)} + 0.27\phi_{\mathrm{O2}(2p''_x)} \quad (8.68)$$

$$\psi_{\mathrm{MO6\alpha}}(\mathrm{O_2}^-) = -0.44\phi_{\mathrm{O1}(2p'_y)} - 0.30\phi_{\mathrm{O1}(2p''_y)} - 0.44\phi_{\mathrm{O2}(2p'_y)} - 0.30\phi_{\mathrm{O2}(2p''_y)} \quad (8.69)$$

$$\psi_{MO6\beta}(O_2^-) = 0.43\phi_{O1(2p'_z)} + 0.31\phi_{O1(2p''_z)} + 0.43\phi_{O2(2p'_z)} + 0.31\phi_{O2(2p''_z)} \tag{8.70}$$

$$\psi_{MO7\alpha}(O_2^-) = -0.44\phi_{O1(2p'_z)} - 0.31\phi_{O1(2p''_z)} - 0.44\phi_{O2(2p'_z)} - 0.31\phi_{O2(2p''_z)} \tag{8.71}$$

$$\psi_{MO7\beta}(O_2^-) = -0.41\phi_{O1(2p'_y)} - 0.33\phi_{O1(2p''_y)} - 0.41\phi_{O2(2p'_y)} - 0.33\phi_{O2(2p''_y)} \tag{8.72}$$

In MO5α and MO5β, there is hybridization between oxygen $2p_x$ and 2s orbitals. The main coefficient is for oxygen $2p_x$ orbital. MO5α and MO5β are paired. There is orbital overlap between O1 and O2 $2p_x$ orbitals. One O1 lobe interacts with one O2 lobe. From chemical bonding rule, it is found that the σ-type covalent bonding is formed between O1 and O2 $2p_x$ orbitals. MO6α and MO7β are paired. In MO6α and MO7β, there is orbital overlap between O1 and O2 $2p_y$ orbitals. Two O1 lobes interact with two O2 lobes. From chemical bonding rule, it is found that the π-type covalent bonding is formed between O1 and O2 $2p_y$ orbitals. MO7α and MO6β are paired. In MO7α and MO6β, there is orbital overlap between O1 and O2 $2p_z$ orbitals. Two O1 lobes interact with two O2 lobes. From chemical bonding rule, it is found that the π-type covalent bonding is formed between O1 and O2 $2p_z$ orbitals.

MO8α, MO8β, MO9α, MO9β, MO10α and MO10β are also for 2p orbital. MO8α, MO8β and MO9α are occupied. The obtained wave-functions of MO8α, MO8β, MO9α, MO9β, MO10α and MO10β are

$$\psi_{MO8\alpha}(O_2^-) = 0.50\phi_{O1(2p'_y)} + 0.40\phi_{O1(2p''_y)} - 0.50\phi_{O2(2p'_y)} - 0.40\phi_{O2(2p''_y)} \tag{8.73}$$

$$\psi_{MO8\beta}(O_2^-) = -0.49\phi_{O1(2p'_z)} - 0.42\phi_{O1(2p''_z)} + 0.49\phi_{O2(2p'_z)} + 0.42\phi_{O2(2p''_z)} \tag{8.74}$$

$$\psi_{MO9\alpha}(O_2^-) = -0.50\phi_{O1(2p'_z)} - 0.41\phi_{O1(2p''_z)} + 0.50\phi_{O2(2p'_z)} + 0.41\phi_{O2(2p''_z)} \tag{8.75}$$

$$\psi_{MO9\beta}(O_2^-) = -0.47\phi_{O1(2p'_y)} - 0.45\phi_{O1(2p''_y)} + 0.47\phi_{O2(2p'_y)} + 0.45\phi_{O2(2p''_y)} \tag{8.76}$$

$$\begin{aligned}\psi_{MO10\alpha}(O_2^-) = &-0.21\phi_{O1(2s')} - 0.65\phi_{O1(2s'')} - 0.49\phi_{O1(2p'_x)} - 0.82\phi_{O1(2p''_x)} \\ &+ 0.21\phi_{O2(2s')} + 0.65\phi_{O2(2s'')} - 0.49\phi_{O2(2p'_x)} - 0.82\phi_{O2(2p''_x)}\end{aligned} \tag{8.77}$$

$$\psi_{MO10\beta}(O_2^-) = 0.21\phi_{O1(2s')} + 0.68\phi_{O1(2s'')} + 0.49\phi_{O1(2p'_x)} + 0.84\phi_{O1(2p''_x)} - 0.21\phi_{O2(2s')} - 0.68\phi_{O2(2s'')} + 0.49\phi_{O2(2p'_x)} + 0.84\phi_{O2(2p''_x)} \quad (8.78)$$

It is found that MO8α is responsible for spin density, due to no paired beta MO. In MO8α, there is orbital overlap between O1 and O2 $2p_y$ orbitals. Two O1 lobes interact with two O2 lobes. The signs of O1 coefficients are opposite to the signs of O2 coefficients. From chemical bonding rule, it is found that the inversion π-type covalent bonding. MO9α and MO8β are paired. In MO9α and MO8β, there is orbital overlap between O1 and O2 $2p_z$ orbitals. Two O1 lobes interact with two O2 lobes. The signs of O1 coefficients are different from the signs of O2 coefficients. From chemical bonding rule, it is found that the inversion π-type covalent bonding is formed between O1 and O2 $2p_z$ orbitals. In triplet oxygen molecule, orbital energies of MO8α and MO9α, which are responsible for spin density, are negative. However, in superoxide, positive orbital energies are given in MO8α, MO9α and MO8β. From orbital energy rule, they are destabilized. It is the reason why superoxide is more reactive.

8.6 Hydrogen Fluoride

Hydrogen fluoride (H–F) exhibits weak acidity in dilute aqueous solution, in spite of the strong electronegativity of fluorine atom. B3LYP/6-31G* calculation is performed for hydrogen fluoride. Seventeen MOs are produced, because hydrogen and fluorine have two and fifteen basis functions in 6-31G* basis set, respectively.

Figure 8.18 depicts the orbital energy diagram and molecular orbitals of hydrogen fluoride at optimized structure. The obtained wave-function of MO1 is

$$\psi_{MO1}(HF) = 0.99\phi_{F(1s)} \quad (8.79)$$

MO1 consists of fluorine 1s orbital. The obtained wave-function of MO2 is

$$\psi_{MO2}(HF) = 0.13\phi_{H(1s')} - 0.23\phi_{F(1s)} + 0.51\phi_{F(2s')} + 0.47\phi_{F(2s'')} - 0.10\phi_{F(2p'_x)} \quad (8.80)$$

In MO2, there is inversion hybridization between fluorine 2s and 1s orbitals, and the fluorine $2p_x$ orbital is also hybridized. The main coefficient is for fluorine 2s orbital. There is orbital overlap between hydrogen 1s and fluorine 2s orbitals. One H lobe interacts with one F lobe. From chemical bonding rule, it is found that σ-type covalent bonding is formed between hydrogen 1s and fluorine 2s orbitals. The obtained wave-function of MO3 is

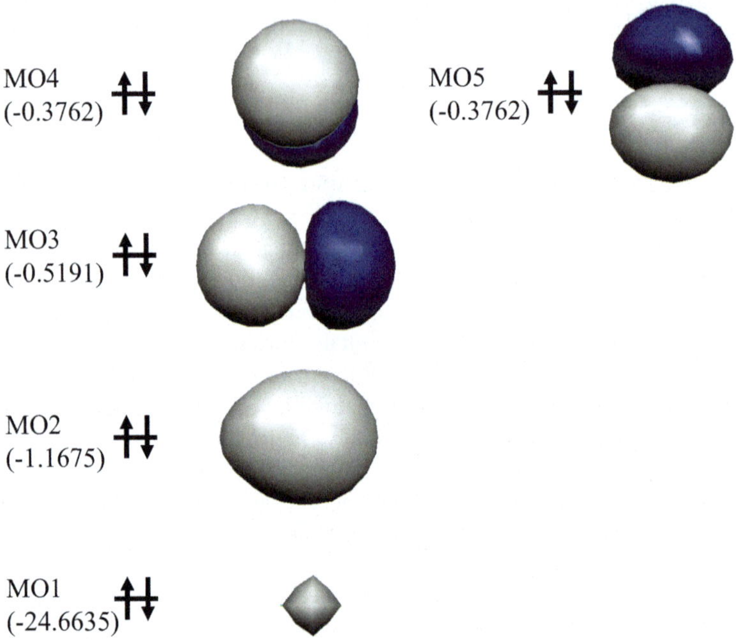

Fig. 8.18 Orbital energy diagram and molecular orbitals of hydrogen fluoride at optimized structure (B3LYP/6-31G*). Hydrogen and fluorine atoms are located at the *left* and *right sides*, respectively. The calculated orbital energy is given in *parentheses*

$$\begin{aligned}\psi_{\mathrm{MO3}}(\mathrm{HF}) = {} & 0.27\phi_{\mathrm{H(1s')}} + 0.15\phi_{\mathrm{H(1s'')}} \\ & - 0.12\phi_{\mathrm{F(2s')}} - 0.36\phi_{\mathrm{F(2s'')}} - 0.55\phi_{\mathrm{F(2p'_x)}} - 0.34\phi_{\mathrm{F(2p''_x)}}\end{aligned} \tag{8.81}$$

In MO3, there is hybridization between fluorine 2s and $2p_x$ orbitals. As shown in Fig. 8.19, there are two orbital overlap patterns between hydrogen and fluorine. One is between hydrogen 1s and fluorine 2s orbitals. The other is between hydrogen 1s and fluorine $2p_x$ orbitals. One H lobe interacts with one F lobe, and the sign of hydrogen 1s coefficient is opposite to the sign of fluorine 2s coefficient. From chemical bonding rule, it is found that the inversion σ-type covalent bonding is formed between hydrogen 1s and fluorine 2s orbitals, and the σ-type covalent bonding is formed between hydrogen 1s and fluorine $2p_x$ orbitals. The latter is more dominative than the former. The obtained wave-functions of MO4 and MO5 are

$$\psi_{\mathrm{MO4}}(\mathrm{HF}) = 0.23\phi_{\mathrm{F(2p'_y)}} + 0.17\phi_{\mathrm{F(2p''_y)}} + 0.62\phi_{\mathrm{F(2p'_z)}} + 0.46\phi_{\mathrm{F(2p''_z)}} \tag{8.82}$$

$$\psi_{\mathrm{MO5}}(\mathrm{HF}) = -0.62\phi_{\mathrm{F(2p'_y)}} - 0.46\phi_{\mathrm{F(2p''_y)}} + 0.23\phi_{\mathrm{F(2p'_z)}} + 0.17\phi_{\mathrm{F(2p''_z)}} \tag{8.83}$$

Fig. 8.19 Two orbital overlap patterns of MO2 and MO3 in hydrogen fluorine

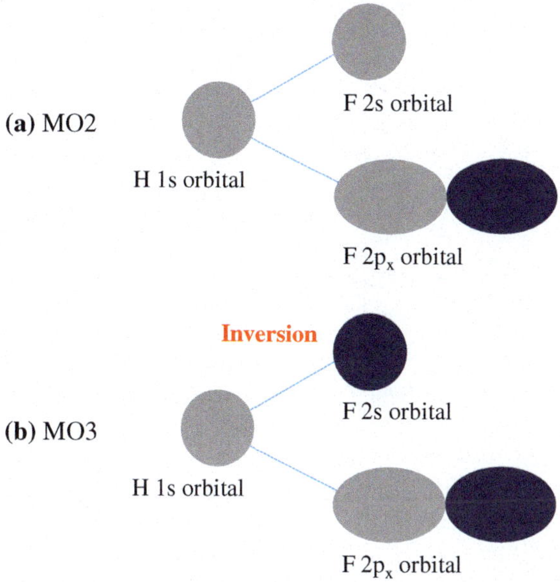

In MO4 and MO5, as the same orbital energy is given, they are degenerated. There is hybridization between fluorine $2p_y$ and $2p_z$ orbitals in fluorine atom. There is no orbital overlap with hydrogen. MO4 and MO5 represent rotated 2p orbital.

Mulliken charge densities of hydrogen and fluorine are 0.465 and −0.465, respectively. As the formal charge of hydrogen is +1.000, it is found that electron exists around hydrogen through covalent bonding formation.

Figure 8.20 shows the potential energy curve of hydrogen fluoride, changing the interatomic H–F distance. The local minimum is given at 0.918 Å, corresponding to H–F distance. Bond dissociation energy can be estimated the total energy difference between the local minimum and completely dissociated point

$$E_{\text{dissociation}}(\text{HF}) = E(\text{H}) + E(\text{F}) - E(\text{HF}) \tag{8.84}$$

The bond dissociation energy is estimated to be 137.3 kcal/mol. It is larger than H_2 molecule. It is because two covalent bonds are formed. The zero-point vibration energy is 5.961 kcal/mol. It is much smaller than bond dissociation energy.

8.7 Hydrogen Chloride

Hydrogen chloride exhibits strong acidity in aqueous solution. The electron configuration of chlorine is $[Ne]3s^2 3p^5$, where 3s and 3p electrons exist as outer shell electron. B3LYP/6-31G* calculation is performed for hydrogen chloride.

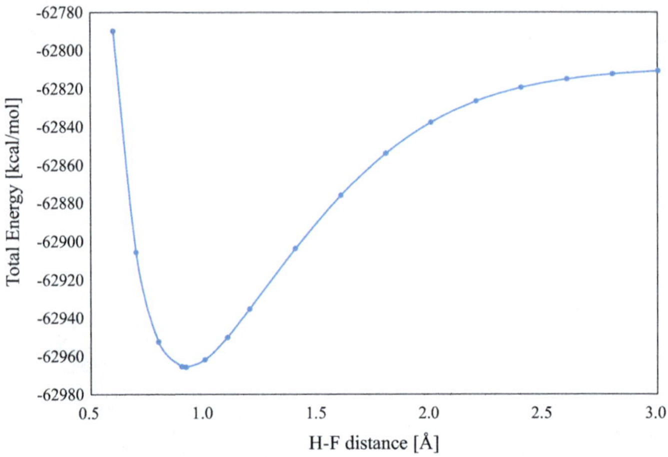

Fig. 8.20 Potential energy curve of hydrogen fluoride, changing the interatomic H–F distance (CCSD/aug-cc-pVTZ)

Twenty-one MOs are produced, because hydrogen and chlorine have two and nineteen basis functions in 6-31G* basis set, respectively.

Figure 8.21 depicts the orbital energy diagram and molecular orbitals of hydrogen chloride at optimized structure. The obtained wave-functions of MO1, MO2, MO3 and MO4 are

$$\psi_{\mathrm{MO1}}(\mathrm{HCl}) = -0.99\phi_{\mathrm{Cl(1s)}} \tag{8.85}$$

$$\psi_{\mathrm{MO2}}(\mathrm{HCl}) = 0.28\phi_{\mathrm{Cl(1s)}} - 1.02\phi_{\mathrm{Cl(2s)}} \tag{8.86}$$

$$\psi_{\mathrm{MO3}}(\mathrm{HCl}) = 0.99\phi_{\mathrm{Cl(2p_x)}} \tag{8.87}$$

$$\psi_{\mathrm{MO4}}(\mathrm{HCl}) = 0.21\phi_{\mathrm{Cl(2p_y)}} + 0.97\phi_{\mathrm{Cl(2p_z)}} \tag{8.88}$$

$$\psi_{\mathrm{MO5}}(\mathrm{HCl}) = -0.97\phi_{\mathrm{Cl(2p_y)}} + 0.21\phi_{\mathrm{Cl(2p_z)}} \tag{8.89}$$

MO1 consists of chlorine 1s orbital. In MO2, though there is inversion hybridization between chlorine 1s and 2s orbitals, the main coefficient is for chlorine 2s orbital. MO3, MO4 and MO5 consist of chlorine 2p orbital in inner L shell. In MO4 and MO5, $2p_y$ and $2p_z$ orbitals are hybridized, implying orbital rotation from standard direction. The obtained wave-function of MO6 is

$$\psi_{\mathrm{MO6}}(\mathrm{HCl}) = 0.16\phi_{\mathrm{H(1s')}} - 0.36\phi_{\mathrm{Cl(2s)}} + 0.72\phi_{\mathrm{Cl(3s')}} + 0.27\phi_{\mathrm{Cl(3s'')}} \tag{8.90}$$

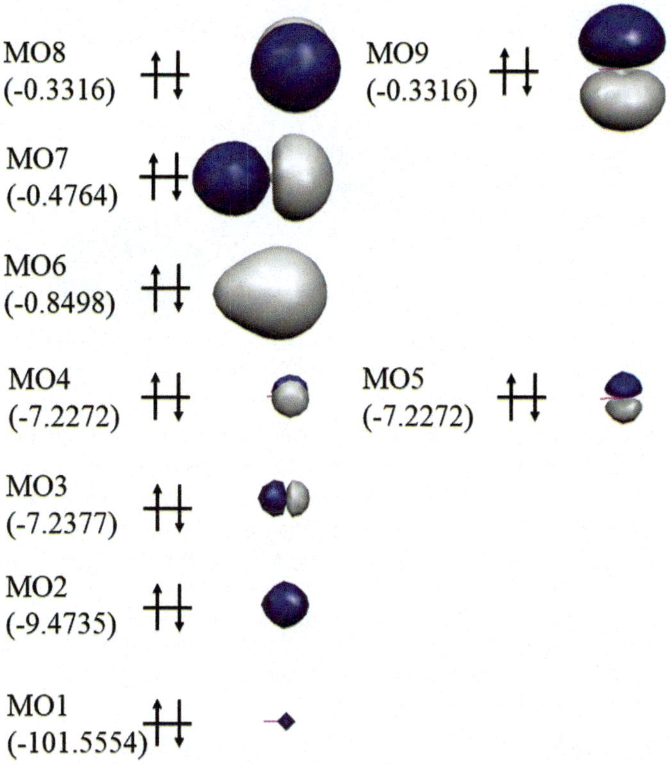

Fig. 8.21 Orbital energy diagram and molecular orbitals of hydrogen chloride at optimized structure (B3LYP/6-31G*). Hydrogen and chlorine atoms are located at the *left* and *right sides*, respectively. The calculated orbital energy is given in *parentheses*

In MO6, though there is inversion hybridization between chlorine 3s and 2s orbitals, the main coefficient of chlorine is 3s orbital. There is orbital overlap between hydrogen 1s and chlorine 3s orbitals. One H lobe interacts with one Cl lobe. From chemical bonding rule, it is found that σ-type covalent bonding is formed between hydrogen 1s and chlorine 3s orbitals. The obtained wave-functions of MO7 is

$$\begin{aligned}\psi_{\mathrm{MO7}}(\mathrm{HCl}) = & -0.29\phi_{\mathrm{H(1s')}} - 0.24\phi_{\mathrm{H(1s'')}} - 0.13\phi_{\mathrm{Cl(2s)}} - 0.23\phi_{\mathrm{Cl(2p_x)}} \\ & + 0.27\phi_{\mathrm{Cl(3s')}} + 0.24\phi_{\mathrm{Cl(3s'')}} + 0.58\phi_{\mathrm{Cl}\left(3p'_x\right)} + 0.17\phi_{\mathrm{Cl}\left(3p''_x\right)}\end{aligned} \tag{8.91}$$

In MO7, there is hybridization between chlorine 3s and $3p_x$ orbitals. As shown in Fig. 8.22, there are two orbital overlap patterns between hydrogen and chlorine. One is between hydrogen 1s and chlorine 3s orbitals. The other is between hydrogen 1s and chlorine $3p_x$ orbitals. In both cases, one H lobe interacts with one

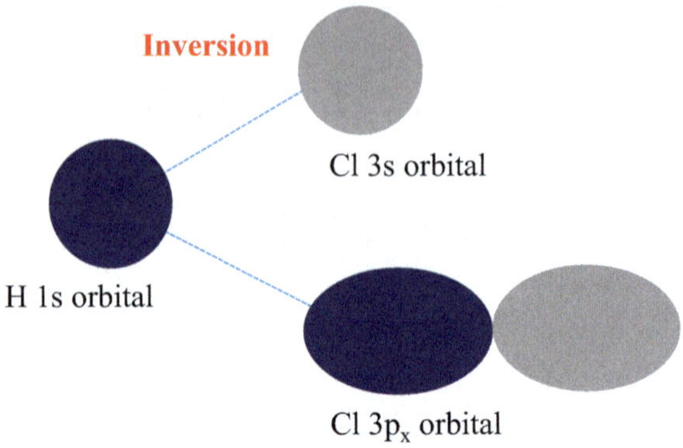

Fig. 8.22 Two orbital overlap patterns of MO7 in hydrogen chloride

Cl lobe. The sign of the coefficient of hydrogen 1s orbital is opposite to the sign of the coefficient of chlorine 3s orbital. From chemical bonding rule, it is found that inversion σ-type covalent bonding is formed between hydrogen 1s and chlorine 3s orbitals, and σ-type covalent bonding is formed between hydrogen 1s and chlorine $3p_x$ orbitals. The obtained wave-functions of MO8 and MO9 are

$$\begin{aligned}\psi_{\mathrm{MO8}}(\mathrm{HCl}) &= 0.27\phi_{\mathrm{Cl}(2p_z)} - 0.22\phi_{\mathrm{Cl}(3p'_y)} - 0.12\phi_{\mathrm{Cl}(3p''_y)} \\ &\quad - 0.70\phi_{\mathrm{Cl}(3p'_z)} - 0.39\phi_{\mathrm{Cl}(3p''_z)}\end{aligned} \tag{8.92}$$

$$\begin{aligned}\psi_{\mathrm{MO9}}(\mathrm{HCl}) &= 0.27\phi_{\mathrm{Cl}(2p_y)} - 0.70\phi_{\mathrm{Cl}(3p'_y)} - 0.39\phi_{\mathrm{Cl}(3p''_y)} \\ &\quad + 0.22\phi_{\mathrm{Cl}(3p'_z)} + 0.12\phi_{\mathrm{Cl}(3p''_z)}\end{aligned} \tag{8.93}$$

In MO8 and MO9, as the same orbital energy is given, they are degenerated. $3p_y$ and $3p_z$ orbitals are hybridized, implying orbital rotation from standard direction.

Figure 8.23 shows the potential energy curve of hydrogen chloride, changing the interatomic H–Cl distance. The local minimum is given at 1.277 Å, corresponding to H–Cl distance. In comparison with HF, the intermolecular distance is larger. It is because more outer 3s and $3p_x$ orbitals overlap with hydrogen 1s orbital.

Bond dissociation energy can be estimated the total energy difference between the local minimum and completely dissociated point

$$E_{\mathrm{dissociation}}(\mathrm{HF}) = E(\mathrm{H}) + E(\mathrm{Cl}) - E(\mathrm{HCl}) \tag{8.94}$$

The bond dissociation energy is estimated to be 103.4 kcal/mol. It is smaller than HF molecule. Hence, HF molecule exhibit weak acidity, compared with HCl molecule. The zero-point vibration energy is 4.309 kcal/mol. It is much smaller than bond dissociation energy.

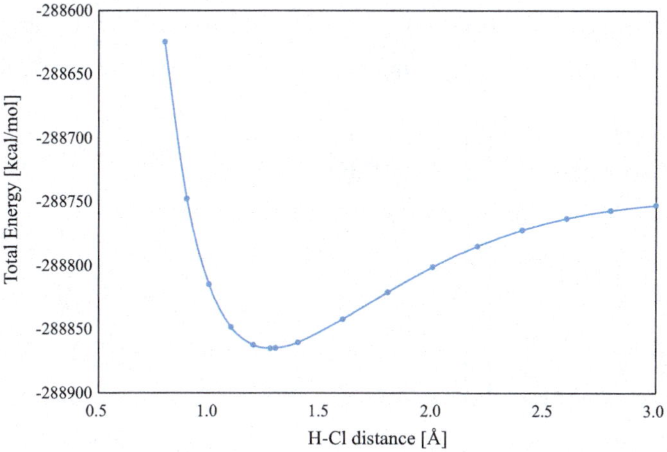

Fig. 8.23 Potential energy curve for hydrogen chloride, changing the intramolecular H–Cl distance (CCSD/aug-cc-pVTZ)

8.8 Hydroxide

8.8.1 Hydroxide

In hydroxide (OH^-), hydrogen and oxygen atoms are bound. The formal charge of hydroxide is −1. As the total number of electrons is as same as hydrogen fluoride, it can be compared with hydrogen fluoride. B3LYP/6-31G* calculation is performed for hydroxide. Seventeen MOs are produced, because hydrogen and oxygen have two and fifteen basis functions in 6-31G* basis set, respectively.

Figure 8.24 depicts the orbital energy diagram and molecular orbitals of hydroxide at optimized structure. The obtained wave-function of MO1 is

$$\psi_{\mathrm{MO1}}(\mathrm{OH}^-) = 0.99\phi_{\mathrm{O(1s)}} \tag{8.95}$$

MO1 consists of oxygen 1s orbital. The obtained wave-functions of MO2 is

$$\psi_{\mathrm{MO2}}(\mathrm{OH}^-) = -0.21\phi_{\mathrm{O(1s)}} + 0.45\phi_{\mathrm{O(2s')}} + 0.50\phi_{\mathrm{O(2s'')}} + 0.11\phi_{\mathrm{O}\left(2\mathrm{p}'_x\right)} + 0.17\phi_{\mathrm{H(1s')}} \tag{8.96}$$

In MO2, there is an inversion hybridization between oxygen 2s and 1s orbitals, and the oxygen $2p_x$ orbital is also hybridized. The main coefficients are for oxygen 2s orbital. There is orbital overlap between oxygen 2s and hydrogen 1s orbitals.

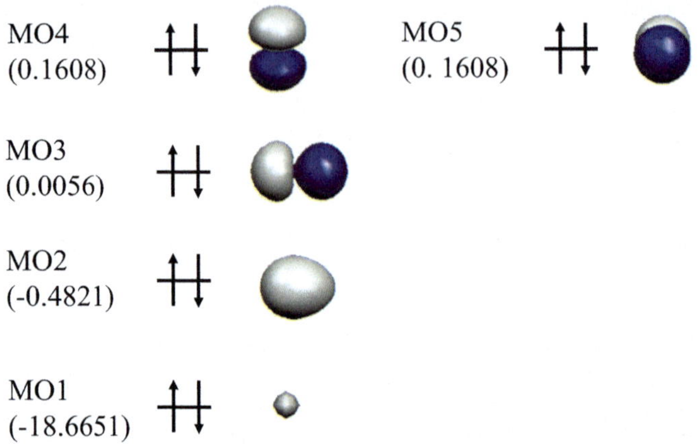

Fig. 8.24 Orbital energy diagram and molecular orbitals of hydroxide at optimized structure (B3LYP/6-31G*). Oxygen and hydrogen atoms are located at the *left* and *right sides*, respectively. The calculated orbital energy is given in *parentheses*

One O lobe interacts with one H lobe. From chemical bonding rule, it is found that σ-type covalent bonding is formed between oxygen 2s and hydrogen 1s orbitals. The obtained wave-functions of MO3 is

$$\begin{aligned}\psi_{\mathrm{MO3}}(\mathrm{OH}^-) = 0.14\phi_{\mathrm{O}(2\mathrm{s}')} + 0.56\phi_{\mathrm{O}(2\mathrm{s}'')} - 0.46\phi_{\mathrm{O}(2\mathrm{p}'_x)} - 0.27\phi_{\mathrm{O}(2\mathrm{p}''_x)} \\ - 0.27\phi_{\mathrm{H}(1\mathrm{s}')} - 0.37\phi_{\mathrm{H}(1\mathrm{s}'')}\end{aligned} \tag{8.97}$$

In MO3, there is hybridization between oxygen 2s and $2\mathrm{p}_x$ orbitals. As shown in Fig. 8.25, there are two orbital overlap patterns between oxygen and hydrogen. One is between oxygen 2s and hydrogen 1s orbitals. The other is between oxygen $2\mathrm{p}_x$ and hydrogen 1s orbitals. In both case, one O lobe interacts with one H lobe. The sign of the coefficient of oxygen 2s orbital is opposite to the signs of the coefficient of hydrogen 1s orbital. From chemical bonding rule, it is found that inversion

Fig. 8.25 Two orbital overlap patterns of MO3 in hydroxide

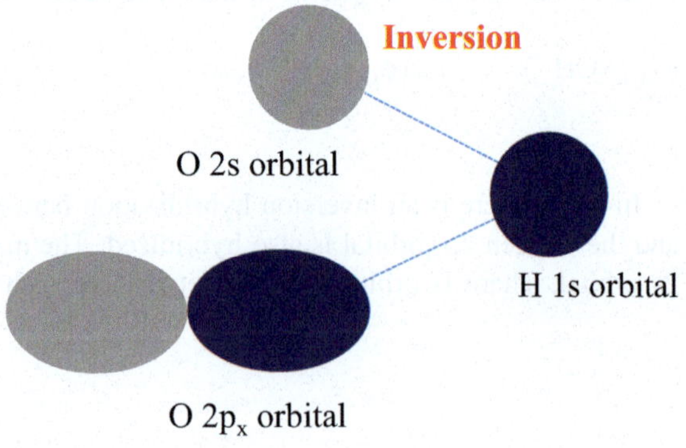

σ-type covalent bonding is formed between oxygen 2s and hydrogen 1s orbitals, and σ-type covalent bonding is formed between oxygen $2p_x$ and hydrogen 1s orbitals. The obtained wave-functions of MO4 and MO5 are

$$\psi_{\mathrm{MO4}}(\mathrm{OH}^-) = 0.58\phi_{\mathrm{F}(2p'_y)} + 0.53\phi_{\mathrm{F}(2p''_y)} + 0.17\phi_{\mathrm{F}(2p'_z)} + 0.16\phi_{\mathrm{F}(2p''_z)} \tag{8.98}$$

$$\psi_{\mathrm{MO5}}(\mathrm{OH}^-) = 0.17\phi_{\mathrm{F}(2p'_y)} + 0.16\phi_{\mathrm{F}(2p''_y)} - 0.58\phi_{\mathrm{F}(2p'_z)} - 0.53\phi_{\mathrm{F}(2p''_z)} \tag{8.99}$$

In MO4 and MO5, as the same orbital energy is given, they are degenerated. $2p_y$ and $2p_z$ orbitals are hybridized, implying orbital rotation from standard direction.

Mulliken charge densities of oxygen and hydrogen are −1.139 and 0.139, respectively. As the formal charge of hydrogen is +1.000, it is found that electron exists around hydrogen through covalent bonding formation.

Figure 8.26 shows the potential energy curve of hydroxide, changing the interatomic O–H distance. The local minimum is given at 0.964 Å, corresponding to O–H distance. Bond dissociation energy can be estimated the total energy difference between the local minimum and completely dissociated point

$$E_{\mathrm{dissociation}}(\mathrm{OH}^-) = E(\mathrm{H}) + E(\mathrm{O}^-) - E(\mathrm{OH}^-) \tag{8.100}$$

Doublet oxygen atom is assumed at completely dissociated point. The bond dissociation energy is estimated to be 111.8 kcal/mol. It is smaller than hydrogen fluoride, though the same electron configuration is given. The zero-point vibration energy is 5.396 kcal/mol. It is much smaller than bond dissociation energy.

In comparison with hydrogen fluoride, the same types of molecular orbitals are given. However, the orbital energies of MO3, MO4 and MO5 are positive. From

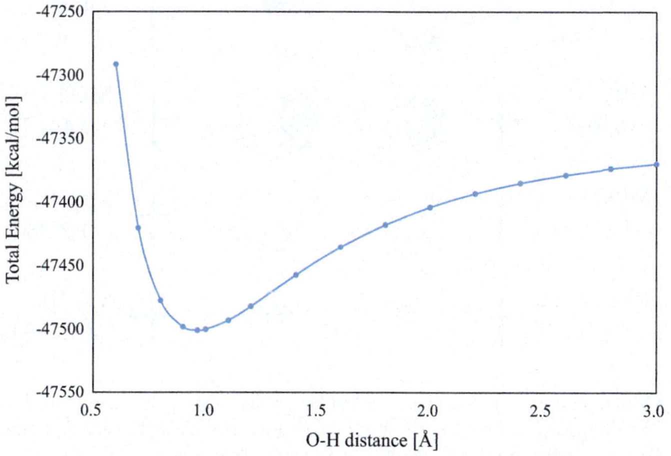

Fig. 8.26 Potential energy curve of hydroxide, changing the interatomic O–H distance (CCSD/aug-cc-pVTZ)

orbital energy rule, it implies that they are destabilized, and more reactive. Hence, hydroxide possesses electron donor property.

8.8.2 *Hydroxide Radical*

Hydroxide radical is open shell system. Five alpha and four beta electrons occupy MOs with doublet spin configuration. B3LYP/6-31G* calculation is performed for hydroxide radical. Seventeen MOs are produced, because hydrogen and oxygen have two and fifteen basis functions in 6-31G* basis set, respectively.

Figure 8.27 depicts the orbital energy diagram and molecular orbitals of hydroxide radical at optimized structure. The obtained wave-functions of MO1α and MO1β are

$$\psi_{\mathrm{MO1}\alpha}(\mathrm{OH}^{\cdot}) = -0.99\phi_{\mathrm{O(1s)}} \tag{8.101}$$

$$\psi_{\mathrm{MO1}\beta}(\mathrm{OH}^{\cdot}) = -0.99\phi_{\mathrm{O(1s)}} \tag{8.102}$$

MO1α and MO1β consist of oxygen 1s orbital. MO1α and MO1β are paired. The obtained wave-functions of MO2α and MO2β are

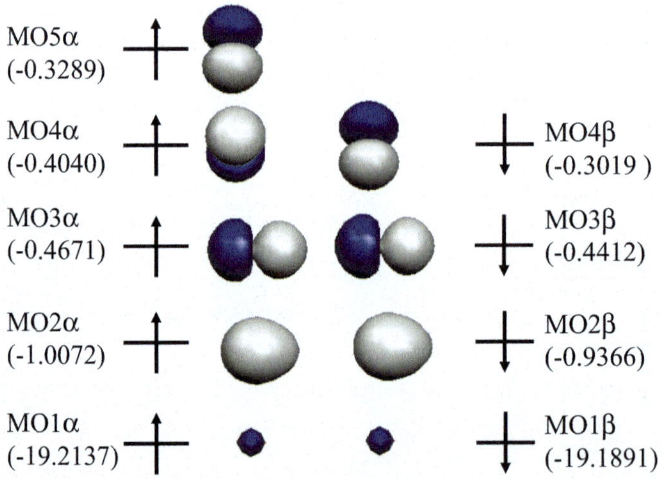

Fig. 8.27 Orbital energy diagram and molecular orbitals of hydroxide radical at optimized structure (B3LYP/6-31G*). Oxygen and hydrogen atoms are located at the *left* and *right sides*, respectively. The calculated orbital energy is given in *parentheses*

$$\psi_{MO2\alpha}(OH^{\cdot}) = -0.22\phi_{O(1s)} + 0.51\phi_{O(2s')} + 0.48\phi_{O(2s'')} + 0.12\phi_{O(2p'_x)} + 0.15\phi_{H(1s')} \tag{8.103}$$

$$\psi_{MO2\beta}(OH^{\cdot}) = -0.21\phi_{O(1s)} + 0.48\phi_{O(2s')} + 0.48\phi_{O(2s'')} + 0.13\phi_{O(2p'_x)} + 0.17\phi_{H(1s')} \tag{8.104}$$

The wave-functions of MO2α and MO2β are qualitatively the same. MO2α and MO2β are paired. In MO2α and MO2β, there is inversion hybridization between oxygen 2s and 1s orbitals, and oxygen $2p_x$ orbital is also hybridized. The main coefficients of oxygen atom are for 2s orbital. There is orbital overlap between oxygen 2s and hydrogen 1s orbitals. One O lobe interacts with one H lobe. From chemical bonding rule, it is found that the σ-type covalent bonding is formed between oxygen 2s and hydrogen 1s orbitals. The obtained wave-functions of MO3α and MO3β are

$$\psi_{MO3\alpha}(OH^{\cdot}) = -0.17\phi_{O(2s')} - 0.35\phi_{O(2s'')} + 0.55\phi_{O(2p'_x)} + 0.30\phi_{O(2p''_x)} + 0.27\phi_{H(1s')} + 0.18\phi_{H(1s'')} \tag{8.105}$$

$$\psi_{MO3\beta}(OH^{\cdot}) = -0.18\phi_{O(2s')} - 0.40\phi_{O(2s'')} + 0.53\phi_{O(2p'_x)} + 0.30\phi_{O(2p''_x)} + 0.27\phi_{H(1s')} + 0.20\phi_{H(1s'')} \tag{8.106}$$

The wave-functions of MO3α and MO3β are qualitatively the same. MO3α and MO3β are paired. In MO3α and MO3β, there is hybridization between oxygen 2s and $2p_x$ orbitals. As shown in Fig. 8.28, there are two orbital overlap patterns

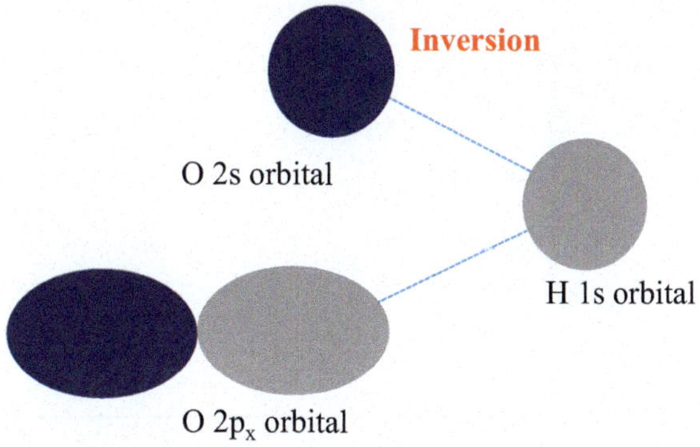

Fig. 8.28 Two orbital overlap patterns of MO3α and MO3β in hydroxide radical

between oxygen and hydrogen. One is between oxygen 2s and hydrogen 1s orbital. The other is between oxygen $2p_x$ and hydrogen 1s orbitals. One O lobe interacts with one H lobe. The sign of the coefficient of oxygen 2s orbital is opposite to the sign of the coefficient of hydrogen 1s orbital. From chemical bonding rule, it is found that the inversion σ-type covalent bonding is formed between oxygen 2s and hydrogen 1s orbitals, and σ-type covalent bonding is formed between oxygen $2p_x$ and hydrogen 1s orbitals. The obtained wave-functions of MO4α, MO4β and MO5α are

$$\psi_{\mathrm{MO4\alpha}}(\mathrm{OH^{\cdot}}) = 0.32\phi_{\mathrm{F}(2p'_y)} + 0.21\phi_{\mathrm{F}(2p''_y)} + 0.61\phi_{\mathrm{F}(2p'_z)} + 0.40\phi_{\mathrm{F}(2p''_z)} \quad (8.107)$$

$$\psi_{\mathrm{MO4\beta}}(\mathrm{OH^{\cdot}}) = -0.58\phi_{\mathrm{F}(2p'_y)} - 0.44\phi_{\mathrm{F}(2p''_y)} + 0.30\phi_{\mathrm{F}(2p'_z)} + 0.23\phi_{\mathrm{F}(2p''_z)} \quad (8.108)$$

$$\psi_{\mathrm{MO5\alpha}}(\mathrm{OH^{\cdot}}) = -0.60\phi_{\mathrm{F}(2p'_y)} - 0.43\phi_{\mathrm{F}(2p''_y)} + 0.31\phi_{\mathrm{F}(2p'_z)} + 0.22\phi_{\mathrm{F}(2p''_z)} \quad (8.109)$$

The wave-functions of MO5α and MO4β are qualitatively the same. MO5α and MO4β are paired. MO4α is responsible for spin density. In MO4α, MO4β and MO5α, $2p_y$ and $2p_z$ orbitals are hybridized, implying orbital rotation from standard direction.

Mulliken charge densities of oxygen and hydrogen are −0.400 and 0.400, respectively. As the formal charge of hydrogen is +1.000, it is found that electron exists around hydrogen through covalent bonding formation.

Figure 8.29 shows the potential energy curve of hydroxide radical, changing the interatomic O–H distance. The local minimum is given at 0.971 Å, corresponding to O–H distance. Bond dissociation energy can be estimated the total energy difference between the local minimum and completely dissociated point

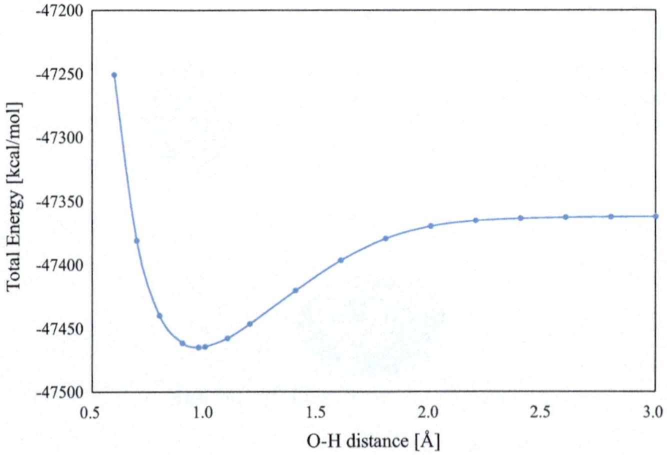

Fig. 8.29 Potential energy curve of hydroxide radical, changing the interatomic O–H distance (CCSD/aug-cc-pVTZ)

$$E_{\text{dissociation}}(\text{OH}\cdot) = E(\text{H}) + E\left(\text{O}_{\text{triplet}}\right) - E(\text{OH}\cdot) \tag{8.110}$$

Triplet oxygen atom is assumed in the completely dissociated point. The bond dissociation energy is estimated to be 103.1 kcal/mol. It is smaller than hydroxide. The zero-point vibration energy is 5.370 kcal/mol. It is much smaller than bond dissociation energy.

In hydroxide radial, no positive orbital energy is given. It is considered that hydroxide radical does not work as electron donor as same as hydroxide. However, it is well known that hydroxide radical acts as reactive oxygen species. The high reactivity of hydroxide radical is due to the existence of unpaired electron.

8.9 Carbon Oxide

Recently, catalysts for oxidation of carbon oxide (CO) have much scientific and industrial interest. Carbon oxide molecule is absorbed on catalyst surface. It is important to know chemical bonding between carbon and oxygen atoms, as the first step. Carbon oxide has seven alpha and seven beta electrons and is closed shell system. B3LYP/6-31G* calculation is performed for carbon oxide. Thirty MOs are produced, because carbon and oxygen have fifteen basis functions in 6-31G* basis set, respectively.

Figure 8.30 depicts the orbital energy diagram and molecular orbitals of carbon oxide at optimized structure. The obtained wave-functions of MO1 and MO2 are

$$\psi_{\text{MO1}}(\text{CO}) = -0.99\phi_{\text{O(1s)}} \tag{8.111}$$

$$\psi_{\text{MO2}}(\text{CO}) = 0.99\phi_{\text{C(1s)}} \tag{8.112}$$

MO1 consists of oxygen 1s orbital, and MO2 consists of carbon 1s orbital. The obtained wave-functions of MO3 and MO4 are

$$\begin{aligned}\psi_{\text{MO3}}(\text{CO}) = {} & 0.12\phi_{\text{C(1s)}} - 0.22\phi_{\text{C(2s')}} - 0.22\phi_{\text{C}\left(2\text{p}'_x\right)} \\ & + 0.20\phi_{\text{O(1s)}} - 0.45\phi_{\text{O(2s')}} - 0.36\phi_{\text{O(2s'')}} + 0.18\phi_{\text{O}\left(2\text{p}'_x\right)}\end{aligned} \tag{8.113}$$

$$\begin{aligned}\psi_{\text{MO4}}(\text{CO}) = {} & -0.14\phi_{\text{C(1s)}} + 0.30\phi_{\text{C(2s')}} + 0.23\phi_{\text{C}\left(2\text{p}'_x\right)} \\ & + 0.12\phi_{\text{O(1s)}} - 0.26\phi_{\text{O(2s')}} - 0.45\phi_{\text{O(2s'')}} - 0.49\phi_{\text{O}\left(2\text{p}'_x\right)} - 0.23\phi_{\text{O}\left(2\text{p}''_x\right)}\end{aligned} \tag{8.114}$$

In MO3 and MO4, there is an inversion hybridization between carbon 2s and 1s orbitals, and between oxygen 2s and 1s orbitals. Carbon and oxygen 2p_x orbitals are also hybridized. As shown in Fig. 8.31, there are four orbital overlap patterns between

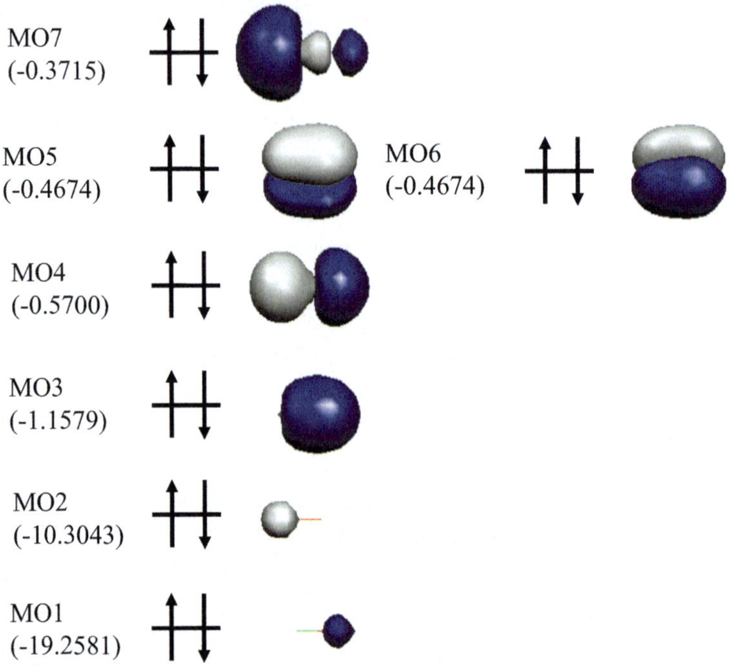

Fig. 8.30 Orbital energy diagram and molecular orbitals of carbon oxide at the optimized structure (B3LYP/6-31G*). Carbon and oxygen atoms are located at the *left* and *right sides*, respectively. The calculated orbital energy is given in *parentheses*

carbon and oxygen: (1) between carbon 2s and oxygen 2s orbitals, (2) between carbon 2s and oxygen $2p_x$ orbitals, (3) between carbon $2p_x$ and oxygen 2s orbitals, (4) between carbon $2p_x$ and oxygen $2p_x$ orbitals. One C lobe interacts with one O lobe. From chemical bonding rule, it is found that σ-type covalent bonding is formed between carbon and oxygen in MO3, and inversion σ-type covalent bonding is formed between carbon and oxygen in MO4. The obtained wave-functions of MO5, MO6 and MO7 are

$$\begin{aligned}\psi_{\mathrm{MO5}}(\mathrm{CO}) = {} & 0.23\phi_{\mathrm{C}(2p'_y)} + 0.22\phi_{\mathrm{C}(2p'_z)} \\ & + 0.41\phi_{\mathrm{O}(2p'_y)} + 0.25\phi_{\mathrm{O}(2p''_y)} + 0.39\phi_{\mathrm{O}(2p'_z)} + 0.23\phi_{\mathrm{O}(2p''_z)}\end{aligned} \tag{8.115}$$

$$\begin{aligned}\psi_{\mathrm{MO6}}(\mathrm{CO}) = {} & 0.22\phi_{\mathrm{C}(2p'_y)} - 0.23\phi_{\mathrm{C}(2p'_z)} \\ & + 0.39\phi_{\mathrm{O}(2p'_y)} + 0.23\phi_{\mathrm{O}(2p''_y)} - 0.41\phi_{\mathrm{O}(2p'_z)} - 0.25\phi_{\mathrm{O}(2p''_z)}\end{aligned} \tag{8.116}$$

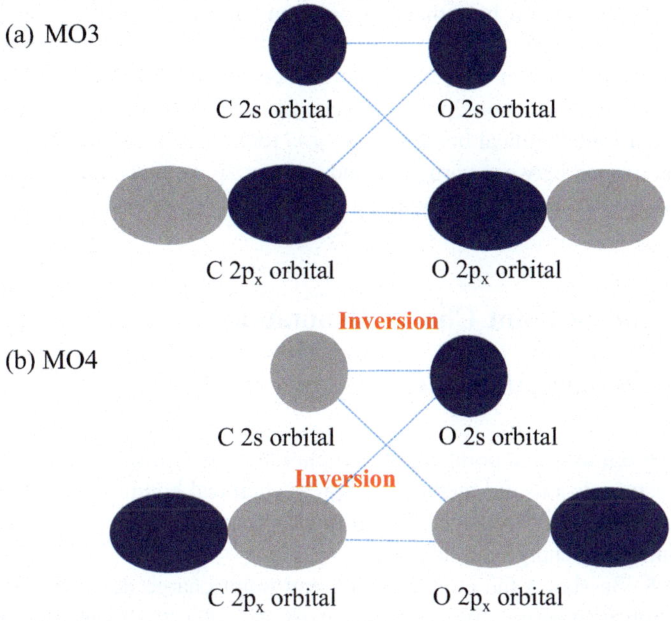

Fig. 8.31 Four orbital overlap patterns of MO3 and MO4 in carbon oxide

$$\psi_{MO7}(CO) = 0.15\phi_{C(1s)} - 0.27\phi_{C(2s')} - 0.62\phi_{C(2s'')} + 0.44\phi_{C(2p'_x)} + 0.14\phi_{C(2p''_x)} - 0.28\phi_{O(2p'_x)} - 0.15\phi_{O(2p''_x)} \quad (8.117)$$

In MO5 and MO6, as the same orbital energy is given, they are degenerated. There are hybridizations between carbon 2p$_y$ and 2p$_z$ orbitals, and between oxygen 2p$_y$ and 2p$_z$ orbitals, implying orbital rotation from standard direction. There is orbital overlap between carbon 2p and oxygen 2p orbitals. Two C lobes interact with two O lobes. From chemical bonding rule, it is found that π-type covalent bonding is formed between carbon 2p and oxygen 2p orbitals. On the other hand, in

Fig. 8.32 Two orbital overlap patterns of MO7 in carbon oxide

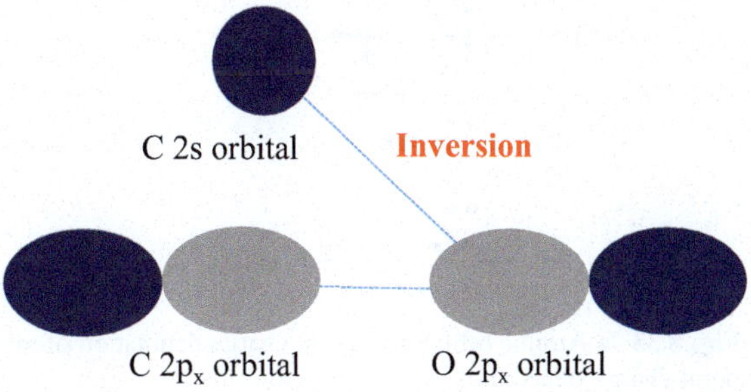

MO7, there is inversion hybridization between carbon 2s and 1s orbital, and $2p_x$ orbitals are also hybridized. As shown in Fig. 8.32, there are two orbital overlap patterns between carbon and oxygen: (1) between carbon 2s and oxygen $2p_x$ orbitals, (2) between carbon $2p_x$ and oxygen $2p_x$ orbitals. One C lobe interacts with one O lobe. From chemical bonding rule, it is found that inversion σ-type covalent bonding is formed between carbon 2s and oxygen $2p_x$ orbitals, and σ-type covalent bonding is formed between carbon $2p_x$ and oxygen $2p_x$ orbitals.

8.10 Limit of Point Charge Denotation

8.10.1 Nitrogen Molecule

It is well recognized that point charge denotation of electrons, which is often called Lewis structure, is a useful method to express chemical bonding formation of outer shell electrons. Let us consider the difference between molecular orbital (atomic orbital) and point charge denotation.

Figure 8.33a shows the atomic orbital and point charge denotation of nitrogen atom. In nitrogen atom, five electrons exist as outer shell electron. Following electron configuration rule, two electrons occupy one 2s orbital, and three 2p electrons occupy three 2p orbitals. Note that three 2p orbitals are half-filled. In point charge denotation, two 2s electrons are shown as paired electrons, and three 2p electrons are shown as unpaired electron. It is found that point charge denotation corresponds to atomic orbital.

Figure 8.33b shows the molecular orbital and point charge denotation of nitrogen molecule. In nitrogen molecule, ten electrons exist as outer shell electron. In MO3 and MO4, covalent bonding is formed between two nitrogen 2s orbitals, and

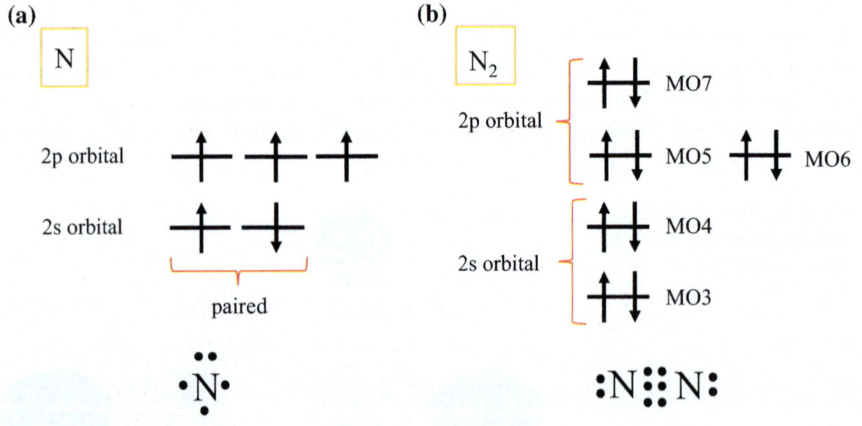

Fig. 8.33 **a** Atomic orbital and point charge denotation of nitrogen atom, **b** molecular orbital and point charge denotation of nitrogen molecule

Fig. 8.34 Corrected point charge denotation of nitrogen molecule

four electrons are occupied with spin pairs. In MO5, MO6 and MO7, as covalent bonding is formed between two nitrogen 2p orbitals, six electrons are allocated with spin pairs. Point charge denotation may be also applicable for nitrogen molecule. Following the present manner of point charge denotation, six electrons are shared between two nitrogen atoms, and two electron pairs are allocated in both nitrogen atoms. On the other hand, in molecular orbital, five covalent bonds are formed. It implies that ten electrons are shared by two nitrogen atoms. The corrected point charge denotation of nitrogen molecule is shown in Fig. 8.34.

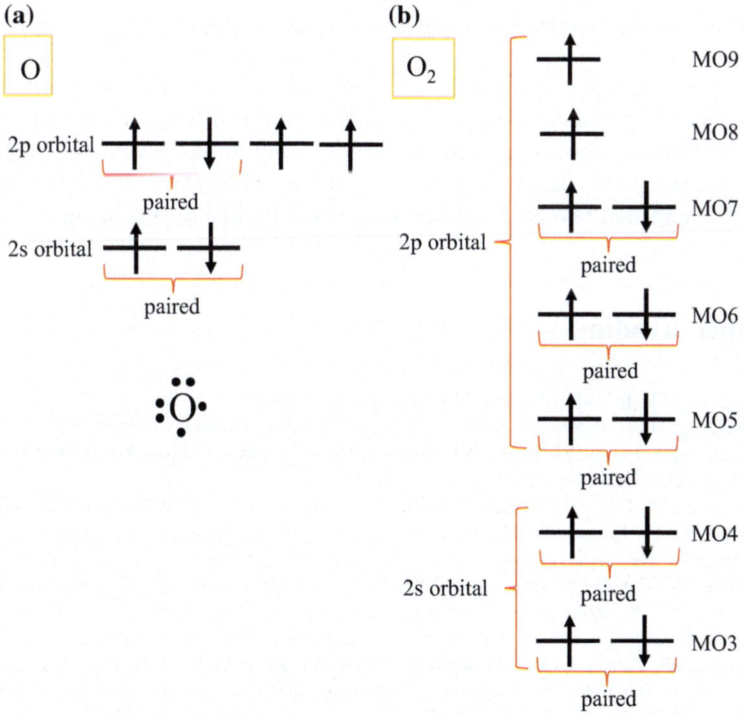

Fig. 8.35 **a** Atomic orbital and point charge denotation of oxygen atom, **b** molecular orbital and point charge denotation of triplet oxygen molecule

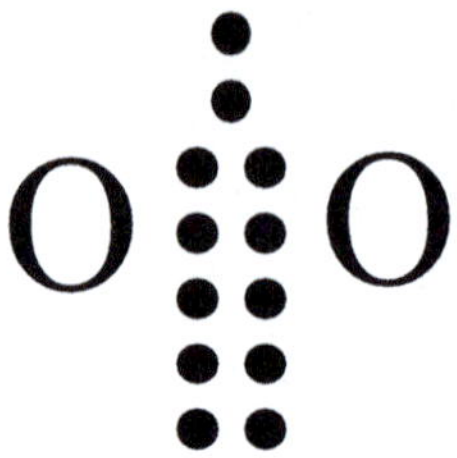

Fig. 8.36 Corrected point charge denotation of triplet oxygen molecule

8.10.2 Oxygen Molecule

Figure 8.35a shows the atomic orbital and point charge denotation of oxygen atom. In oxygen atom, six electrons exist as outer shell electron. Following electron configuration rule, two electrons occupy one 2s orbital, and four 2p electrons occupy three 2p orbitals with triplet electron configuration. In point charge denotation, two 2s electrons are shown as paired electrons. Two 2p electrons are shown as paired electrons, and two 2p electrons are shown as unpaired electron. It is found that point charge denotation corresponds to atomic orbital.

Figure 8.35b shows the molecular orbital of triplet oxygen molecule. In triplet oxygen molecule, twelve electrons exist as outer shell electron. In MO3 and MO4, covalent bonding is formed between two oxygen 2s orbitals, and four electrons are occupied with spin pairs. In MO5, MO6 and MO7, as covalent bonding is formed between two nitrogen 2p orbitals, six electrons are allocated with spin pairs. In MO8 and MO9, inversion covalent bonding is formed between two nitrogen 2p orbitals. Two electrons occupy MO8 and MO9 with triplet electron configuration. The corrected point charge denotation of nitrogen molecule is shown in Fig. 8.36. Note that unpaired two electrons are also shared by two oxygen atoms.

Further Readings

1. Dunning TH Jr (1989) J Chem Phys 90(2):1007–1023
2. Dunning TH Jr, Peterson KA, Woon DE (1999) Encycl Comput Chem 88–115
3. Francl MM, Pietro WJ, Hehre WJ, Binkley JS, Gordon MS, DeFrees DJ, Pople JA (1982) J Chem Phys 77(7):3654–3665
4. Gaussian 09, Frisch MJ, Trucks GW, Schlegel HB, Scuseria GE, Robb MA, Cheeseman JR, Scalmani G, Barone V, Mennucci B, Petersson GA, Nakatsuji H, Caricato M, Li X, Hratchian HP, Izmaylov AF, Bloino J, Zheng G, Sonnenberg JL, Hada M, Ehara M, Toyota K, Fukuda R, Hasegawa J, Ishida M, Nakajima T, Honda Y, Kitao O, Nakai H, Vreven T, Montgomery JA Jr, Peralta JE, Ogliaro F, Bearpark M, Heyd JJ, Brothers E, Kudin KN, Staroverov VN, Kobayashi R, Normand J, Raghavachari K, Rendell A, Burant JC, Iyengar SS, Tomasi J, Cossi M, Rega N, Millam JM, Klene M, Knox JE, Cross JB, Bakken V, Adamo C, Jaramillo J, Gomperts R, Stratmann RE, Yazyev O, Austin AJ, Cammi R, Pomelli C, Ochterski JW, Martin RL, Morokuma K, Zakrzewski VG, Voth GA, Salvador P, Dannenberg JJ, Dapprich S, Daniels AD, Farkas Ö, Foresman JB, Ortiz JV, Cioslowski J, Fox DJ, Gaussian, Inc., Wallingford CT, 2009

5. Hariharan PC, Pople JA (1973) Theor Chim Acta 28:213–222
6. Rassolov VA, Pople JA, Ratner MA, Windus TL (1998) J Chem Phys 109(4):1223–1229
7. Rassolov VA, Ratner MA, Pople JA, Redfern PC, Curtiss LA (2001) J Comput Chem 22 (9):976–984
8. Schmidt MW, Baldridge KK, Boatz JA, Elbert ST, Gordon MS, Jensen JH, Koseki S, Matsunaga N, Nguyen KA, Su S, Windus TL, Dupuis M, Montgomery JA (1993) J Comput Chem 14:1347–1363
9. Woon DE, Dunning TH Jr (1994) J Chem Phys 100(4):2975–2988
10. Woon DE, Dunning TH Jr (1994) J Chem Phys 98(2):1358–1371
11. Varetto U, <MOLEKEL 4.3.>; Swiss National Supercomputing Centre. Manno, Switzerland

Part III
Theoretical Background of Inorganic Chemistry

Chapter 9
Model Construction

Abstract When performing molecular orbital calculation, model construction is required. In small molecule, the real molecule corresponds to calculation model. The situation changes in solid. Within solid, the same unit structures are continuously allocated. To represent an electronic structure of unit structure in ideal solid, a minimum cluster model corresponding to unit structure is favourable. When constructing a minimum cluster model, three conditions are required: (1) no neutral condition; (2) no geometry optimization; (3) experimental interatomic distance. On the other hand, in larger cluster model including many unit structures, the equality of unit structure is not kept. The difference between molecular orbital and band structure are also explained. It is often recognized that molecular orbitals of infinite cluster model should correspond to band structure. The breakdown of the idea is also explained in comparison with molecular orbital and band structure. A geometric structure of solid is determined by a short-range chemical bonding and long-range ionic interaction. In a minimum cluster model, a long-range ionic interaction is incorporated by the use of experimental lattice distance. Finally, two useful indices such as ionic radius and tolerance factor are introduced.

Keywords Solid · Cluster model construction · Long-range ionic interaction · Ionic radius · Tolerance factor

9.1 Solid and Cluster Model Construction

As shown in Fig. 9.1, the same unit structures, which possess the same electronic configuration and geometric structure, are continuously allocated within ideal solid. On the other hand, a variety of electron configurations and geometric structures are considered near solid surface. It is because boundary condition, which implies the same units are continuously allocated, is not applied there. The distortion of a unit structure may be observed. In this chapter, a unit structure is focused to investigate a solid state property.

T. Onishi, *Quantum Computational Chemistry*,
DOI 10.1007/978-981-10-5933-9_9

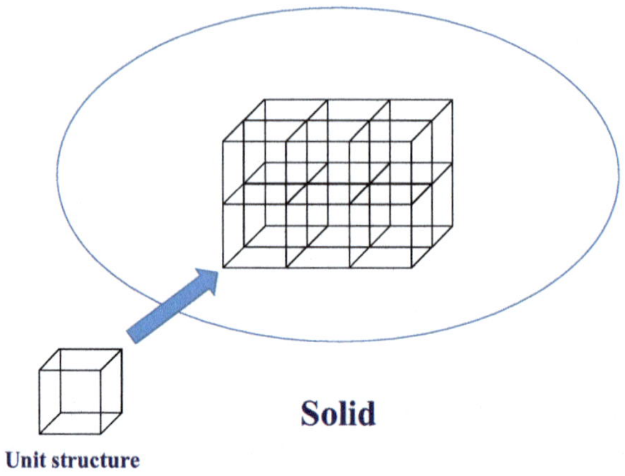

Fig. 9.1 A unit structure within ideal solid

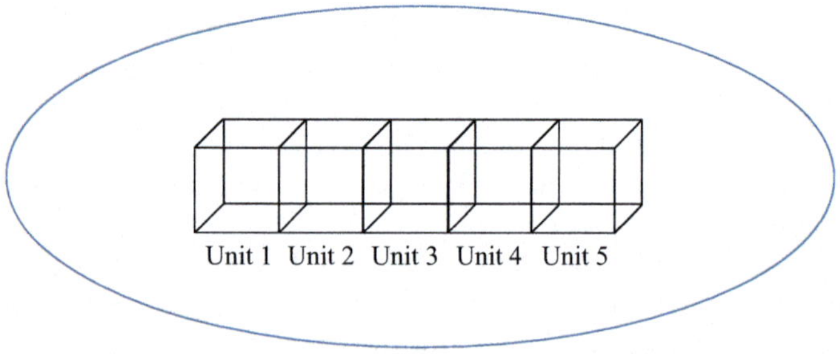

Fig. 9.2 Same electronic and geometric structures of unit structures in ideal solid

To perform molecular orbital calculation for solid, cluster model construction is required. For example, the electronic and geometric structures of the units 1, 2, 3, 4 and 5 must be the same in an ideal solid (see Fig. 9.2). A unit structure near solid surface may be distorted, due to the breakdown of boundary condition. The wave-functions of units 1, 2, 3, 4 and 5 must be the same in ideal solid, though the wave-function of the distorted unit structure near solid surface may be different.

When considering the large cluster model containing the units 1, 2, 3, 4 and 5, what does it represent? In large cluster model, electrons of each unit are not equally treated. It corresponds to one big molecule containing the five units. For example, the electronic state of unit 1 is different from unit 2, unit 3 and unit 4, though the same electronic state is given in unit 5, due to the symmetry. To represent an electronic structure of unit structure in ideal solid, a minimum cluster model corresponding to unit structure is favourable. In nanoparticle, contrarily, larger cluster model is favourable. It is because the size of unit structure may be changeable or different unit structures may be mixed. When constructing a minimum cluster model, the following conditions are required, as shown in Fig. 9.3.

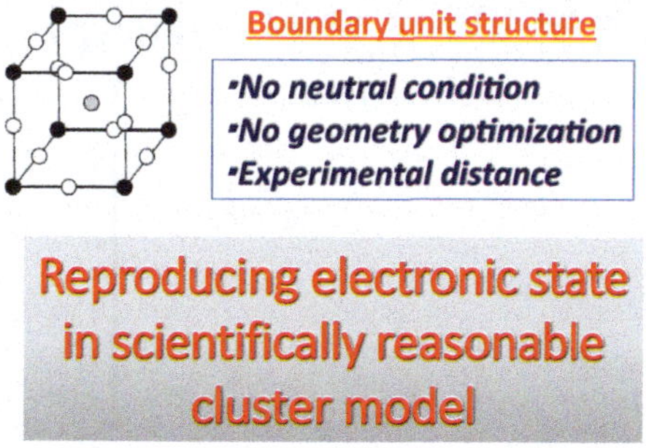

Fig. 9.3 Three required conditions for cluster model construction

Three conditions

(1) **No neutral condition**

A minimum cluster model is not neutral molecule but a part of solid. Neutral condition is not required. The total charge is estimated as the summation of formal charges of all atoms.

(2) **No geometry optimization**

If performing geometry optimization, it corresponds that a minimum cluster model is treated as molecule or nanoparticle. The largely distorted structure will be given.

(3) **Experimental interatomic distance**

Instead of geometry optimization, experimental interatomic distances are applied.

9.2 Molecular Orbital Versus Band

Let us consider the relationship between molecular orbital and band structure. In band structure, electrons are allocated in not real space but momentum space. Figure 9.4 depicts the schematic drawing of the difference between molecular orbital and band structure. It is difficult to characterize the position of electron in momentum space, in comparison with molecular orbital.

One may think that molecular orbitals of infinite cluster model correspond to band structure. Figure 9.5 depicts the schematic drawing of molecular orbital in cluster model extension. Let us consider molecular orbitals in a large cluster model, assuming that two electrons are allocated per a unit structure. Note that N denotes the number of molecular orbitals. In $N = 2$, two different molecular orbitals, which have different orbital energies, are given. When N is very large number, many

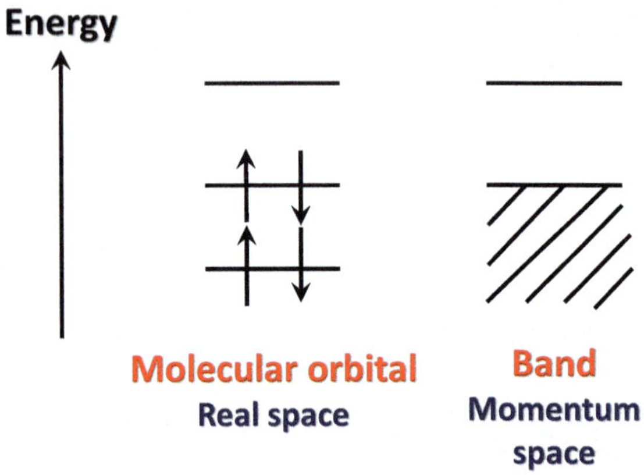

Fig. 9.4 Schematic drawing of the difference between molecular orbital and band structure

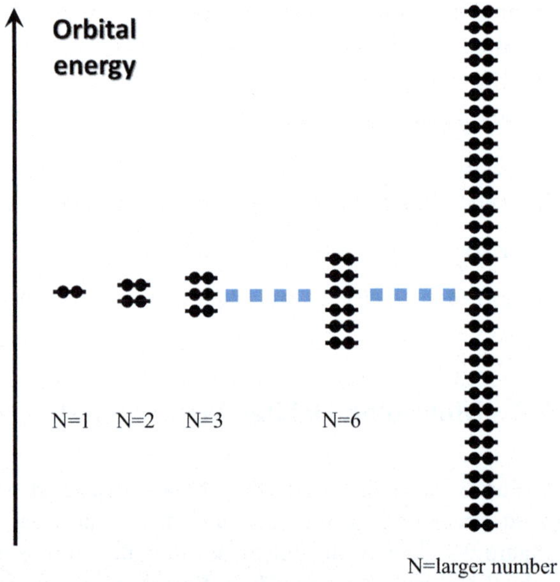

Fig. 9.5 Schematic drawing of molecular orbital in cluster model extension

molecular orbitals with different orbital energies are given. Even if degenerated molecular orbitals are given, orbital directions are different. Cluster model extension breaks the equality of unit structure. A minimum cluster model is desirable, due to the equality of unit structures within solid. Instead, a larger cluster model is favourable in nanoparticle.

9.3 Long-Range Ionic Interaction

A geometric structure of solid is determined by short-range chemical bonding and long-range ionic interaction. Chemical bonding such as covalent bonding and ionic bonding are considered as a short-range interaction. In addition, a long-range ionic interaction is combined.

Let us consider transition metal oxide, for example. Transition metal is directly bound with oxygen, combined with a long-range ionic interaction between positively charged transition metal and negatively charged oxygen. If performing geometry optimization of a minimum cluster model directly, the latter will be neglected. Instead, the experimental lattice distance is used without geometry optimization.

Embedding point charges around unit structure (minimum cluster model) is another solution to take long-range ionic interaction into account (see Fig. 9.6). However, as both charge transfer and orbital overlap are not represented between atom and point charge, geometry optimization for a minimum cluster model embedding point charges cannot be universal manner. In addition, the magnitude of point charge must be arbitrarily determined. The reasonable value is different from a formal charge.

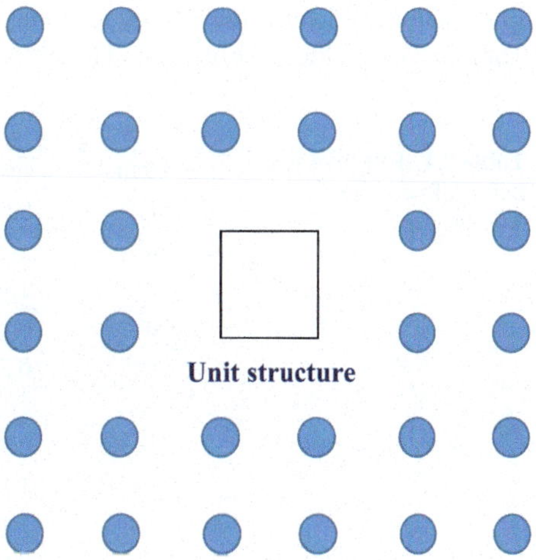

Fig. 9.6 Embedding point charges around unit structure. *Blue dot* denotes point charge

9.4 Useful Index

9.4.1 Ionic Radius

In transition metal compound, chemical bonding is formed between transition metal cation and anion. Shannon empirically determined the effective ionic radii of cation and anion (see Table 9.1). In most cases, though covalent bonding is combined, the index is useful to predict crystal structure before molecular orbital calculation. For example, in a cubic perovskite, it can be predicted whether counter cation can be allocated or not. The details will be shown in Part 4.

9.4.2 Tolerance Factor

In AMX_3 perovskite, tolerance factor (t) is empirically defined to express the stability of cubic structure. It is given by

$$t = \frac{(r_A + r_X)}{\sqrt{2}(r_M + r_X)} \tag{9.1}$$

where r_A, r_M and r_X denote the empirical ionic radii for A, M and X, respectively. For example, using the effective ionic radii of octahedral coordination (see Table 9.1), t values is obtained. The empirical prediction rule is below.

Table 9.1 Effective ionic radius of octahedral coordination

Atom	Ionic radius (Å)
Al^{3+}	0.675
Ba^{2+}	1.49, 1.56 (eight-coordination)
Cl^{-}	1.67
Co^{2+}	0.885 (high spin state), 0.79 (low spin state)
Cu^{2+}	0.87
F^{-}	1.19
Fe^{2+}	0.92 (high spin state), 0.75 (low spin state)
K^{+}	1.52, 1.65 (eight-coordination)
La^{3+}	1.17, 1.30 (eight-coordination)
Li^{+}	0.90, 1.06 (eight-coordination)
Mn^{2+}	0.97 (high spin state), 0.81 (low spin state)
Na^{+}	1.16, 1.32 (eight-coordination)
O^{2-}	1.26
OH^{-}	1.23
Sr^{2+}	1.32, 1.40 (eight-coordination)
Ti^{4+}	0.745
Zr^{4+}	0.86

Tolerance factor

$t = 1.0$: **Ideal cubic structure**
$0.89 < t < 1.0$: **Cubic structure**
$0.8 < t < 0.89$: **Distorted structure**

When the ionic radii of eight-coordination are used for counter cation, t values of $KMnF_3$, $KCoF_3$, $SrTiO_3$ and $BaTiO_3$ are 0.93, 0.97, 0.94 and 0.99, respectively. The values are within $0.89 < t < 1.0$, predicting cubic structure. Tolerance factor is an empirical indication, because ionic radius is also an empirical value. In fact, as covalent bonding between transition metal and anion is combined, tolerance factor is a rough estimate.

How can we use it? For example, let us consider barium-doping at counter cation site in $KMnF_3$ perovskite. From Table 9.1, it is found that the ionic radius of barium is close to potassium. The t value of $BaMnF_3$ unit is 0.90. When substituting potassium for barium, it is predicted that Ba-doped $KMnF_3$ perovskite keeps a cubic structure.

Further Readings

1. Onishi T (2012) Adv Quant Chem 64:31–81
2. Onishi T (2015) AIP Conf Proc 1702:090002
3. Shannon RD (1969) Acta Cryst B25:925–946
4. Shannon RD (1976) Acta Cryst A32:751–767
5. Müller U (1992) Inorganic structural chemistry. Wiley, Singapore, pp 197–201
6. Shriver DF, Atkins PW (1999) Inorganic chemistry, 3rd edn. Oxford University Press, pp 23–24
7. Goldsmidt VM (1926) Geochemische Verteilungsgesetze Der Elemente VII Die Gesetze Der Krystallocheme, pp 1–117

Chapter 10
Superexchange Interaction

Abstract In linear MXM system, where M is transition metal and X is bridge-anion, the magnetic interaction is often antiferromagnetic. In Kanamori-Goodenough rule, the magnetic interaction can be predicted based on slight charge transfer from ligand anion to transition metal. As the magnetic interaction occurs between transition metal atoms via ligand anion, it is called “superexchange interaction”. In this chapter, superexchange interaction is reconsidered, from the viewpoint of molecular orbital theory. In fact, there are two direct interactions between transition metal and ligand anion. One is charge transfer, and the other is orbital overlap. Kanamori-Goodenough rule is revised (“superexchange rule”). In MnFMn, Mn_4F_4 and KMn_8F_{12} models, the mechanism of superexchange interaction is explained according to superexchange rule. In Cu_2F_2 model, slight σ-type superexchange interaction occurs in bent CuFCu. Finally, two-atom bridge superexchange interaction is explained in MnCNMn model.

Keywords Kanamori-Goodenough rule · Superexchange rule · $KMnF_3$ perovskite · Covalent bonding · Orbital overlap

10.1 Kanamori-Goodenough Rule

Let us consider the magnetic interaction in linear MXM model, where M and X denote transition metal and ligand anion, respectively. It is assumed that the left and right transition metals have α and β spins, respectively. Kanamori-Goodenough rule predicts the mechanism of the magnetic interaction between transition metals. This rule is based on charge transfer from ligand anion to transition metal, at Heiter-London approximation level.

<u>**Kanamori-Goodenough rule**</u>

***Charge transfer**

In Kanamori-Goodenough rule, the direct interaction between transition metal and ligand anion is explained by charge transfer. As shown in Fig. 10.1, slight charge transfer occurs from ligand anion to the right transition metal, and slight charge

T. Onishi, *Quantum Computational Chemistry*,
DOI 10.1007/978-981-10-5933-9_10

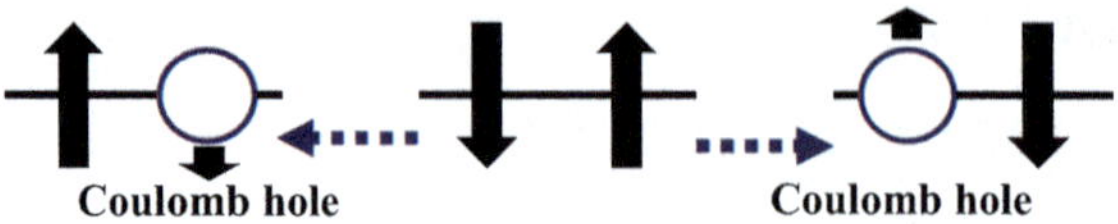

Fig. 10.1 Schematic drawing of Kanamori-Goodenough rule in MXM model. *M* and *X* denote transition metal and ligand anion, respectively

Table 10.1 Typical charge transfer patterns in MXM model

From	$2p_x$	$2p_y$	$2p_y$
To	$3d_{x^2-y^2}$ $(3d_{3x^2-r^2})$	$3d_{xy}$	$3d_{xz}$

*MXM model is allocated along *x* axis. Charge transfer occurs from ligand 2p electron to transition metal 3d electron

transfer occurs from ligand anion to the left transition metal. The charge transfer patterns are shown in Table 10.1.

***Spin**

Ligand anion has no spin, due to two charge transfer patterns. α and β spins remain in left and right transition metals, respectively. As the result, antiferromagnetic interaction occurs between transition metals via ligand anion. It is called "superexchange interaction".

10.2 Superexchange Rule

Let us reconsider the mechanism of superexchange interaction, from the viewpoint of molecular orbital (MO) theory. Due to electron correlation effect, two direct interactions exist between transition metal and ligand anion. One is charge transfer from ligand anion to transition metal, as mentioned Kanamori-Goodenough rule. The other is electron spread between transition metal and ligand anion. In MO method, it is called "orbital overlap". Although Kanamori-Goodenough rule predicts superexchange interaction correctly, orbital overlap is not fully taken into account. To include the effect of electron spread between transition metal and ligand anion, Kanamori-Goodenough rule is revised. It is called "superexchange rule".

Superexchange rule

***Orbital overlap**

Figure 10.2 depicts the schematic drawing of the superexchange rule in MXM model. There are two orbital overlap patterns: (1) between transition metal α orbital and ligand anion α orbital in MOα; (2) between transition metal β orbital and ligand anion β orbital in MOβ. Table 10.2 shows the typical orbital overlap patterns in MXM model.

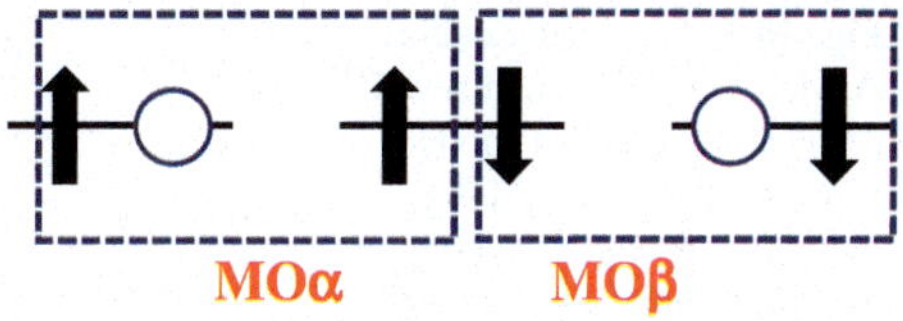

Fig. 10.2 Schematic drawing of superexchange rule in MXM model. *M* and *X* denote transition metal and ligand anion, respectively

Table 10.2 Typical orbital overlap patterns in MXM model

Covalent bonding	σ-type	π-type	π-type
Transition metal 3d orbital	$2p_x$	$2p_y$	$2p_y$
Ligand anion 2p orbital	$3d_{x^2-y^2}$ $(3d_{3x^2-r^2})$	$3d_{xy}$	$3d_{xz}$

*MXM model is allocated along *x* axis

*Spin

In MOα and MOβ, ligand anion α spin is cancelled out with ligand anion β spin. Hence, left transition metal α spin remains in MOα, and right transition metal β spin remains in MOβ. As the result, antiferromagnetic interaction occurs between transition metals via ligand anion.

10.3 Cluster Model of Superexchange System

Figure 10.3 shows the geometric structures of A_2MX_4 and AMX_3 perovskites. Transition metal (*M*) coordinates with six ligand anions (*X*). In A_2MX_4 perovskite, the two-dimensional layers are alternately stacked. Figure 10.4 shows three cluster models are constructed for A_2MX_4 and AMX_3 perovskites. *A*, *M* and *X* denote

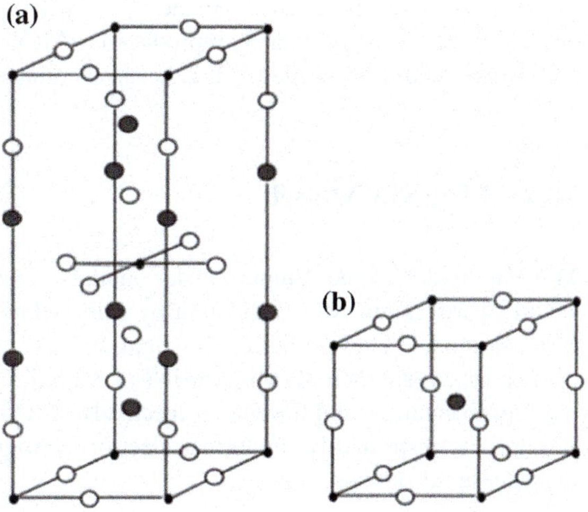

Fig. 10.3 Geometric structures of **a** A_2MX_4 perovskite and **b** AMX_3 perovskite. *Black dot*, *white dot* and *grey dot* denote transition metal (*M*), ligand anion (*X*) and counter cation (*A*), respectively

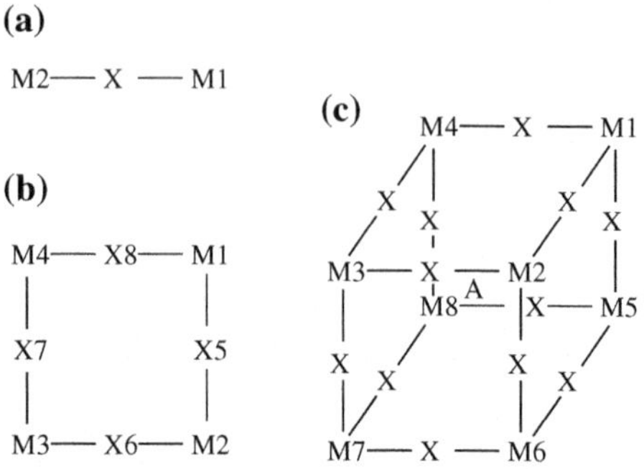

Fig. 10.4 Cluster models for A_2MX_4 perovskite and AMX_3 perovskite: **a** MXM model, **b** M_4X_4 model, **c** AM_8X_{12} model. *A*, *M* and *X* denote transition metal ligand anion and counter cation, respectively

counter cation, transition metal and ligand anion, respectively. In both perovskites, the minimum unit structure is MXM. The details are explained below.

Linear chain MXM model

MXM model is the minimum cluster model of AMX_3 perovskite and A_2MX_4 perovskite. From our calculation results, it was found that orbital overlap is overestimated, though superexchange interaction is qualitatively reproduced in MXM model. In this book, this model is used for the simplicity and qualitative discussion.

Two-dimensional M_4X_4 model

M_4X_4 is the best cluster model for $A_2M_4X_4$ perovskite. Two-dimensional orbital overlap is approximately reproduced. In most cubic perovskites, though counter cation plays an important role in stabilizing cubic structure, it does not affect superexchange interaction directly. However, there is the case that counter cation forms chemical bonding with conductive ion. In that case, counter cation must be included in M_4X_4 model. The details will be explained in Part 4.

Three-dimensional AM_8X_{12} model

AM_8X_{12} is the best cluster model for AMX_3 perovskite. Three-dimensional orbital overlap is approximately reproduced. AM_8X_{12} model is more favourable, in comparison with M_8X_{12} model. It is because counter cation affects lattice distance.

10.4 MnFMn Model

MnFMn model is the simple cluster model of antiferromagnetic K_2MnF_4 and $KMnF_3$ perovskites. MO calculation by using BHHLYP method is performed for MnFMn model. The site number of right and left manganese atoms are Mn1 and Mn2, respectively (Mn2-F-Mn1). MINI (5.3.3.3/5.3/4.1) and 6-31G* basis sets are used for manganese and fluorine, respectively. The formal charges of Mn and F are +2 and −1, respectively. In formal electron configuration, five electrons occupy manganese 3d orbitals.

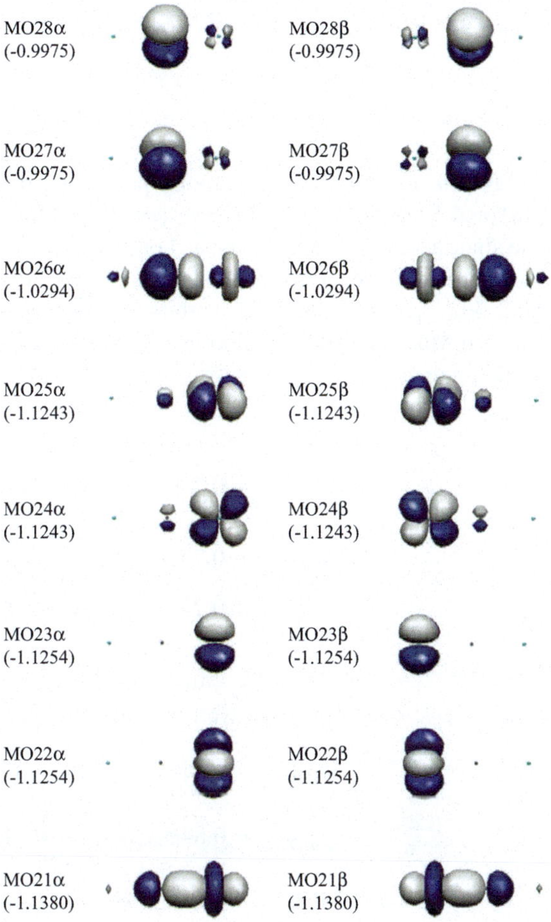

Fig. 10.5 Selected molecular orbitals of MnFMn model (BHHLYP method)

Mulliken charge densities of manganese and fluorine are 1.870 and −0.740, respectively. Spin densities of Mn1 and Mn2 spins are 4.967 and −4.967, respectively. Figure 10.5 depicts the selected molecular orbitals of MnFMn model. The obtained wave-functions of MO21α and MO21β are

$$\psi_{\mathrm{MO21}\alpha} = 0.74\phi_{\mathrm{Mn1}\left(3dx^{2\prime}\right)} + 0.27\phi_{\mathrm{Mn1}\left(3dx^{2\prime\prime}\right)} - 0.37\phi_{\mathrm{Mn1}\left(3dy^{2\prime}\right)} - 0.13\phi_{\mathrm{Mn1}\left(3dy^{2\prime\prime}\right)}$$
$$- 0.37\phi_{\mathrm{Mn1}\left(3dz^{2\prime}\right)} - 0.13\phi_{\mathrm{Mn1}\left(3dz^{2\prime\prime}\right)} + 0.26\phi_{\mathrm{F}(2px^{\prime})} + 0.19\phi_{\mathrm{F}(2px^{\prime\prime})} \tag{10.1}$$

$$\psi_{MO21\beta} = 0.74\phi_{Mn2(3dx^{2\prime})} + 0.27\phi_{Mn2(3dx^{2\prime\prime})} - 0.37\phi_{Mn2(3dy^{2\prime})} - 0.13\phi_{Mn2(3dy^{2\prime\prime})} - 0.37\phi_{Mn2(3dz^{2\prime})} - 0.13\phi_{Mn2(3dz^{2\prime\prime})} - 0.26\phi_{F(2px^{\prime})} - 0.19\phi_{F(2px^{\prime\prime})} \tag{10.2}$$

MO21α and MO21β are partially paired in $2p_x$ orbital. α and β spins exist in Mn1 and Mn2 $3d_{3x^2-r^2}$ orbitals, respectively. One manganese lobe interacts with one fluorine lobe. From chemical bonding rule, it is found that σ-type covalent bonding is formed between Mn1 $3d_{3x^2-r^2}$ and fluorine $2p_x$ orbitals, and between Mn2 $3d_{3x^2-r^2}$ and fluorine $2p_x$ orbitals. σ-type superexchange interaction occurs between Mn1 and Mn2 via fluorine. The obtained wave-functions of MO22α and MO22β are

$$\psi_{MO22\alpha} = -0.73\phi_{Mn1(3dy^{2\prime})} - 0.27\phi_{Mn1(3dy^{2\prime\prime})} + 0.73\phi_{Mn1(3dz^{2\prime})} + 0.27\phi_{Mn1(3dz^{2\prime\prime})} \tag{10.3}$$

$$\psi_{MO22\beta} = -0.73\phi_{Mn2(3dy^{2\prime})} - 0.27\phi_{Mn2(3dy^{2\prime\prime})} + 0.73\phi_{Mn2(3dz^{2\prime})} + 0.27\phi_{Mn2(3dz^{2\prime\prime})} \tag{10.4}$$

In MO22α and MO22β, Mn1 and Mn2 $3d_{y^2-z^2}$ orbitals are represented, respectively. The obtained wave-functions of MO23α and MO23β are

$$\psi_{MO23\alpha} = 0.84\phi_{Mn1(3dyz^{\prime})} + 0.31\phi_{Mn1(3dyz^{\prime\prime})} \tag{10.5}$$

$$\psi_{MO23\beta} = 0.84\phi_{Mn2(3dyz^{\prime})} + 0.31\phi_{Mn2(3dyz^{\prime\prime})} \tag{10.6}$$

In MO23α and MO23β, Mn1 and Mn2 $3d_{yz}$ orbitals are represented, respectively. The obtained wave-functions of MO24α and MO24β are

$$\psi_{MO24\alpha} = -0.79\phi_{Mn1(3dxy^{\prime})} - 0.29\phi_{Mn1(3dxy^{\prime\prime})} - 0.26\phi_{Mn1(3dxz^{\prime})} - 0.09\phi_{Mn1(3dxz^{\prime\prime})} + 0.09\phi_{F(2py^{\prime})} + 0.07\phi_{F(2py^{\prime\prime})} + 0.03\phi_{F(2pz^{\prime})} + 0.02\phi_{F(2pz^{\prime\prime})} \tag{10.7}$$

$$\psi_{MO24\beta} = 0.74\phi_{Mn2(3dxy^{\prime})} + 0.27\phi_{Mn2(3dxy^{\prime\prime})} + 0.37\phi_{Mn2(3dxz^{\prime})} + 0.14\phi_{Mn2(3dxz^{\prime\prime})} + 0.08\phi_{F(2py^{\prime})} + 0.06\phi_{F(2py^{\prime\prime})} + 0.04\phi_{F(2pz^{\prime})} + 0.03\phi_{F(2pz^{\prime\prime})} \tag{10.8}$$

MO24α and MO24β are partially paired in hybridized fluorine 2p orbital. α and β spins exist in hybridized Mn1 and Mn2 3d orbital, respectively. Two manganese lobes interact with two fluorine lobes. From chemical bonding rule, it is found that

π-type covalent bonding is formed between Mn1 hybridized 3d and hybridized fluorine 2p orbitals, and between Mn2 hybridized 3d and hybridized fluorine 2p orbitals. π-type superexchange interaction occurs between Mn1 and Mn2 via fluorine. The obtained wave-functions of MO25α and MO25β are

$$\begin{aligned}\psi_{\mathrm{MO25\alpha}} = & -0.26\phi_{\mathrm{Mn1(3d}xy')} - 0.09\phi_{\mathrm{Mn1(3d}xy'')} + 0.79\phi_{\mathrm{Mn1(3d}xz')} + 0.29\phi_{\mathrm{Mn1(3d}xz'')} \\ & + 0.03\phi_{\mathrm{F(2py')}} + 0.02\phi_{\mathrm{F(2py'')}} - 0.09\phi_{\mathrm{F(2pz')}} - 0.07\phi_{\mathrm{F(2pz'')}}\end{aligned} \tag{10.9}$$

$$\begin{aligned}\psi_{\mathrm{MO25\beta}} = & \, 0.37\phi_{\mathrm{Mn2(3d}xy')} + 0.13\phi_{\mathrm{Mn2(3d}xy'')} - 0.74\phi_{\mathrm{Mn2(3d}xz')} - 0.27\phi_{\mathrm{Mn2(3d}xz'')} \\ & + 0.04\phi_{\mathrm{F(2py')}} + 0.03\phi_{\mathrm{F(2py'')}} - 0.08\phi_{\mathrm{F(2pz')}} - 0.06\phi_{\mathrm{F(2pz'')}}\end{aligned} \tag{10.10}$$

MO25α and MO25β are partially paired in hybridized fluorine 2p orbital. α and β spins exist in Mn1 and Mn2 hybridized 3d orbital, respectively. Two manganese lobes interact with two fluorine lobes. From chemical bonding rule, it is found that π-type covalent bonding is formed between Mn1 hybridized 3d and hybridized fluorine 2p orbitals, and between Mn2 hybridized 3d and hybridized fluorine 2p orbitals. π-type superexchange interaction occurs between Mn1 and Mn2 via fluorine. In comparison with σ-type superexchange interaction, the contribution of fluorine 2p coefficients is slight in π-type superexchange interaction. It shows the weak superexchange interaction. The obtained wave-functions of MO26α and MO26β are

$$\begin{aligned}\psi_{\mathrm{MO26\alpha}} = & -0.38\phi_{\mathrm{Mn1}(3\mathrm{d}x^{2\prime})} - 0.10\phi_{\mathrm{Mn1}(3\mathrm{d}x^{2\prime\prime})} + 0.23\phi_{\mathrm{Mn1}(3\mathrm{d}y^{2\prime})} + 0.12\phi_{\mathrm{Mn1}(3\mathrm{d}y^{2\prime\prime})} \\ & + 0.23\phi_{\mathrm{Mn1}(3\mathrm{d}z^{2\prime})} + 0.12\phi_{\mathrm{Mn1}(3\mathrm{d}z^{2\prime\prime})} + 0.53\phi_{\mathrm{F(2p}x')} + 0.45\phi_{\mathrm{F(2p}x'')}\end{aligned} \tag{10.11}$$

$$\begin{aligned}\psi_{\mathrm{MO26\beta}} = & -0.38\phi_{\mathrm{Mn2}(3\mathrm{d}x^{2\prime})} - 0.10\phi_{\mathrm{Mn2}(3\mathrm{d}x^{2\prime\prime})} + 0.23\phi_{\mathrm{Mn2}(3\mathrm{d}y^{2\prime})} + 0.12\phi_{\mathrm{Mn2}(3\mathrm{d}y^{2\prime\prime})} \\ & + 0.23\phi_{\mathrm{Mn2}(3\mathrm{d}z^{2\prime})} + 0.12\phi_{\mathrm{Mn2}(3\mathrm{d}z^{2\prime\prime})} - 0.53\phi_{\mathrm{F(2p}x')} - 0.45\phi_{\mathrm{F(2p}x'')}\end{aligned} \tag{10.12}$$

MO26α and MO26β are partially paired in fluorine $2p_x$ orbital. α and β spins exist in Mn1 and Mn2 hybridized 3d orbitals, respectively. Onc manganese lobe interacts with one fluorine lobe. There are nodes between Mn1 and F, and between Mn2 and F. From chemical bonding rule, it is found that inversion σ-type covalent bonding is formed between Mn1 hybridized 3d and fluorine $2p_x$ orbitals, and between Mn2 hybridized 3d and fluorine $2p_x$ orbitals. Inversion σ-type superexchange interaction occurs between Mn1 and Mn2 via fluorine. MO26α and MO26β are inversion σ-type covalent bonding to MO21α and MO21β, respectively. The obtained wave-functions of MO27α, MO27β, MO28α and MO28β are

$$\psi_{\mathrm{MO27\alpha}} = 0.10\phi_{\mathrm{Mn1(3d}xy')} - 0.12\phi_{\mathrm{Mn1(3d}xz')} + 0.41\phi_{\mathrm{F(2py')}} + 0.32\phi_{\mathrm{F(2py'')}} - 0.49\phi_{\mathrm{F(2pz')}} - 0.38\phi_{\mathrm{F(2pz'')}} \tag{10.13}$$

$$\psi_{\mathrm{MO27\beta}} = -0.14\phi_{\mathrm{Mn2(3d}xy')} + 0.08\phi_{\mathrm{Mn2(3d}xz')} + 0.55\phi_{\mathrm{F(2py')}} + 0.42\phi_{\mathrm{F(2py'')}} - 0.33\phi_{\mathrm{F(2pz')}} - 0.25\phi_{\mathrm{F(2pz'')}} \tag{10.14}$$

$$\psi_{\mathrm{MO28\beta}} = 0.12\phi_{\mathrm{Mn1(3d}xy')} + 0.10\phi_{\mathrm{Mn1(3d}xz')} + 0.49\phi_{\mathrm{F(2py')}} + 0.38\phi_{\mathrm{F(2py'')}} + 0.41\phi_{\mathrm{F(2pz')}} + 0.32\phi_{\mathrm{F(2pz'')}} \tag{10.15}$$

$$\psi_{\mathrm{MO28\beta}} = -0.08\phi_{\mathrm{Mn2(3d}xy')} - 0.14\phi_{\mathrm{Mn2(3d}xz')} + 0.33\phi_{\mathrm{F(2py')}} + 0.25\phi_{\mathrm{F(2py'')}} + 0.55\phi_{\mathrm{F(2pz')}} + 0.42\phi_{\mathrm{F(2pz'')}} \tag{10.16}$$

In MO27α, MO27β, MO28α and MO28β, two manganese lobes interact with two fluorine lobes. There are nodes between Mn1 and F, and between Mn2 and F. From chemical bonding rule, it is found that inversion π-type covalent bonding is formed between Mn1 hybridized 3d and hybridized fluorine 2p orbitals, and between Mn2 hybridized 3d and hybridized fluorine 2p orbitals. They are inversion π-type covalent bonding to MO24α, MO24β, MO25α and MO25β.

In MnFMn model, two types of superexchange interaction are reproduced. One is σ-type superexchange interaction in MO21α, MO21β, MO26α and MO26β. The other is π-type superexchange interaction in MO24α, MO24β, MO25α, MO25β, MO27α, MO27β, MO28α and MO28β.

10.5 Mn_4F_4 Model

MO calculation by using BHHLYP method is performed for two-dimensional Mn_4F_4 model. MINI (5.3.3.3/5.3/4.1) and 6-31G* basis sets are used for manganese and fluorine, respectively. The formal charges of Mn and F are +2 and −1, respectively. In formal electron configuration, five electrons occupy manganese 3d orbitals.

Mulliken charge densities for manganese and fluorine are 1.759 and −0.759. Spin densities of manganese with α and β spins are 4.925 and −4.925, respectively. Figure 10.6 depicts the selected molecular orbitals of Mn_4F_4 model. The obtained wave-functions of MO45α and MO45β are

Fig. 10.6 Selected molecular orbitals of Mn_4F_4 model (BHHLYP method)

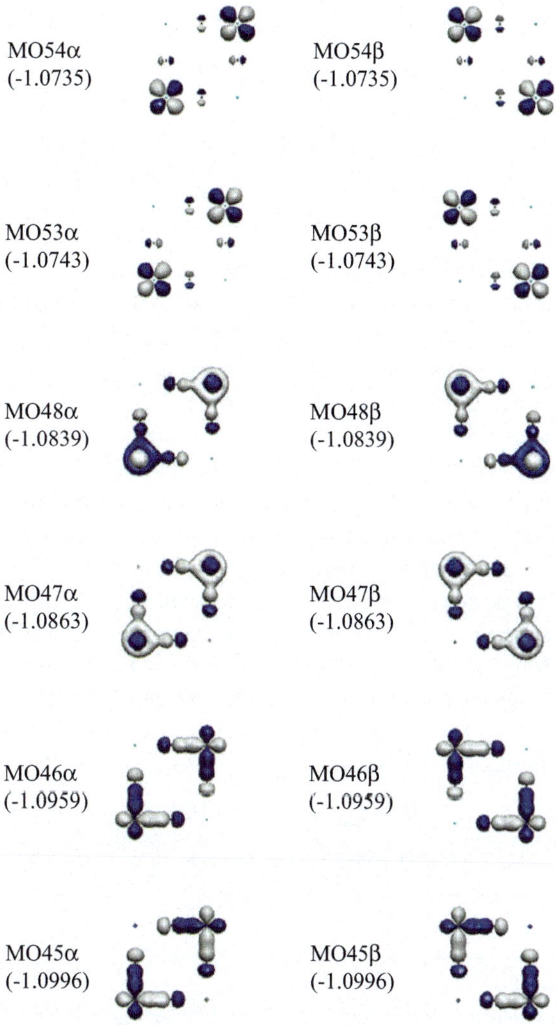

$$\begin{aligned}\psi_{\mathrm{MO45\alpha}} = & -0.42\phi_{\mathrm{Mn1}\left(3dx^{2\prime}\right)} - 0.16\phi_{\mathrm{Mn1}\left(3dx^{2\prime\prime}\right)} + 0.42\phi_{\mathrm{Mn1}\left(3dy^{2\prime}\right)} + 0.16\phi_{\mathrm{Mn1}\left(3dy^{2\prime\prime}\right)} \\ & + 0.42\phi_{\mathrm{Mn3}\left(3dx^{2\prime}\right)} + 0.16\phi_{\mathrm{Mn3}\left(3dx^{2\prime\prime}\right)} - 0.42\phi_{\mathrm{Mn3}\left(3dy^{2\prime}\right)} - 0.16\phi_{\mathrm{Mn3}\left(3dy^{2\prime\prime}\right)} \\ & + 0.16\phi_{\mathrm{F5(2py^{\prime})}} + 0.12\phi_{\mathrm{F5(2py^{\prime\prime})}} - 0.16\phi_{\mathrm{F6(2px^{\prime})}} - 0.12\phi_{\mathrm{F6(2px^{\prime\prime})}} \\ & + 0.16\phi_{\mathrm{F7(2py^{\prime})}} + 0.12\phi_{\mathrm{F7(2py^{\prime\prime})}} - 0.16\phi_{\mathrm{F8(2px^{\prime})}} - 0.12\phi_{\mathrm{F8(2px^{\prime\prime})}}\end{aligned} \tag{10.17}$$

$$\begin{aligned}\psi_{\mathrm{MO45\beta}} = {} & 0.42\phi_{\mathrm{Mn2}\left(3\mathrm{d}x^{2'}\right)} + 0.16\phi_{\mathrm{Mn2}\left(3\mathrm{d}x^{2''}\right)} - 0.42\phi_{\mathrm{Mn2}\left(3\mathrm{d}y^{2'}\right)} - 0.16\phi_{\mathrm{Mn2}\left(3\mathrm{d}y^{2''}\right)} \\ & - 0.42\phi_{\mathrm{Mn4}\left(3\mathrm{d}x^{2'}\right)} - 0.16\phi_{\mathrm{Mn4}\left(3\mathrm{d}x^{2''}\right)} + 0.42\phi_{\mathrm{Mn4}\left(3\mathrm{d}y^{2'}\right)} + 0.16\phi_{\mathrm{Mn4}\left(3\mathrm{d}y^{2''}\right)} \\ & + 0.16\phi_{\mathrm{F5(2p}y')} + 0.12\phi_{\mathrm{F5(2p}y'')} + 0.16\phi_{\mathrm{F6(2p}x')} + 0.12\phi_{\mathrm{F6(2p}x'')} \\ & + 0.16\phi_{\mathrm{F7(2p}y')} + 0.12\phi_{\mathrm{F7(2p}y'')} + 0.16\phi_{\mathrm{F8(2p}x')} + 0.12\phi_{\mathrm{F8(2p}x'')}\end{aligned} \tag{10.18}$$

MO45α and MO45β are paired partially in F5 $2p_y$, F6 $2p_x$, F7 $2p_y$ and F8 $2p_x$ orbitals. α spins exist in $3d_{x^2-y^2}$ orbitals of Mn1 and Mn3, and β spins exist in $3d_{x^2-y^2}$ orbitals of Mn2 and Mn4. In MO45α, Mn1 lobe interacts with F5 lobe and one F8 lobe, and Mn3 lobe interacts with F6 lobe and F7 lobe. From chemical bonding rule, it is found that Mn1 $3d_{x^2-y^2}$ orbital forms σ-type covalent bonding with F5 $2p_y$ orbital and F8 $2p_x$ orbital, and Mn3 $3d_{x^2-y^2}$ orbital forms σ-type covalent bonding with F6 $2p_x$ orbital and F7 $2p_y$ orbital. In MO45β, Mn2 lobe interacts with F5 lobe and one F6 lobe, and Mn3 lobe interacts with F7 lobe and F8 lobe. From chemical bonding rule, it is found that Mn2 $3d_{x^2-y^2}$ orbital forms σ-type covalent bonding with F5 $2p_y$ orbital and F6 $2p_x$ orbital, and Mn4 $3d_{x^2-y^2}$ orbital forms σ-type covalent bonding with F7 $2p_y$ orbital and F8 $2p_x$ orbital. σ-type superexchange interaction occurs between Mn1 and Mn2 via F5, between Mn2 and Mn3 via F6, between Mn3 and Mn4 via F7, and between Mn4 and Mn1 via F8. The obtained wave-functions of MO46α and MO46β are

$$\begin{aligned}\psi_{\mathrm{MO46\alpha}} = {} & 0.43\phi_{\mathrm{Mn1}\left(3\mathrm{d}x^{2'}\right)} + 0.16\phi_{\mathrm{Mn1}\left(3\mathrm{d}x^{2''}\right)} - 0.43\phi_{\mathrm{Mn1}\left(3\mathrm{d}y^{2'}\right)} - 0.16\phi_{\mathrm{Mn1}\left(3\mathrm{d}y^{2''}\right)} \\ & + 0.43\phi_{\mathrm{Mn3}\left(3\mathrm{d}x^{2'}\right)} + 0.16\phi_{\mathrm{Mn3}\left(3\mathrm{d}x^{2''}\right)} - 0.43\phi_{\mathrm{Mn3}\left(3\mathrm{d}y^{2''}\right)} - 0.16\phi_{\mathrm{Mn3}\left(3\mathrm{d}y^{2''}\right)} \\ & - 0.15\phi_{\mathrm{F5(2p}y')} - 0.11\phi_{\mathrm{F5(2p}y'')} - 0.15\phi_{\mathrm{F6(2p}x')} - 0.11\phi_{\mathrm{F6(2p}x'')} \\ & + 0.15\phi_{\mathrm{F7(2p}y')} + 0.11\phi_{\mathrm{F7(2p}y'')} + 0.15\phi_{\mathrm{F8(2p}x')} + 0.11\phi_{\mathrm{F8(2p}x'')}\end{aligned} \tag{10.19}$$

$$\begin{aligned}\psi_{\mathrm{MO46\beta}} = {} & 0.43\phi_{\mathrm{Mn2}\left(3\mathrm{d}x^{2'}\right)} + 0.16\phi_{\mathrm{Mn2}\left(3\mathrm{d}x^{2''}\right)} - 0.43\phi_{\mathrm{Mn2}\left(3\mathrm{d}y^{2'}\right)} - 0.16\phi_{\mathrm{Mn2}\left(3\mathrm{d}y^{2''}\right)} \\ & + 0.43\phi_{\mathrm{Mn4}\left(3\mathrm{d}x^{2'}\right)} + 0.16\phi_{\mathrm{Mn4}\left(3\mathrm{d}x^{2''}\right)} - 0.43\phi_{\mathrm{Mn4}\left(3\mathrm{d}y^{2'}\right)} - 0.16\phi_{\mathrm{Mn4}\left(3\mathrm{d}y^{2''}\right)} \\ & + 0.15\phi_{\mathrm{F5(2p}y')} + 0.11\phi_{\mathrm{F5(2p}y'')} + 0.15\phi_{\mathrm{F6}\left(2\mathrm{p}x'\right)} + 0.11\phi_{\mathrm{F6(2p}x'')} \\ & - 0.15\phi_{\mathrm{F7(2p}y')} - 0.11\phi_{\mathrm{F7(2p}y'')} - 0.15\phi_{\mathrm{F8}\left(2\mathrm{p}x'\right)} - 0.11\phi_{\mathrm{F8(2p}x'')}\end{aligned} \tag{10.20}$$

In MO46α and MO46β, σ-type superexchange interaction occurs between Mn1 and Mn2 via F5, between Mn2 and Mn3 via F6, between Mn3 and Mn4 via F7, and between Mn4 and Mn1 via F8. There are inversion interactions between F5 and F6, and between F7 and F8. The orbital energies of MO46α and MO46β (−1.0959 au) are slightly higher than MO45α and MO45β (−1.0996 au). The obtained wave-functions of MO47α and MO47β are

$$\begin{aligned}\psi_{MO47\alpha} = {} & 0.26\phi_{Mn1(3dx^{2\prime})} + 0.10\phi_{Mn1(3dx^{2\prime\prime})} + 0.26\phi_{Mn1(3dy^{2\prime})} + 0.10\phi_{Mn1(3dy^{2\prime\prime})} \\ & - 0.51\phi_{Mn1(3dz^{2\prime})} - 0.18\phi_{Mn1(3dz^{2\prime\prime})} + 0.26\phi_{Mn3(3dx^{2\prime})} + 0.10\phi_{Mn3(3dx^{2\prime\prime})} \\ & + 0.26\phi_{Mn3(3dy^{2\prime})} + 0.10\phi_{Mn3(3dy^{2\prime\prime})} - 0.51\phi_{Mn3(3dz^{2\prime})} - 0.18\phi_{Mn3(3dz^{2\prime\prime})} \\ & + 0.14\phi_{F5(2py')} + 0.10\phi_{F5(2py'')} - 0.14\phi_{F6(2px')} - 0.10\phi_{F6(2px'')} \\ & - 0.14\phi_{F7(2py')} - 0.10\phi_{F7(2py'')} + 0.14\phi_{F8(2px')} + 0.10\phi_{F8(2px'')}\end{aligned} \tag{10.21}$$

$$\begin{aligned}\psi_{MO47\beta} = {} & 0.26\phi_{Mn2(3dx^{2\prime})} + 0.10\phi_{Mn2(3dx^{2\prime\prime})} + 0.26\phi_{Mn2(3dy^{2\prime})} + 0.10\phi_{Mn2(3dy^{2\prime\prime})} \\ & - 0.51\phi_{Mn2(3dz^{2\prime})} - 0.18\phi_{Mn2(3dz^{2\prime\prime})} + 0.26\phi_{Mn4(3dx^{2\prime})} + 0.10\phi_{Mn4(3dx^{2\prime\prime})} \\ & + 0.26\phi_{Mn4(3dy^{2\prime})} + 0.10\phi_{Mn4(3dy^{2\prime\prime})} - 0.51\phi_{Mn4(3dz^{2\prime})} - 0.18\phi_{Mn4(3dz^{2\prime\prime})} \\ & - 0.14\phi_{F5(2py')} - 0.10\phi_{F5(2py'')} + 0.14\phi_{F6(2px')} + 0.10\phi_{F6(2px'')} \\ & + 0.14\phi_{F7(2py')} + 0.10\phi_{F7(2py'')} - 0.14\phi_{F8(2px')} - 0.10\phi_{F8(2px'')}\end{aligned} \tag{10.22}$$

MO47α and MO47β are partially paired in F5 $2p_y$, F6 $2p_x$, F7 $2p_y$ and F8 $2p_x$ orbitals. α spins exist in $3d_{3z^2-r^2}$ orbitals of Mn1 and Mn3, and β spins exist in $3d_{3z^2-r^2}$ orbitals of Mn2 and Mn4. In MO47α, Mn1 lobe interacts with F5 lobe and F8 lobe, and Mn3 lobe interacts with F6 lobe and F7 lobe. From chemical bonding rule, it is found that Mn1 $3d_{3z^2-r^2}$ orbital forms σ-type covalent bonding with F5 $2p_y$ orbital and F8 $2p_x$ orbital, and Mn3 $3d_{3z^2-r^2}$ orbital forms σ-type covalent bonding with F6 $2p_x$ orbital and F7 $2p_y$ orbital. In MO47β, Mn2 lobe interacts with F5 lobe and F6 lobe, and Mn4 interacts with F7 lobe and F8 lobe. From chemical bonding rule, it is found that Mn2 $3d_{3z^2-r^2}$ orbital forms σ-type covalent bonding with F5 $2p_y$ orbital and F6 $2p_x$ orbital, and Mn4 $3d_{3z^2-r^2}$ orbital forms σ-type covalent bonding with F7 $2p_y$ orbital and F8 $2p_x$ orbital. σ-type superexchange interaction occurs between Mn1 and Mn2 via F5, between Mn2 and Mn3 via F6, between Mn3 and Mn4 via F7, and between Mn4 and Mn1 via F8. The obtained wave-functions of MO48α and MO48β are

$$\begin{aligned}\psi_{MO48\alpha} = {} & 0.27\phi_{Mn1(3dx^{2\prime})} + 0.11\phi_{Mn1(3dx^{2\prime\prime})} + 0.27\phi_{Mn1(3dy^{2\prime})} + 0.11\phi_{Mn1(3dy^{2\prime\prime})} \\ & - 0.53\phi_{Mn1(3dz^{2\prime})} - 0.19\phi_{Mn1(3dz^{2\prime\prime})} - 0.27\phi_{Mn3(3dx^{2\prime})} - 0.11\phi_{Mn3(3dx^{2\prime\prime})} \\ & - 0.27\phi_{Mn3(3dy^{2\prime})} - 0.11\phi_{Mn3(3dy^{2\prime\prime})} + 0.53\phi_{Mn3(3dz^{2\prime})} + 0.19\phi_{Mn3(3dz^{2\prime\prime})} \\ & + 0.12\phi_{F5(2py')} + 0.09\phi_{F5(2py'')} + 0.12\phi_{F6(2px')} + 0.09\phi_{F6(2px'')} \\ & + 0.12\phi_{F7(2py')} + 0.09\phi_{F7(2py'')} + 0.12\phi_{F8(2px')} + 0.09\phi_{F8(2px'')}\end{aligned} \tag{10.23}$$

$$\begin{aligned}\psi_{\text{MO48}\beta} = &-0.27\phi_{\text{Mn2}(3dx^{2\prime})} - 0.10\phi_{\text{Mn2}(3dx^{2\prime\prime})} - 0.27\phi_{\text{Mn2}(3dy^{2\prime})} - 0.10\phi_{\text{Mn2}(3dy^{2\prime\prime})} \\ &+0.51\phi_{\text{Mn2}(3dz^{2\prime})} + 0.18\phi_{\text{Mn2}(3dz^{2\prime\prime})} + 0.27\phi_{\text{Mn4}(3dx^{2\prime})} + 0.10\phi_{\text{Mn4}(3dx^{2\prime\prime})} \\ &+0.27\phi_{\text{Mn4}(3dy^{2\prime})} + 0.10\phi_{\text{Mn4}(3dy^{2\prime\prime})} - 0.51\phi_{\text{Mn4}(3dz^{2\prime})} - 0.18\phi_{\text{Mn4}(3dz^{2\prime\prime})} \\ &+0.12\phi_{\text{F5}(2py')} + 0.09\phi_{\text{F5}(2py'')} - 0.12\phi_{\text{F6}(2px')} - 0.09\phi_{\text{F6}(2px'')} \\ &+0.12\phi_{\text{F7}(2py')} + 0.09\phi_{\text{F7}(2py'')} - 0.12\phi_{\text{F8}(2px')} - 0.09\phi_{\text{F8}(2px'')}\end{aligned} \tag{10.24}$$

In MO48α and MO48β, σ-type superexchange interaction occurs between Mn1 and Mn2 via F5, between Mn2 and Mn3 via F6, between Mn3 and Mn4 via F7, and between Mn4 and Mn1 via F8. There are inversion interactions between F5 and F6, and between F7 and F8. The orbital energies of MO48α and MO48β (−1.0839 au) are slightly higher than MO47α and MO47β (−1.0863 au). The obtained wave-functions of MO53α and MO53β are

$$\begin{aligned}\psi_{\text{MO53}\alpha} = &\ 0.56\phi_{\text{Mn1}(3dxy')} + 0.21\phi_{\text{Mn1}(3dxy'')} + 0.56\phi_{\text{Mn3}(3dxy')} + 0.21\phi_{\text{Mn3}(3dxy'')} \\ &-0.08\phi_{\text{F5}(2px')} - 0.07\phi_{\text{F5}(2px'')} + 0.08\phi_{\text{F6}(2py')} + 0.07\phi_{\text{F6}(2py'')} \\ &+0.08\phi_{\text{F7}(2px')} + 0.07\phi_{\text{F7}(2px'')} - 0.08\phi_{\text{F8}(2py')} - 0.07\phi_{\text{F8}(2py'')}\end{aligned} \tag{10.25}$$

$$\begin{aligned}\psi_{\text{MO53}\beta} = &-0.56\phi_{\text{Mn2}(3dxy')} - 0.21\phi_{\text{Mn2}(3dxy'')} - 0.56\phi_{\text{Mn4}(3dxy')} - 0.21\phi_{\text{Mn4}(3dxy'')} \\ &-0.08\phi_{\text{F5}(2px')} - 0.07\phi_{\text{F5}(2px'')} + 0.08\phi_{\text{F6}(2py')} + 0.07\phi_{\text{F6}(2py'')} \\ &+0.08\phi_{\text{F7}(2px')} + 0.07\phi_{\text{F7}(2px'')} - 0.08\phi_{\text{F8}(2py')} - 0.07\phi_{\text{F8}(2py'')}\end{aligned} \tag{10.26}$$

MO53α and MO53β are partially paired in F5 $2p_x$, F6 $2p_y$, F7 $2p_x$ and F8 $2p_y$ orbitals. α spins exist in $3d_{xy}$ orbitals of Mn1 and Mn3, and β spins exist in $3d_{xy}$ orbitals of Mn2 and Mn4. In MO53α, two Mn1 lobes interact with two F5 lobes and two F8 lobes, and two Mn3 lobes interact with two F6 lobes and two F7 lobes. From chemical bonding rule, it is found that Mn1 $3d_{xy}$ orbital forms π-type covalent bonding with F5 $2p_x$ orbital and F8 $2p_y$ orbital, and Mn3 $3d_{xy}$ orbital forms π-type covalent bonding with F6 $2p_y$ orbital and F7 $2p_x$ orbital. In MO53β, two Mn2 lobes interact with two F5 lobes and two F6 lobes, and two Mn4 lobes interact with two F7 lobes and two F8 lobes. From chemical bonding rule, it is found that Mn2 $3d_{xy}$ orbital forms π-type covalent bonding with F5 $2p_x$ orbital and F6 $2p_y$ orbital, and Mn4 $3d_{xy}$ orbital forms π-type covalent bonding with F7 $2p_x$ orbital and F8 $2p_y$ orbital. π-type superexchange interaction occurs between Mn1 and Mn2 via F5, between Mn2 and Mn3 via F6, between Mn3 and Mn4 via F7, and between Mn4 and Mn1 via F8. The obtained wave-functions of MO54α and MO54β are

$$\psi_{MO54\alpha} = 0.57\phi_{Mn1(3dxy')} + 0.21\phi_{Mn1(3dxy'')} - 0.57\phi_{Mn3(3dxy')} - 0.21\phi_{Mn3(3dxy'')} - 0.07\phi_{F5(2px')} - 0.06\phi_{F5(2px'')} - 0.07\phi_{F6(2py')} - 0.06\phi_{F6(2py'')} - 0.07\phi_{F7(2px')} - 0.06\phi_{F7(2px'')} - 0.07\phi_{F8(2py')} - 0.06\phi_{F8(2py'')} \quad (10.27)$$

$$\psi_{MO54\beta} = -0.57\phi_{Mn2(3dxy')} - 0.21\phi_{Mn2(3dxy'')} + 0.57\phi_{Mn4(3dxy')} + 0.21\phi_{Mn4(3dxy'')} - 0.07\phi_{F5(2px')} - 0.06\phi_{F5(2px'')} + 0.07\phi_{F6(2py')} + 0.06\phi_{F6(2py'')} - 0.07\phi_{F7(2px')} - 0.06\phi_{F7(2px'')} + 0.07\phi_{F8(2py')} + 0.06\phi_{F8(2py'')} \quad (10.28)$$

In MO54α and MO54β, π-type superexchange interaction occurs between Mn1 and Mn2 via F5, between Mn2 and Mn3 via F6, between Mn3 and Mn4 via F7, and between Mn4 and Mn1 via F8. There are inversion interactions between F5 and F6, and between F7 and F8. The orbital energies of MO54α and MO54β (−1.0735 au) are slightly higher than MO53α and MO53β (−1.0743 au).

In Mn_4F_4 model, two types of two-dimensional superexchange interaction are reproduced. One is σ-type superexchange interaction in MO45α, MO45β, MO46α, MO46β, MO47α, MO47β, MO48α and MO48β. The other is π-type superexchange interaction is represented through MO53α, MO53β, MO54α and MO54β. Mn_4F_4 model reproduces well two-dimensional superexchange interaction within MnF_2 layer in K_2MnF_4. Note that σ-type superexchange interaction in z direction is not reproduced, though $3d_{z^2}$ component participates in molecular orbitals.

10.6 KMn_8X_{12} Model

MO calculation by using BHHLYP method is performed for three-dimensional KMn_8F_{12} model. MINI (5.3.3.3/5.3/4.1), 6-31G* and MINI(4.3.3.3/4.3) basis sets are used for manganese, fluorine and potassium, respectively. The formal charges of Mn, F and K are +2, −1 and +1, respectively. In formal electron configuration, five electrons occupy manganese 3d orbitals.

Mulliken charge densities for manganese and fluorine are 1.692 and −0.778. The spin densities of manganese with α and β spins are 4.891 and −4.891, respectively. Table 10.3 shows the population analysis of alpha and beta orbitals of KMn_8F_{12} model. Note that spin densities of manganese 3d orbitals are expressed by two components, because they are expressed by two basis functions. For example, $3d_{x^2}$ orbital is represented by two $3d_{x^{2\prime}}$ and $3d_{x^{2\prime\prime}}$ components. It is considered that twelve superexchange interactions occur between two neighbouring manganese atoms via fluorine, due to α spins of Mn1, Mn3, Mn6 and Mn8, and β spins of Mn2, Mn4, Mn5 and Mn7.

Table 10.3 Population analysis of alpha and beta electrons of KMn_8F_{12} model (BHHLYP method)

Site	Component	Alpha	Beta
Mn1, Mn3, Mn6, Mn8	$3d_{x^{2\prime}}$, $3d_{y^{2\prime}}$, $3d_{z^{2\prime}}$	0.5416	
	$3d_{x^{2\prime\prime}}$, $3d_{y^{2\prime\prime}}$, $3d_{z^{2\prime\prime}}$	0.1446	
	$3d_{xy'}$, $3d_{yz'}$, $3d_{xz'}$	0.7994	
	$3d_{xy''}$, $3d_{yz''}$, $3d_{xz''}$	0.2037	
Mn2, Mn4, Mn5, Mn7	$3d_{x^{2\prime}}$, $3d_{y^{2\prime}}$, $3d_{z^{2\prime}}$		0.5416
	$3d_{x^{2\prime\prime}}$, $3d_{y^{2\prime\prime}}$, $3d_{z^{2\prime\prime}}$		0.1446
	$3d_{xy'}$, $3d_{yz'}$, $3d_{xz'}$		0.7994
	$3d_{xy''}$, $3d_{yz''}$, $3d_{xz''}$		0.2037

10.7 Bent Superexchange Interaction: Cu_2F_2 Model

Let us investigate superexchange interaction in bent CuFCu. Figure 10.7 depicts Cu_2F_2 model and expected covalent bonds in 90°-bent CuFCu. Note that it is assumed that the formal charge of copper is +2, and Cu1 and Cu2 have spin density on $3d_{x^2-y^2}$ orbital. It is expected that Cu1 $3d_{x^2-y^2}$ orbital forms covalent bonding with F3 $2p_x$ and F4 $2p_y$ orbitals, and Cu2 $3d_{x^2-y^2}$ orbital forms covalent bonding with F3 $2p_y$ and F4 $2p_x$ orbitals. Following superexchange rule, as two orbitals remain as spin source, ferromagnetic interaction is expected.

BHHLYP calculation is performed for Cu_2F_2 model. MINI (5.3.3.3/5.3/5) and 6-31G* are used for copper and fluorine, respectively. At the geometry optimization structure, the Cu–F–Cu angle is different from 90°. Antiferromagnetic spin state is stabilized, and spin densities of Cu1 and Cu2 are 0.996 and −0.996, respectively.

Figure 10.8 depicts the selected molecular orbitals of Cu_2F_2 model. The obtained wave-functions of MO23α and MO23β are

$$\psi_{\mathrm{MO23\alpha}} = 0.96\phi_{\mathrm{Cu1(3dx^2)}} - 0.67\phi_{\mathrm{Cu1(3dy^2)}} - 0.30\phi_{\mathrm{Cu1(3dz^2)}} \\ - 0.02\phi_{\mathrm{F3(2s')}} - 0.03\phi_{\mathrm{F3(2s'')}} - 0.02\phi_{\mathrm{F4(2s')}} - 0.03\phi_{\mathrm{F4(2s'')}} \tag{10.29}$$

$$\psi_{\mathrm{MO23\beta}} = 0.96\phi_{\mathrm{Cu2(3dx^2)}} - 0.67\phi_{\mathrm{Cu2(3dy^2)}} - 0.30\phi_{\mathrm{Cu2(3dz^2)}} \\ - 0.02\phi_{\mathrm{F3(2s')}} - 0.03\phi_{\mathrm{F3(2s'')}} - 0.02\phi_{\mathrm{F4(2s')}} - 0.03\phi_{\mathrm{F4(2s'')}} \tag{10.30}$$

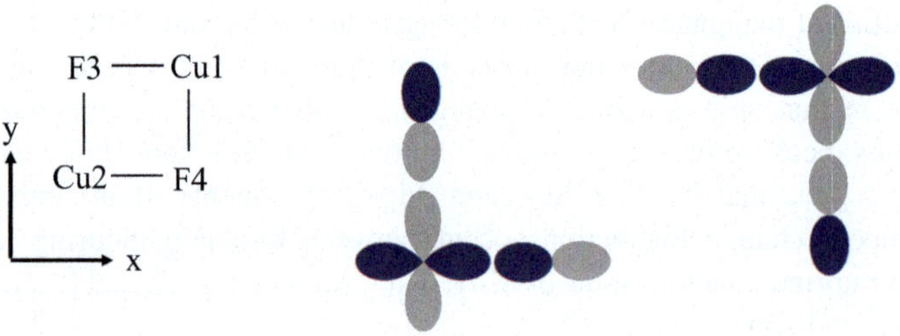

Fig. 10.7 Cubic Cu_2F_2 model and expected covalent bonding

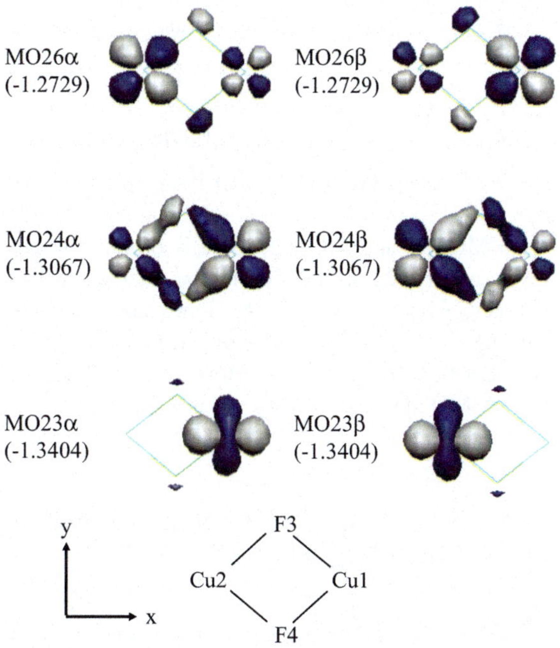

Fig. 10.8 Selected molecular orbitals of Cu_2F_2 model (BHHLYP method)

In MO23α, there are slight orbital overlap between Cu1 hybridized 3d and F3 2s orbitals, and Cu1 hybridized 3d and F4 2s orbitals. In MO23β, Cu2 has the same types of orbital overlaps. One Cu1 lobe interacts with one F3 lobe, and one F4 lobe. From chemical bonding rule, it is found that slight σ-type covalent bonding is formed in MO23α and MO24β. Slight σ-type superexchange interaction occurs between Cu1 and Cu2 via F3 and F4. The obtained wave-functions of MO24α and MO24β are

$$\begin{aligned}\psi_{\mathrm{MO24\alpha}} = {} & 0.89\phi_{\mathrm{Cu1(3d}xy)} + 0.36\phi_{\mathrm{Cu2(3d}xy)} \\ & - 0.10\phi_{\mathrm{F3(2p}x')} - 0.06\phi_{\mathrm{F3(2p}x'')} + 0.10\phi_{\mathrm{F4(2p}x')} + 0.06\phi_{\mathrm{F4(2p}x'')}\end{aligned} \tag{10.31}$$

$$\begin{aligned}\psi_{\mathrm{MO24\beta}} = {} & 0.36\phi_{\mathrm{Cu1(3d}xy)} + 0.89\phi_{\mathrm{Cu2(3d}xy)} \\ & - 0.10\phi_{\mathrm{F3(2p}x')} - 0.06\phi_{\mathrm{F3(2p}x'')} + 0.10\phi_{\mathrm{F4(2p}x')} + 0.06\phi_{\mathrm{F4(2p}x'')}\end{aligned} \tag{10.32}$$

In MO24α and MO24β, there is orbital overlap between Cu1 $3d_{xy}$, Cu2 $3d_{xy}$, F3 $2p_x$ and F4 $2p_x$ orbitals. Cu1 lobe interacts with F2 lobe and F3 lobe, and Cu2 interacts with F2 lobe and F3 lobe. From chemical bonding rule, it is found that σ-type covalent bonding is formed. However, as MO24α and MO24β are paired, MO24α and MO24β are not spin source. The obtained wave-function of MO26α and MO26β are

$$\psi_{MO26\alpha} = 0.40\phi_{Cu1(3dxy)} - 0.90\phi_{Cu2(3dxy)} + 0.04\phi_{F3(2s')} + 0.10\phi_{F3(2s'')} - 0.04\phi_{F4(2s')} - 0.10\phi_{F4(2s'')} \quad (10.33)$$

$$\psi_{MO26\beta} = -0.90\phi_{Cu1(3dxy)} + 0.40\phi_{Cu2(3dxy)} - 0.04\phi_{F3(2s')} - 0.10\phi_{F3(2s'')} + 0.04\phi_{F4(2s')} + 0.10\phi_{F4(2s'')} \quad (10.34)$$

In MO26α and MO26β, there is orbital overlap between Cu1 $3d_{xy}$, Cu2 $3d_{xy}$, F3 2s and F4 2s orbitals. Cu1 lobe interacts with F2 lobe and F3 lobe, and Cu2 interacts with F2 lobe and F3 lobe. There are nodes between Cu1-F3, Cu1-F4, Cu2-F3 and Cu2-F4. From chemical bonding rule, it is found that inversion σ-type covalent bonding is formed. However, as MO26α and MO26β are approximately paired, MO24α and MO24β are not spin source.

10.8 Two-Atom Bridge Superexchange Interaction: MnCNMn Model

In $KMnF_3$ perovskite, superexchange interaction atoms occur between manganese atoms via one fluorine bridge. Let us consider two-atom bridge superexchange interaction. In Prussian blue and its analogues, manganese atoms are bound with cyano-ligand (CN). MO calculation by using BHHLYP method is performed for MnCNMn model. Note that carbon and nitrogen are allocated at the right and left sides, respectively. MINI (5.3.3.3/5.3/5) basis set is used for manganese, combined with 6-31G* basis set for carbon and nitrogen. In formal electron configuration, five electrons occupy manganese 3d orbitals.

Mulliken charge densities for Mn1, Mn2, C and N are 1.821, 1.885, −0.067 and −0.638, respectively. The spin densities of Mn1 and Mn2 are 4.950 and −4.983, respectively. There exist small spin densities of carbon (−0.151) and nitrogen (0.183). In comparison with MnFMn model, charge and spin densities are non-symmetric. Figure 10.9 depicts the molecular orbitals of MnCNMn model. The obtained wave-functions of MO22α and MO22β are

$$\psi_{MO22\alpha} = 0.62\phi_{Mn1(3dx^2)} - 0.29\phi_{Mn1(3dy^2)} - 0.29\phi_{Mn1(3dz^2)} - 0.13\phi_{C(1s)} + 0.25\phi_{C(2s')} + 0.16\phi_{C(2s'')} + 0.18\phi_{C(2px')} + 0.23\phi_{C(2px'')} - 0.18\phi_{N(2s')} + 0.15\phi_{N(2px')} + 0.20\phi_{N(2px'')} \quad (10.35)$$

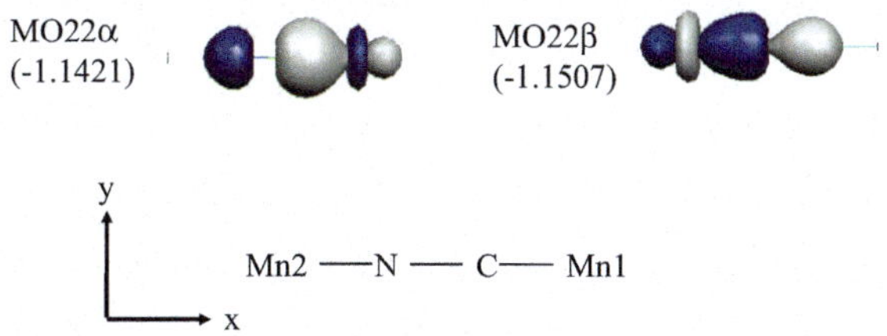

Fig. 10.9 Molecular orbitals of MnCNMn model (BHHLYP method)

$$\begin{aligned}\psi_{\mathrm{MO22\beta}} = & -0.75\phi_{\mathrm{Mn2(3d}x^2)} + 0.35\phi_{\mathrm{Mn2(3d}y^2)} + 0.35\phi_{\mathrm{Mn2(3d}z^2)} \\ & + 0.19\phi_{\mathrm{C(2s')}} + 0.10\phi_{\mathrm{C(2s'')}} + 0.12\phi_{\mathrm{C(2p}x'')} \\ & - 0.15\phi_{\mathrm{C(2s')}} - 0.13\phi_{\mathrm{C(2p}x')} + 0.25\phi_{\mathrm{N(2p}x')} + 0.19\phi_{\mathrm{N(2p}x'')}\end{aligned} \quad (10.36)$$

MO22α and MO22β are partially paired in cyano-ligand: carbon 2s, carbon $2p_x$, nitrogen 2s and nitrogen $2p_x$ orbitals. In MO22α, one Mn1 lobe interacts with one carbon lobe. In MO23β, one Mn2 lobe interacts with nitrogen lobe. From chemical bonding rule, it is found that Mn1 and Mn2 form σ-type covalent bonding with cyano-ligand. σ-type superexchange interaction occurs between Mn1 and Mn2 via cyano-ligand. It is called two-atom bridge superexchange interaction.

Further Readings

1. Buijse MA (1991) Electron correlation Fermi and Coulomb holes dynamical and nondynamical correlation, pp 9–10
2. Kanamori J (1956) J Phys Chem Solids 10:p87–p98
3. Kanamori J (1960) J Appl Phys 31(5):14S–23S
4. Onishi T (2014) J Comput Chem Jpn 13(6):p319–p320
5. Onishi T (2012) Adv Quant Chem 64:p36–p39
6. Granovsky AA, Firefly version 8, http://classic.chem.msu.su/gran/firefly/index.html
7. Schmidt MW, Baldridge KK, Boatz JA, Elbert ST, Gordon MS, Jensen JH, Koseki S, Matsunaga N, Nguyen KA, Su S, Windus TL, Dupuis M, Montgomery JA (1993) J Comput Chem 14:1347–1363
8. Varetto U <MOLEKEL 4.3.>; Swiss National Supercomputing Centre. Manno, Switzerland
9. Huzinaga S, Andzelm J, Radzio-Andzelm E, Sakai Y, Tatewaki H, Klobukowski M (1984) Gaussian basis sets for molecular calculations. Elsevier, Amsterdam
10. Hariharan PC, Pople JA (1973) Theor Chim Acta 28:213–222
11. Francl MM, Pietro WJ, Hehre WJ, Binkley JS, Gordon MS, DeFrees DJ, Pople JA (1982) J Chem Phys 77(7):3654–3665
12. Rassolov VA, Pople JA, Ratner MA, Windus TL (1998) J Chem Phys 109(4):1223–1229
13. Rassolov VA, Ratner MA, Pople JA, Redfern PC, Curtiss LA (2001) J Comput Chem 22 (9):976–984

Chapter 11
Ligand Bonding Effect

Abstract In ligand field theory, electron configuration of transition metal is empirically predicted based on Coulomb repulsion between transition metal and ligand anion. In octahedral coordination, transition metal 3d orbitals are split into two e_g $\left(3d_{3z^2-r^2},\ 3d_{x^2-y^2}\right)$ and t_{2g} ($3d_{xy}$, $3d_{yz}$, $3d_{xz}$) orbitals. However, it does not always predict correct electronic structure. It is because quantum effects of charge transfer and orbital overlap are missing. The alternate copper $3d_{z^2-x^2}$ type orbital ordering occurs in K_2CuF_4 perovskite. From molecular orbital calculation, it is found that the elongation and shrink of Cu–F distance occur. The electron configuration of transition metal is determined by quantum effect and structural distortion. The effect is called ligand bonding effect. In $KCoF_3$ perovskite, Co^{2+} has the degree of freedom in cobalt electron configuration. Two spin states such as quartet and doublet spin state are compared. Finally, in ideal FeF_6 model, the relationship between Fe–F distance and total energy is discussed.

Keywords Ligand field effect · Ligand bonding effect · K_2CuF_4 perovskite · Alternate $3d_{z^2-x^2}$ type orbital ordering · Electron configuration

11.1 Ligand Field Theory

In transition metal compounds, the orbital energies of 3d orbitals are split. Ligand field theory predicts the orbital energy splitting, based on Coulomb interaction between transition metal and ligand anion. If transition metal is isolated, 3d orbitals are degenerated.

In octahedral coordination, it is well known that transition metal 3d orbitals are split into two e_g $\left(3d_{3z^2-r^2},\ 3d_{x^2-y^2}\right)$ and t_{2g} ($3d_{xy}$, $3d_{yz}$, $3d_{xz}$) orbitals (see Fig. 11.1). e_g orbitals are destabilized, compared with t_{2g} orbitals. It is because Coulomb repulsion between e_g electron and ligand anion is larger. However, ligand field theory does not always predict a correct electronic structure, because of pseudo-quantum

T. Onishi, *Quantum Computational Chemistry*,
DOI 10.1007/978-981-10-5933-9_11

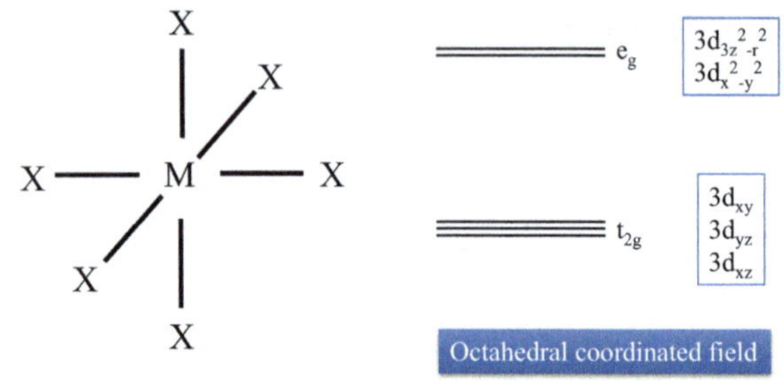

Fig. 11.1 Orbital energy splitting of transition metal 3d orbitals in octahedral coordinated field. *M* and *X* denote transition metal and ligand anion, respectively

level. For example, the effect of orbital overlap between transition metal and ligand anion is missing.

11.2 Ligand Bonding Effect

In transition metal compounds, orbital energies of transition metal 3d orbitals are split, due to Coulomb repulsion, charge transfer and orbital overlap between transition metal and ligand anion. In addition, the elongation or shrink of transition metal-ligand anion distance is combined. The effect is called "ligand bonding effect". The magnitudes of elongation and shrink depend on the type of covalent bonding. For example, in octahedral coordination, t_{2g} orbitals have σ-type orbital overlap with ligand anion, and e_g orbitals have π-type orbital overlap with ligand anion.

11.3 K_2CuF_4 Perovskite

In K_2CuF_4 perovskite, magnetic CuF_2 layer is separated by two non-magnetic KF layers, as shown in Fig. 10.3. The formal charges of copper and fluorine are +2 and −1, respectively. The change of copper electron configuration from the conventional octahedral coordination occurs, combined with displacements of fluorine anions on CuF_2 layer.

It was reported that apical and equatorial Cu–F distances are 1.95 Å and 2.08 Å, respectively. Note that the average Cu–F distance is observed in experiment. The further displacements are connected, in relation to Q_2 vibration mode (see Fig. 11.2). In one copper atom, as Cu–F distance along *x* and *y* axes shrinks and is elongated, respectively, the orbital energy of $3d_{z^2-x^2}$ orbital becomes higher. In neighbouring copper atom, as Cu–F distance along *x* and *y* axes is elongated and shrinks, respectively, the orbital energy of $3d_{z^2-y^2}$ orbital becomes higher. As the result, alternate $3d_{z^2-x^2}$ type orbital ordering is caused in K_2CuF_4 perovskite.

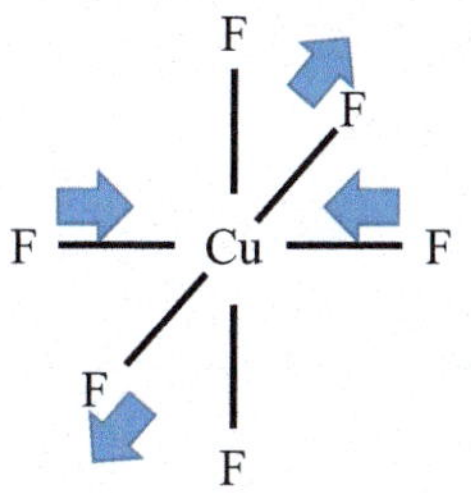

Fig. 11.2 Schematic drawing of Q_2 normal vibration mode. The *arrows* denote the directions of elongation and shrink

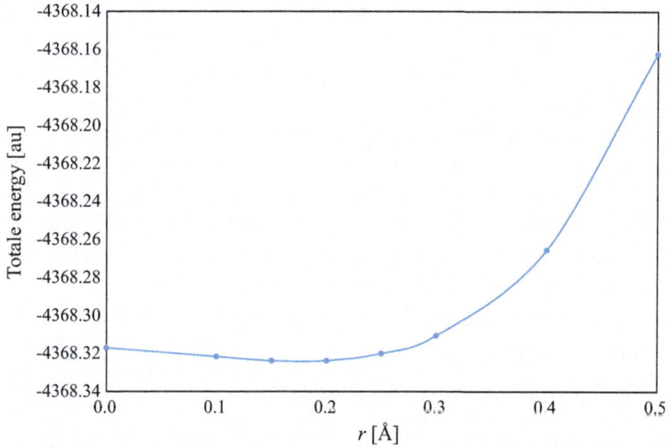

Fig. 11.3 Potential energy curve of $F_5CuFCuF_5$ model, displacing fluorine following Q_2 mode. The displaced distance (*r*) is defined from the lattice position. (BHHLYP method)

BHHLYP calculation is performed for two-nuclear $F_5CuFCuF_5$ model. MINI (5.3.3.3/5.3/5) and 6-31G* basis sets are used for copper and fluorine, respectively.

Figure 11.3 shows the potential energy curve of $F_5CuFCuF_5$ model, displacing fluorine following Q_2 mode. In Q_2 mode, all apical Cu–F distances are fixed (1.95 Å), and equatorial Cu–F distances change. When $r = 0.00$ Å, equatorial Cu–F distance corresponds to experimental Cu–F distance (2.08 Å). The local minimum is given at around $r = 0.15$ Å. It is found that fluorine anions are displaced, following Q_2 vibrational mode.

Mulliken charge densities of Cu1 and Cu2 are 1.73 and 1.74, respectively. Spin densities of Cu1, Cu2 and surrounding fluorine are 0.96, 0.97 and 0.00, respectively. Figure 11.4 depicts the selected molecular orbitals, which are related to alternate $3d_{z^2-x^2}$ type orbital ordering. The obtained wave-functions of MO41α and MO42α are

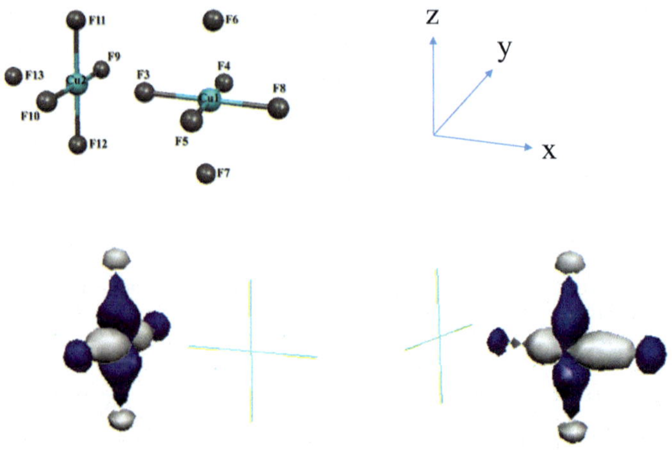

MO41α
(0.2933au)

MO42α
(0.3015au)

Fig. 11.4 Selected molecular orbitals related to alternate $3d_{z^2-x^2}$ type orbital ordering in $F_5CuFCuF_5$ model. The orbital energy is shown in parenthesis. (BHHLYP method)

$$\begin{aligned}\psi_{MO41\alpha} = &-0.78\phi_{Cu2(3dz^2)} + 0.82\phi_{Cu2(3dy^2)}\\ &- 0.11\phi_{F9(2py')} - 0.07\phi_{F9(2py'')} + 0.11\phi_{F10(2py')} + 0.07\phi_{F10(2py'')}\\ &+ 0.10\phi_{F11(2pz')} + 0.06\phi_{F11(2pz'')} - 0.10\phi_{F12(2pz')} - 0.06\phi_{F12(2pz'')}\end{aligned} \tag{11.1}$$

$$\begin{aligned}\psi_{MO42\alpha} = &-0.77\phi_{Cu1(3dz^2)} + 0.81\phi_{Cu1(3dx^2)}\\ &+ 0.08\phi_{F3(2px')} + 0.04\phi_{F3(2px'')} - 0.15\phi_{F8(2px')} - 0.11\phi_{F8(2px'')}\\ &+ 0.10\phi_{F6(2pz')} + 0.06\phi_{F6(2pz'')} - 0.10\phi_{F7(2pz')} - 0.06\phi_{F7(2pz'')}\end{aligned} \tag{11.2}$$

In MO41α, as there is hybridization between copper $3d_{z^2}$ and $3d_{y^2}$ orbitals, it is found that copper $3d_{z^2-y^2}$ orbital has orbital overlap with fluorine 2 orbitals. One copper lobe interacts with one fluorine lobe. From chemical bonding rule, it is found that σ-type covalent bonding is formed between Cu2 $3d_{z^2-y^2}$ and fluorine 2p orbitals. On the other hand, in MO42α, there is hybridization between copper $3d_{z^2}$ and $3d_{x^2}$ orbitals. σ-type covalent bonding is formed between Cu2 $3d_{z^2-x^2}$ and fluorine 2p orbitals. As MO41α and MO42α are spin source, it is found that alternate $3d_{z^2-x^2}$ type orbital ordering is caused, and no superexchange interaction occurs between copper atoms.

11.4 $KCoF_3$ Perovskite

At room temperature, $KCoF_3$ perovskite has the cubic structure, where the Co–F distance is 2.035 Å. As shown in Fig. 11.5, there are two possible electron configurations in Co^{2+} (see Fig. 11.5). BHHLYP calculation is performed for two-nuclear $F_5CoFCoF_5$ model, under consideration of two cobalt electron configurations (quartet and doublet). MINI(5.3.3.3/5.3/5) and 6-31G* basis sets are used for cobalt and fluorine, respectively.

Table 11.1 summarizes the population analysis of cobalt alpha and beta electrons, and spin densities. It is found that quartet and doublet electron configurations are reproduced. In quartet electron configuration, one spin is delocalized over three t_{2g} orbitals, and two spins are delocalized over two e_g orbitals. On the other hand, in doublet electron configuration, one spin is delocalized over three t_{2g} orbitals. The total energies of quartet and doublet cobalt electron configurations are −3860.02731 au and −3859.93008 au, respectively. It is found that quartet cobalt electron configuration is more stabilized than doublet cobalt electron configuration.

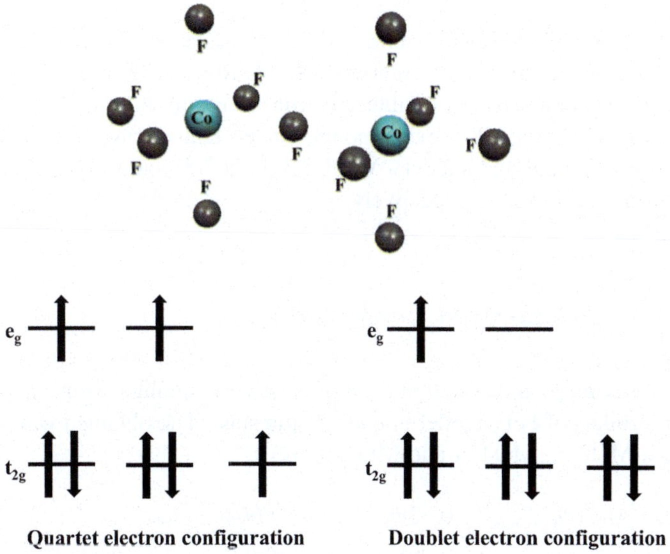

Fig. 11.5 Cluster model of $KCoF_3$ perovskite ($F_5CoFCoF_5$ model), and two possible cobalt electron configuration

Table 11.1 Population analysis of alpha and beta electrons, and spin density in $F_5CoFCoF_5$ model by BHHLYP calculation

Quartet					Doublet				
Site		Alpha	Beta	Spin	Site		Alpha	Beta	Spin
Co1	$3d_{x^2}$	0.6830	0.0997	0.5833	Co1	$3d_{x^2}$	0.6790	0.0601	0.6190
	$3d_{y^2}$	0.6814	0.1003	0.5811		$3d_{y^2}$	0.2650	0.0824	0.1826
	$3d_{z^2}$	0.6814	0.1003	0.5811		$3d_{z^2}$	0.2285	0.0843	0.1442
	$3d_{xy}$	1.0020	0.6449	0.3572		$3d_{xy}$	1.0014	1.0012	0.0002
	$3d_{yz}$	1.0020	0.6449	0.3572		$3d_{yz}$	1.0014	1.0012	0.0002
	$3d_{xz}$	1.0020	0.6525	0.3496		$3d_{xz}$	1.0017	1.0014	0.0003
Co2	$3d_{x^2}$	0.0997	0.6830	−0.5833	Co2	$3d_{x^2}$	0.0601	0.6790	−0.6190
	$3d_{y^2}$	0.1003	0.6814	−0.5811		$3d_{y^2}$	0.0824	0.2650	−0.1826
	$3d_{z^2}$	0.1003	0.6814	−0.5811		$3d_{z^2}$	0.0843	0.2285	−0.1442
	$3d_{xy}$	0.6449	1.0020	−0.3572		$3d_{xy}$	1.0012	1.0014	−0.0002
	$3d_{yz}$	0.6449	1.0020	−0.3572		$3d_{yz}$	1.0012	1.0014	−0.0002
	$3d_{xz}$	0.6525	1.0020	−0.3496		$3d_{xz}$	1.0014	1.0017	−0.0003

11.5 FeF_6 Model

To investigate the ligand bonding effect of iron fluorides, let us consider ideal FeF_6 model for the simplicity. The formal charges of iron and fluorine are +2 and −1, respectively. There are two possible electron configurations in Fe^{2+}. BHHLYP calculation is performed for FeF_6 model, under considering two iron electron configurations (quintet and singlet). MINI(5.3.3.3/5.3/5) and 6-31G* basis sets are used for iron and fluorine, respectively.

11.5.1 Quintet Electron Configuration

Figure 11.6 depicts the electron configuration of quintet iron, and selected molecular orbitals of FeF_6 model in quintet spin state. The obtained wave-functions of MO22α, MO23α, MO43α and MO44α are

$$\begin{aligned}\psi_{\mathrm{MO22}\alpha} &= 0.31\phi_{\mathrm{Fe1(3dx^2)}} + 0.32\phi_{\mathrm{Fe1(3dy^2)}} - 0.66\phi_{\mathrm{Fe1(3dz^2)}} \\ &\quad + 0.28\phi_{\mathrm{F6(2pz')}} + 0.22\phi_{\mathrm{F6(2pz'')}} - 0.28\phi_{\mathrm{F7(2pz')}} - 0.22\phi_{\mathrm{F7(2pz'')}}\end{aligned} \tag{11.3}$$

$$\begin{aligned}\psi_{\mathrm{MO23}\alpha} &= -0.51\phi_{\mathrm{Fe1(3dx^2)}} + 0.51\phi_{\mathrm{Fe1(3dy^2)}} \\ &\quad + 0.23\phi_{\mathrm{F2(2px')}} + 0.18\phi_{\mathrm{F2(2px'')}} + 0.24\phi_{\mathrm{F3(2py')}} + 0.19\phi_{\mathrm{F3(2py'')}} \\ &\quad - 0.23\phi_{\mathrm{F4(2px')}} - 0.18\phi_{\mathrm{F4(2px'')}} - 0.24\phi_{\mathrm{F5(2py')}} - 0.19\phi_{\mathrm{F5(2py'')}}\end{aligned} \tag{11.4}$$

Fig. 11.6 Electron configuration of quintet iron, and selected molecular orbitals of FeF_6 model in quintet spin state. The orbital energy is shown in parenthesis

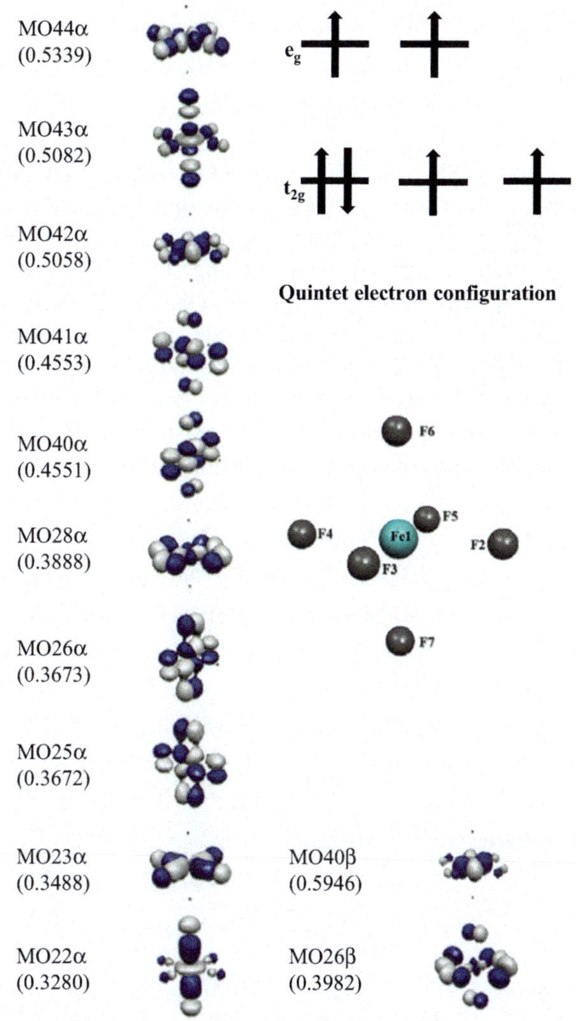

$$\begin{aligned}\psi_{\mathrm{MO43}\alpha} = {} & 0.37\phi_{\mathrm{Fe1(3dx^2)}} + 0.40\phi_{\mathrm{Fe1(3dy^2)}} - 0.78\phi_{\mathrm{Fe1(3dz^2)}} \\ & + 0.14\phi_{\mathrm{F2(2px')}} + 0.12\phi_{\mathrm{F2(2px'')}} - 0.15\phi_{\mathrm{F3(2py')}} - 0.13\phi_{\mathrm{F3(2py'')}} \\ & - 0.14\phi_{\mathrm{F4(2px')}} - 0.12\phi_{\mathrm{F4(2px'')}} + 0.15\phi_{\mathrm{F5(2py')}} + 0.13\phi_{\mathrm{F5(2py'')}} \\ & - 0.27\phi_{\mathrm{F6(2pz')}} - 0.23\phi_{\mathrm{F6(2pz'')}} + 0.27\phi_{\mathrm{F7(2pz')}} + 0.23\phi_{\mathrm{F7(2pz'')}}\end{aligned} \tag{11.5}$$

$$\psi_{\mathrm{MO44}\alpha} = -0.73\phi_{\mathrm{Fe1(3dx^2)}} + 0.71\phi_{\mathrm{Fe1(3dy^2)}} - 0.22\phi_{\mathrm{F2(2px')}} - 0.19\phi_{\mathrm{F2(2px'')}} - 0.22\phi_{\mathrm{F3(2py')}} - 0.19\phi_{\mathrm{F3(2py'')}} + 0.22\phi_{\mathrm{F4(2px')}} + 0.19\phi_{\mathrm{F4(2px'')}} + 0.22\phi_{\mathrm{F5(2py')}} + 0.19\phi_{\mathrm{F5(2py'')}} \quad (11.6)$$

In MO22α, iron $3d_{3z^2-r^2}$ orbital has orbital overlap with fluorine 2p orbitals. One iron 3d lobe interacts with one fluorine 2p lobe. From chemical bonding rule, it is found that σ-type covalent bonding is formed. MO43α represents corresponding inversion σ-type covalent bonding. In MO23α, iron $3d_{x^2-y^2}$ orbital has orbital overlap with fluorine 2p orbitals. One iron 3d lobe interacts with one fluorine 2p lobe. From chemical bonding rule, it is found that σ-type covalent bonding is formed. MO44α represents corresponding inversion σ-type covalent bonding. The spin population of iron $3d_{x^2}$, $3d_{y^2}$ and $3d_{z^2}$ orbitals is 0.634, 0.636 and 0.630, respectively. As the whole, there exist about two spins in iron $3d_{3z^2-r^2}$ and $3d_{x^2-y^2}$ orbitals. The obtained wave-functions of MO25α, MO26α, MO40α and MO41α are

$$\psi_{\mathrm{MO25}\alpha} = 0.61\phi_{\mathrm{Fe1(3dxz)}} + 0.21\phi_{\mathrm{F2(2pz')}} + 0.18\phi_{\mathrm{F2(2pz'')}} - 0.21\phi_{\mathrm{F4(2pz')}} - 0.18\phi_{\mathrm{F4(2pz'')}} + 0.24\phi_{\mathrm{F6(2px')}} + 0.21\phi_{\mathrm{F6(2px'')}} - 0.24\phi_{\mathrm{F7(2px')}} - 0.21\phi_{\mathrm{F7(2px'')}} \quad (11.7)$$

$$\psi_{\mathrm{MO26}\alpha} = 0.61\phi_{\mathrm{Fe1(3dyz)}} - 0.21\phi_{\mathrm{F3(2pz')}} - 0.18\phi_{\mathrm{F3(2pz'')}} + 0.21\phi_{\mathrm{F5(2pz')}} + 0.18\phi_{\mathrm{F5(2pz'')}} + 0.24\phi_{\mathrm{F6(2py')}} + 0.21\phi_{\mathrm{F6(2py'')}} - 0.24\phi_{\mathrm{F7(2py')}} - 0.21\phi_{\mathrm{F7(2py'')}} \quad (11.8)$$

$$\psi_{\mathrm{MO40}\alpha} = 0.78\phi_{\mathrm{Fe1(3dyz)}} + 0.24\phi_{\mathrm{F3(2pz')}} + 0.21\phi_{\mathrm{F3(2pz'')}} - 0.24\phi_{\mathrm{F5(2pz')}} - 0.21\phi_{\mathrm{F5(2pz'')}} - 0.17\phi_{\mathrm{F6(2py')}} - 0.15\phi_{\mathrm{F6(2py'')}} + 0.17\phi_{\mathrm{F7(2py')}} + 0.15\phi_{\mathrm{F7(2py'')}} \quad (11.9)$$

$$\psi_{\mathrm{MO41}\alpha} = -0.13\phi_{\mathrm{Fe1(3dxy)}} + 0.77\phi_{\mathrm{Fe1(3dxz)}} - 0.24\phi_{\mathrm{F2(2pz')}} - 0.20\phi_{\mathrm{F2(2pz'')}} + 0.24\phi_{\mathrm{F4(2pz')}} + 0.20\phi_{\mathrm{F4(2pz'')}} - 0.17\phi_{\mathrm{F6(2px')}} - 0.14\phi_{\mathrm{F6(2px'')}} + 0.17\phi_{\mathrm{F7(2px')}} + 0.14\phi_{\mathrm{F7(2px'')}} \quad (11.10)$$

In MO25α, iron $3d_{3xz}$ orbital has orbital overlap with fluorine 2p orbitals. Two iron 3d lobes interact with two fluorine 2p lobes. From chemical bonding rule, it is found that π-type covalent bonding is formed. MO41α represents corresponding inversion σ-type covalent bonding. In MO26α, iron $3d_{yz}$ orbital has orbital overlap with fluorine 2p orbitals. Two iron 3d lobes interact with two fluorine 2p lobes. From chemical bonding rule, it is found that π-type covalent bonding is formed. MO40α represents corresponding inversion π-type covalent bonding. The obtained wave-functions of MO28α, MO26β, MO42α and MO40β are

$$
\begin{aligned}
\psi_{\mathrm{MO28\alpha}} = {} & 0.39\phi_{\mathrm{Fe1(3d}xy)} \\
& + 0.27\phi_{\mathrm{F2(2p}y')} + 0.24\phi_{\mathrm{F2(2p}y'')} - 0.27\phi_{\mathrm{F3(2p}x')} - 0.24\phi_{\mathrm{F3(2p}x'')} \\
& - 0.27\phi_{\mathrm{F4(2p}y')} - 0.24\phi_{\mathrm{F4(2p}y'')} + 0.27\phi_{\mathrm{F5(2p}x')} + 0.24\phi_{\mathrm{F5(2p}x'')}
\end{aligned} \tag{11.11}
$$

$$
\begin{aligned}
\psi_{\mathrm{MO26\beta}} = {} & 0.19\phi_{\mathrm{Fe1(3d}xy)} \\
& + 0.25\phi_{\mathrm{F2(2p}y')} + 0.22\phi_{\mathrm{F2(2p}y'')} + 0.14\phi_{\mathrm{F2(2p}z')} + 0.12\phi_{\mathrm{F2(2p}z'')} \\
& - 0.25\phi_{\mathrm{F3(2p}x')} - 0.22\phi_{\mathrm{F3(2p}x'')} \\
& - 0.25\phi_{\mathrm{F4(2p}y')} - 0.22\phi_{\mathrm{F4(2p}y'')} - 0.14\phi_{\mathrm{F4(2p}z')} - 0.12\phi_{\mathrm{F4(2p}z'')} \\
& + 0.25\phi_{\mathrm{F5(2p}x')} + 0.22\phi_{\mathrm{F5(2p}x'')} \\
& + 0.18\phi_{\mathrm{F6(2p}x')} + 0.16\phi_{\mathrm{F6(2p}x'')} - 0.18\phi_{\mathrm{F7(2p}x')} - 0.16\phi_{\mathrm{F7(2p}x'')}
\end{aligned} \tag{11.12}
$$

$$
\begin{aligned}
\psi_{\mathrm{MO42\alpha}} = {} & -0.91\phi_{\mathrm{Fe1(3d}xy)} - 0.16\phi_{\mathrm{Fe1(3d}xz)} \\
& + 0.14\phi_{\mathrm{F2(2p}y')} + 0.11\phi_{\mathrm{F2(2p}y'')} - 0.14\phi_{\mathrm{F3(2p}x')} - 0.11\phi_{\mathrm{F3(2p}x'')} \\
& - 0.14\phi_{\mathrm{F4(2p}y')} - 0.11\phi_{\mathrm{F4(2p}y'')} + 0.14\phi_{\mathrm{F5(2p}x')} + 0.11\phi_{\mathrm{F5(2p}x'')}
\end{aligned} \tag{11.13}
$$

$$
\begin{aligned}
\psi_{\mathrm{MO40\beta}} = {} & -0.96\phi_{\mathrm{Fe1(3d}xy)} - 0.17\phi_{\mathrm{Fe1(3d}xz)} \\
& + 0.09\phi_{\mathrm{F2(2p}y')} + 0.07\phi_{\mathrm{F2(2p}y'')} - 0.09\phi_{\mathrm{F3(2p}x')} - 0.07\phi_{\mathrm{F3(2p}x'')} \\
& - 0.09\phi_{\mathrm{F4(2p}y')} - 0.07\phi_{\mathrm{F4(2p}y'')} + 0.09\phi_{\mathrm{F5(2p}x')} + 0.07\phi_{\mathrm{F5(2p}x'')}
\end{aligned} \tag{11.14}
$$

MO28α and MO26β are approximately paired. In MO28α and MO26β, iron 3d$_{xy}$ orbital has orbital overlap with fluorine 2p orbitals. Two iron 3d lobes interact with two fluorine 2p lobes. From chemical bonding rule, it is found that π-type covalent bonding is formed. MO42α and MO40β are also approximately paired. They are corresponding inversion π-type covalent bonding.

11.5.2 Singlet Electron Configuration

Figure 11.7 depicts the electron configuration of singlet iron and selected molecular orbitals of FeF$_6$ model in singlet spin state. Degenerated MO23, MO24 and MO25 represent hybridized t$_{2g}$ orbitals. Two iron lobes interact with two fluorine lobes. From chemical bonding rule, it is found that π-type covalent bonding is formed. Degenerated MO40, MO41 and MO42 correspond to inversion π-type covalent bonding.

Figure 11.8 shows the potential energy curve of FeF$_6$ model, changing Fe–F distance. Local minima are given around 2.2 Å in quartet electron configuration, and 2.1 Å in singlet electron configuration. In all regions, quintet electron configuration is more stable than singlet electron configuration. In FeF$_6$ model, potential energy curves are not crossed. However, if two potential energy curves are crossed between

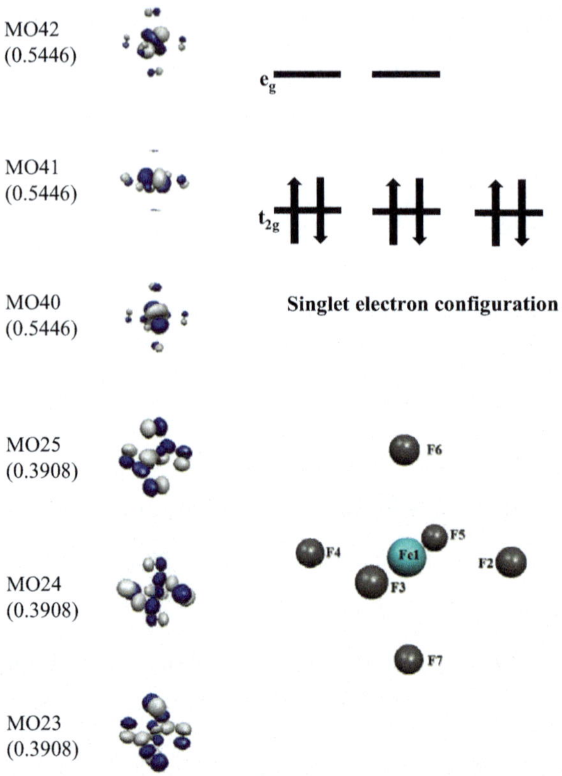

Fig. 11.7 Electron configuration of singlet iron, and molecular orbitals of FeF_6 model in singlet spin state. The orbital energy is shown in parenthesis

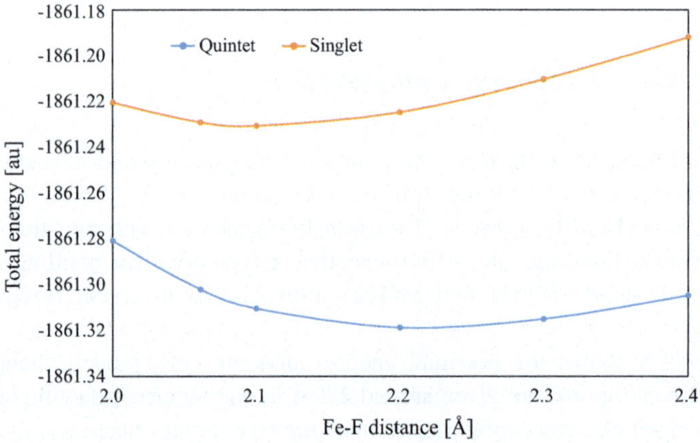

Fig. 11.8 Potential energy curve of FeF_6 model, changing Fe–F distance

different electron configurations, spin transition occurs between different electron configurations. The phenomenon is called spin crossover. In fact, spin crossover is observed in Prussian blue. The spin transition occurs between quartet and singlet electron configurations by changing Fe–CN distance. The conditions of spin crossover are very sensitive and complex, depending on patterns of Coulomb repulsion, charge transfer, orbital overlap, superexchange interaction, etc.

Further Readings

1. Onishi T, Yamaguchi K (2009) Polyhedron 28:1972–1976
2. Kanamori J (1960) J Appl Phys 31(5):14S–23S
3. Onishi T, Yoshioka Y, e-J Surf Sci Natotech (2007) 5: p17–19
4. Okazaki A, Suemune Y, Fuchikami T (1959) J Phys Soc Jpn 14:1823–1824
5. Sato O (2003) Acc Chem Res 36:692–700
6. Onishi T (2012) Adv Quant Chem 64:39–43
7. Granovsky AA, Firefly version 8. http://classic.chem.msu.su/gran/firefly/index.html
8. Schmidt MW, Baldridge KK, Boatz JA, Elbert ST, Gordon MS, Jensen JH, Koseki S, Matsunaga N, Nguyen KA, Su S, Windus TL, Dupuis M, Montgomery JA (1993) J Comput Chem 14:1347–1363
9. Varetto U <MOLEKEL 4.3.>; Swiss National Supercomputing Centre. Manno, Switzerland
10. Huzinaga S, Andzelm J, Radzio-Andzelm E, Sakai Y, Tatewaki H, Klobukowski M (1984) Gaussian basis sets for molecular calculations. Elsevier, Amsterdam
11. Hariharan PC, Pople JA (1973) Theoret Chim Acta 28:213–222
12. Francl MM, Pietro WJ, Hehre WJ, Binkley JS, Gordon MS, DeFrees DJ, Pople JA (1982) J Chem Phys 77(7):3654–3665
13. Rassolov VA, Pople JA, Ratner MA, Windus TL (1998) J Chem Phys 109(4):1223–1229
14. Rassolov VA, Ratner MA, Pople JA, Redfern PC, Curtiss LA (2001) J Compt Chem 22 (9):976–984

Part IV
Advanced Inorganic Materials

Chapter 12
Photocatalyst

Abstract $SrTiO_3$ perovskite has been utilized as photocatalyst. The bandgap (3.27 eV) corresponds to the wave-length of ultraviolet light. In general, virtual molecular orbital does not represent excited electronic structure. However, in $SrTiO_3$ perovskite, the reliable LUMO is given, due to the inclusion of electron correlation effect and stable crystal structure. Bandgap can be estimated as HOMO–LUMO energy gap. To enhance visible light response, nitrogen-doping and carbon-doping at oxygen site are performed to decrease bandgap, corresponding to the wave-length of visible light. From the viewpoint of energetics and bonding, the mechanism of bandgap change is explained. In nitrogen-doping, combined oxygen vacancy disturbs visible light response. Instead, in carbon-doping, visible light response is enhanced, due to no oxygen vacancy.

Keywords Bandgap · HOMO–LUMO energy gap · Hybrid-DFT · Photocatalyst · $SrTiO_3$ perovskite

12.1 Bandgap

As is explained in Chap. 4, virtual (unoccupied) molecular orbitals (MOs) are produced as the result of the introduction of basis function. In general, virtual MO does not represent excited electronic structure. However, in $SrTiO_3$ perovskite, the reliable lowest unoccupied molecular orbital (LUMO) is given, due to the inclusion of the electron correlation effect in LUMO, and stable crystal structure. The excitation energy of solid is called “bandgap”. As shown in Fig. 12.1, in molecular orbital, bandgap corresponds to HOMO–LUMO energy gap.

T. Onishi, *Quantum Computational Chemistry*,
DOI 10.1007/978-981-10-5933-9_12

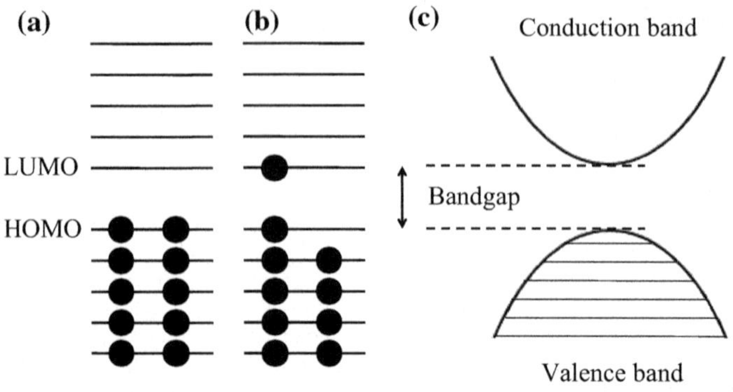

Fig. 12.1 Schematic drawing of the relationship between HOMO–LUMO energy gap and bandgap: **a** ground state of molecular orbital; **b** excited state of molecular orbital; **c** band structure. Ref. [1] by permission from Elsevier

12.2 Bandgap Estimation in $SrTiO_3$ Perovskite

BHHLYP calculation is performed for $SrTi_8O_{12}$ model of $SrTiO_3$ perovskite, where Ti-O-Ti distance is 3.91Å (see Fig. 12.2). Basis sets used for titanium, oxygen and strontium are MINI(5.3.3.3/5.3/5), 6-31G* and MINI(4.3.3.3.3/4.3.3/4), respectively. The formal charges of titanium and oxygen are +4 and −2, respectively. It implies that titanium formally has no 3d electron.

Figure 12.3 depicts the orbital energy diagram and molecular orbitals of $SrTi_8O_{12}$ model. The obtained wave-function of HOMO is

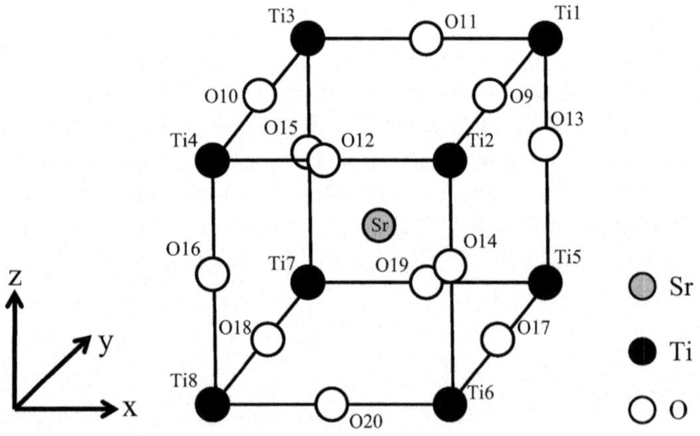

Fig. 12.2 $SrTi_8O_{12}$ model of $SrTiO_3$ perovskite

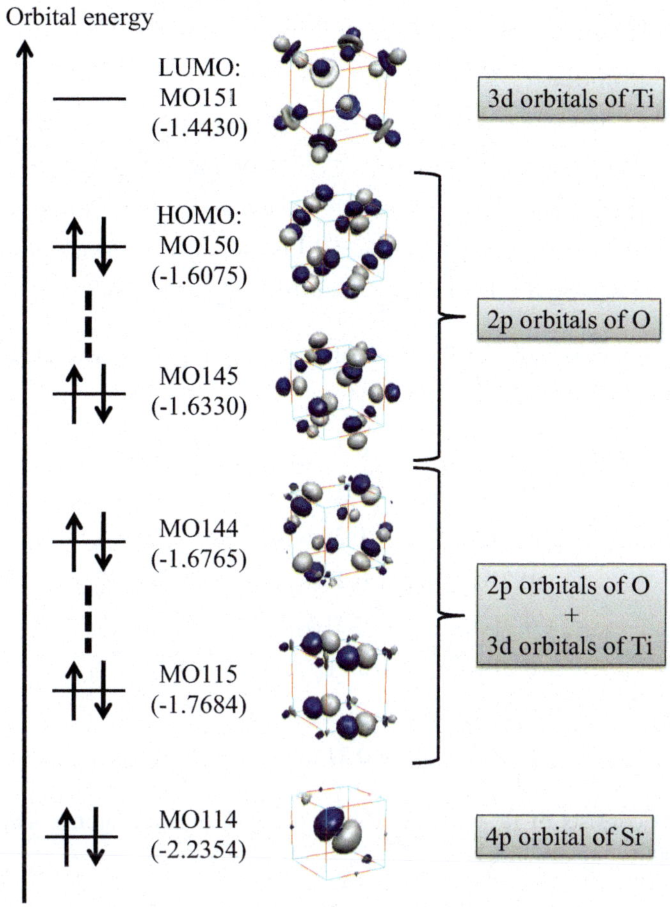

Fig. 12.3 Orbital energy diagram and molecular orbitals of $SrTi_8O_{12}$ model. (BHHLYP method) Ref. [1] by permission from Elsevier

$$
\begin{aligned}
\psi_{\mathrm{HOMO}} = & -0.14\phi_{O9(2px')} - 0.10\phi_{O9(2px'')} + 0.14\phi_{O9(2pz')} + 0.10\phi_{O(2pz'')} \\
& + 0.14\phi_{O10(2px')} + 0.10\phi_{O10(2px'')} + 0.14\phi_{O10(2pz')} + 0.10\phi_{O10(2pz'')} \\
& + 0.14\phi_{O11(2py')} + 0.10\phi_{O11(2py'')} - 0.14\phi_{O11(2pz')} - 0.10\phi_{O11(2pz'')} \\
& - 0.14\phi_{O12(2py')} - 0.10\phi_{O12(2py'')} - 0.14\phi_{O12(2pz')} - 0.10\phi_{O12(2pz'')} \\
& + 0.14\phi_{O13(2px')} + 0.10\phi_{O13(2px'')} - 0.14\phi_{O13(2py')} - 0.10\phi_{O13(2py'')}
\end{aligned}
$$

$$
\begin{aligned}
&+0.14\phi_{O14(2px')}+0.10\phi_{O14(2px'')}+0.14\phi_{O14(2py')}+0.10\phi_{O14(2py'')}\\
&-0.14\phi_{O15(2px')}-0.10\phi_{O15(2px'')}-0.14\phi_{O15(2py')}-0.10\phi_{O15(2py'')}\\
&-0.14\phi_{O16(2px')}-0.10\phi_{O16(2px'')}+0.14\phi_{O16(2py')}+0.10\phi_{O16(2py'')}\\
&-0.14\phi_{O17(2px')}-0.10\phi_{O17(2px'')}-0.14\phi_{O17(2pz')}-0.10\phi_{O17(2pz'')}\\
&+0.14\phi_{O18(2px')}+0.10\phi_{O18(2px'')}-0.14\phi_{O18(2pz')}-0.10\phi_{O18(2pz'')}\\
&+0.14\phi_{O19(2py')}+0.10\phi_{O19(2py'')}+0.14\phi_{O19(2pz')}+0.10\phi_{O19(2pz'')}\\
&-0.14\phi_{O20(2py')}-0.10\phi_{O20(2py'')}+0.14\phi_{O19(2pz')}+0.10\phi_{O19(2pz'')}
\end{aligned}
\tag{12.1}
$$

HOMO consists of 2p orbitals of twelve oxygen atoms. The obtained wave-function of LUMO is

$$
\begin{aligned}
\psi_{LOMO}=&-0.21\phi_{Ti1(3dxy)}-0.21\phi_{Ti1(3dxz)}-0.21\phi_{Ti1(3dyz)}\\
&-0.21\phi_{Ti2(3dxy)}+0.21\phi_{Ti2(3dxz)}-0.21\phi_{Ti2(3dyz)}\\
&-0.21\phi_{Ti3(3dxy)}-0.21\phi_{Ti3(3dxz)}+0.21\phi_{Ti3(3dyz)}\\
&-0.21\phi_{Ti4(3dxy)}+0.21\phi_{Ti4(3dxz)}+0.21\phi_{Ti4(3dyz)}\\
&+0.21\phi_{Ti5(3dxy)}-0.21\phi_{Ti5(3dxz)}-0.21\phi_{Ti5(3dyz)}\\
&+0.21\phi_{Ti6(3dxy)}+0.21\phi_{Ti6(3dxz)}-0.21\phi_{Ti6(3dyz)}\\
&+0.21\phi_{Ti7(3dxy)}-0.21\phi_{Ti7(3dxz)}+0.21\phi_{Ti7(3dyz)}\\
&+0.21\phi_{Ti8(3dxy)}+0.21\phi_{Ti8(3dxz)}+0.21\phi_{Ti8(3dyz)}
\end{aligned}
\tag{12.2}
$$

LUMO consists of t_{2g}-type 3d orbitals of eight titanium atoms. The obtained wave-function of MO114 is

$$
\begin{aligned}
\psi_{MO114}=&\,0.10\phi_{Sr(2px)}-0.31\phi_{Sr(3px)}+0.72\phi_{Sr(4px)}\\
&+0.19\phi_{Sr(3py)}-0.43\phi_{Sr(4py)}+0.28\phi_{Sr(3pz)}-0.63\phi_{Sr(4pz)}
\end{aligned}
\tag{12.3}
$$

MO114 consists of strontium 4p orbitals, which are outer shell orbitals. It is found that strontium is isolated as counter cation.

The orbital overlap between titanium 3d and oxygen 2p orbitals is observed in between MO115 and MO144. Although LUMO consists of titanium 3d orbital, the electron correlation effect between titanium and oxygen is taken into account. MO145–MO150 consist of oxygen 2p orbitals. Thus, it is found that charge transfer occurs from oxygen to titanium 3d electron, and orbital overlap occurs between titanium 3d and oxygen 2p orbitals, due to the electron correlation effect.

Bandgap depends on the magnitudes of charge transfer and orbital overlap. It is well known that bandgap is underestimated by pure DFT methods such as LDA and GGA. It is closely related to the fact that pure DFT overestimates the magnitudes of charge transfer and orbital overlap (delocalization effect). To solve the problem, hybrid-DFT is utilized to incorporate the localization effect by Hartree-Fock exchange functional. In hybrid-DFT, the exchange and correlation energy is expressed as

$$E_{XC} = c_1 E_X^{HF} + c_2 E_X^{Slater} + c_3 E_X^{Becke} + c_4 E_C^{VWN} + c_5 E_C^{LYP} \quad (12.4)$$

where E_X^{HF} E_X^{Slater} and E_X^{Becke} denote Hartree-Fock, Slater and Becke exchange energies, respectively; E_C^{VWN} and E_C^{LYP} denote Vosko-Wilk-Nusair and Lee-Yang-Parr correlation energies, respectively. The coefficients of Hartree-Fock exchange energy are 1.0, 0.5, 0.2 and 0.0 for Hartree-Fock, BHHLYP, B3LYP and BLYP methods, respectively.

Figure 12.4 shows the variation of bandgap by changing Hartree-Fock exchange coefficient in $SrTi_8O_{12}$ model. Bandgap approximately increases, in proportion to Hartree-Fock coefficient. The experimental $SrTiO_3$ bandgap (3.27 eV) is reproduced by the Hartree-Fock coefficient between BHHLYP and B3LYP.

Figure 12.5 shows the variations of Mulliken charge densities of titanium, oxygen and strontium by changing Hartree-Fock exchange coefficient in $SrTi_8O_{12}$ model. Charge density of titanium monotonously increases, and charge density of oxygen monotonously decreases, in proportion to Hartree-Fock coefficient. On the other hand, charge density of strontium is unchanged. It is concluded that bandgap depends on the magnitude of Hartree-Fock coefficient. Here, the scaling factor (k) can be applicable as substitution for determining the best Hartree-Fock coefficient.

$$\Delta E = k \Delta E_{BHHLYP} \quad (12.5)$$

The k value is 0.73 in $SrTiO_3$ perovskite. The corrected bandgap (ΔE) can be estimated from the calculated one by BHHLYP (ΔE_{BHHLYP}).

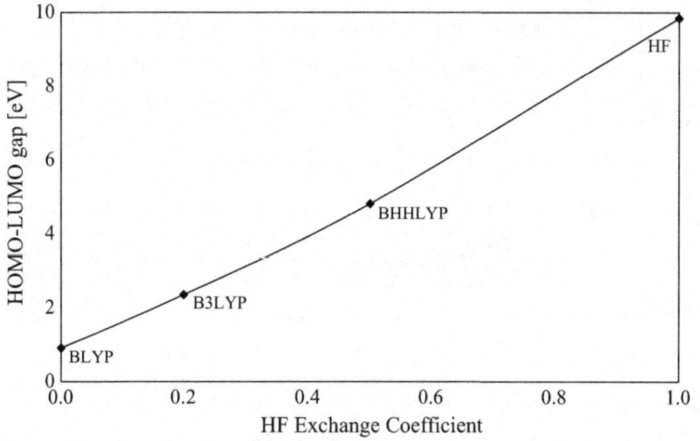

Fig. 12.4 Variation of corrected bandgap, changing the Hartree-Fock exchange coefficient in $SrTi_8O_{12}$ model. (BHHLYP method) Ref. [2] by permission from Wiley

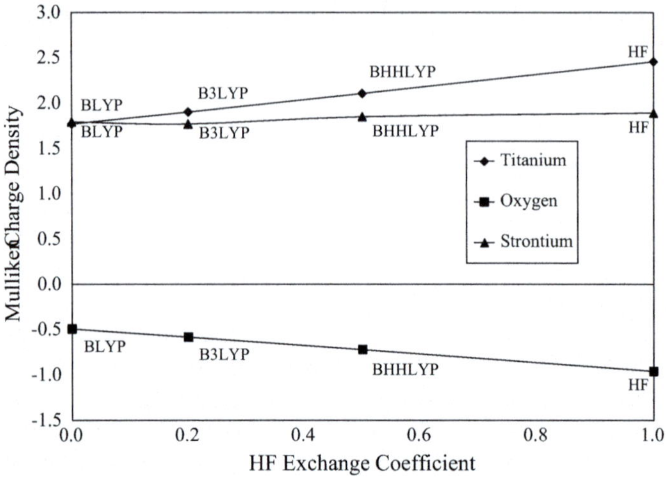

Fig. 12.5 Variation of Mulliken charge densities of titanium, oxygen and strontium, changing Hartree-Fock exchange coefficient in $SrTi_8O_{12}$ model. (BHHLYP method) Ref. [2] by permission from Wiley

12.3 Photocatalytic Activity of $SrTiO_3$ Perovskite

12.3.1 Introduction of Photocatalyst

Titanium oxides such as $SrTiO_3$ and TiO_2 are widely utilized as photocatalyst under ultraviolet light. About 40% of sunlight belongs to visible light, though ultraviolet light is less than 5% of sunlight. For the effective use of sunlight, photocatalyst with visible light response has been explored. $SrTiO_3$ bandgap (3.27 eV) corresponds to wave-length of ultraviolet light. To enhance a visible light response, the bandgap must be decreased, corresponding to wave-length of visible light (see Fig. 12.6).

Figure 12.7 depicts the schematic drawing of orbital energy diagram and photocatalytic reactions. When electron is excited by sunlight, electron hole (h) is

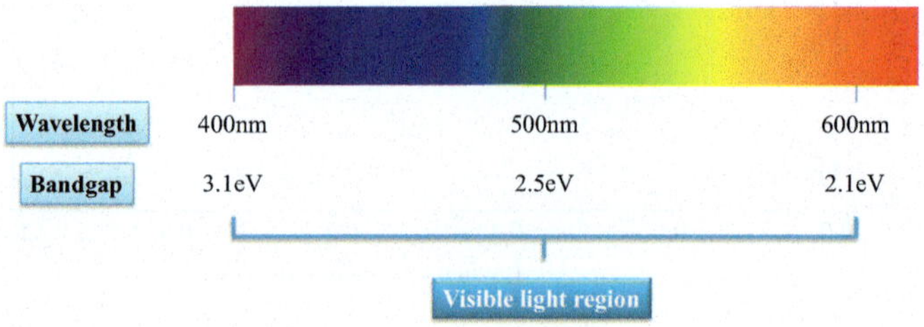

Fig. 12.6 Relationship between wave-length and bandgap

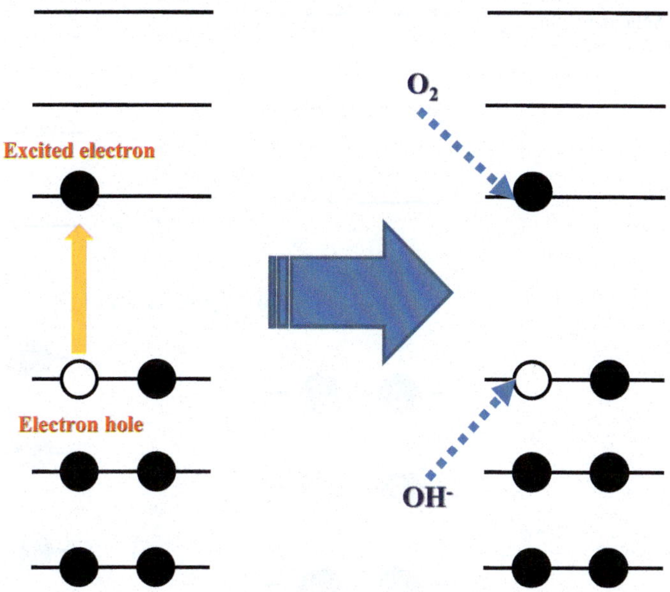

Fig. 12.7 Schematic drawing of orbital energy diagram and photocatalytic reactions

produced within occupied molecular orbital. Let us explain the possible major reactions. One is the reaction between excited electron and oxygen molecule:

$$e^- + O_2 \rightarrow O_2^- \tag{12.6}$$

Then, superoxide reacts with proton:

$$2O_2^- + 2H^+ \rightarrow H_2O_2 + O_2 \tag{12.7}$$

The other is the reaction between electron hole (h) and hydroxyl group (OH^-)

$$h + OH^- \rightarrow \cdot OH \tag{12.8}$$

It is known that the produced active species on the surface are closely related to photocatalytic reactions such as water oxidation, decomposition, etc. Though several reactions on surface are proposed, the details are still unclear.

12.3.2 *Nitrogen-Doping*

It was reported that bandgap decreases by dopings of nitrogen, carbon and sulphur, and transition metals (see Fig. 12.8). Here, the effect of nitrogen-doping at oxygen

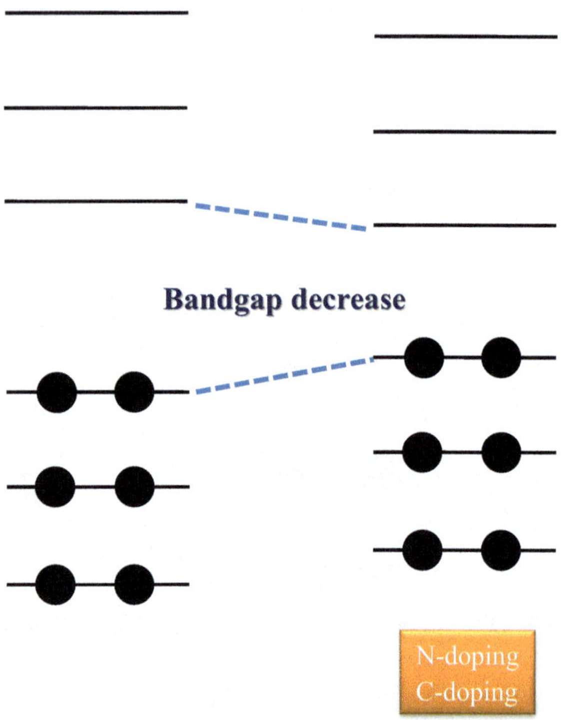

Fig. 12.8 Schematic drawing of bandgap decrease by doping

site on photocatalytic activity is investigated, from the viewpoint of energetics and bonding.

As shown in Fig. 12.9a, one-nitrogen-doped $SrTi_8O_{11}N$ model is constructed. The formal charges of oxygen and nitrogen are −2.0 and −3.0, respectively. One oxygen vacancy is produced per two-nitrogen-doping, due to the neutral condition as the whole. To investigate the effect of oxygen vacancy on bandgap change, $SrTi_8O_{10}N$ model is also constructed (see Fig. 12.9b).

BHHLYP calculation is performed for $SrTi_8O_{11}N$ and $SrTi_8O_{10}N$ models. Basis sets for titanium, oxygen, nitrogen and strontium are MINI(5.3.3.3/5.3/4.1), 6-311 + G*, 6-311 + G* and MINI(4.3.3.3.3/4.3.3/4), respectively. Due to the smaller formal charge of nitrogen, Coulomb interaction between titanium and nitrogen is larger than between titanium and oxygen. Hence, the shrink of Ti-N-Ti bond is taken into account as partially structural relaxation. Titanium is displaced from the cubic corner towards nitrogen of Ti-N-Ti bond.

Figure 12.10 shows the potential energy curve, when displacing titanium. The local minimum is given between 0.15 and 0.20Å. Figure 12.11 shows the variation of corrected bandgap, when displacing titanium. The corrected bandgap near the local minimum (between 0.15 and 0.20Å) is between 3.00 and 3.18 eV. It is found that nitrogen-doping enhances visible light response.

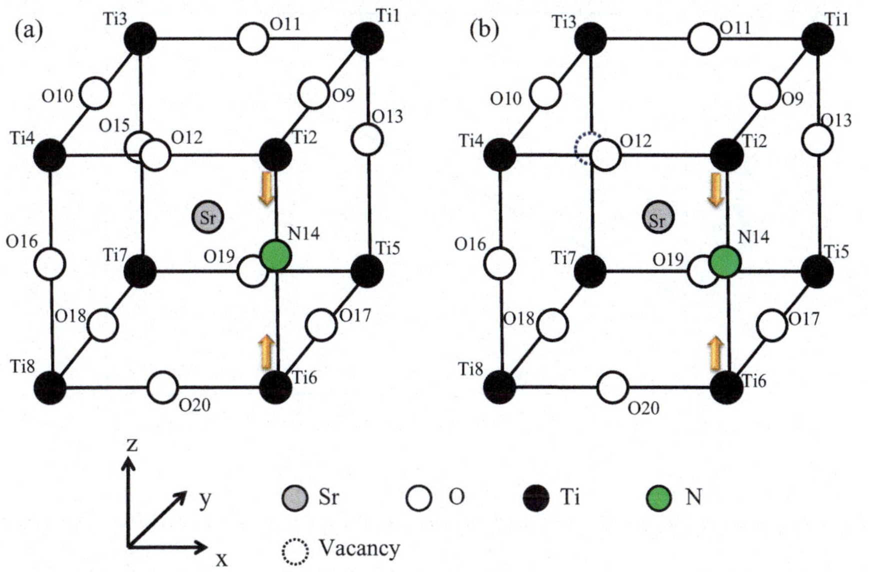

Fig. 12.9 Cluster models of nitrogen-doped $SrTiO_3$ perovskite: **a** $SrTi_8O_{11}N$ model; **b** $SrTi_8O_{10}N$ model. The *arrows* depicts a titanium displacement direction

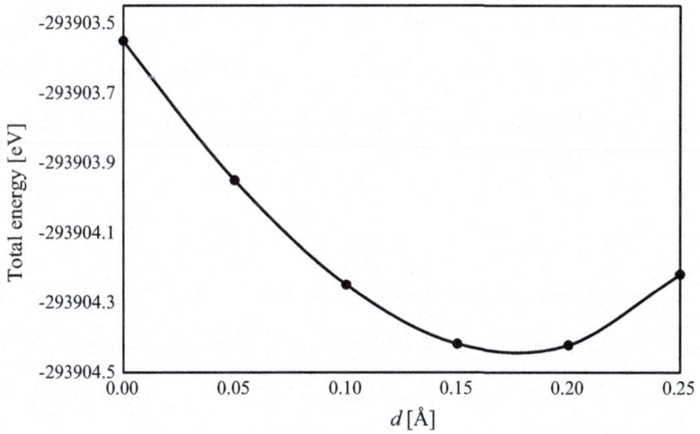

Fig. 12.10 Potential energy curve of $SrTi_8O_{11}N$ model, when displacing titanium towards nitrogen in Ti-N-Ti bond. *d* is the displacement distance from the cubic corner. (BHHLYP method)

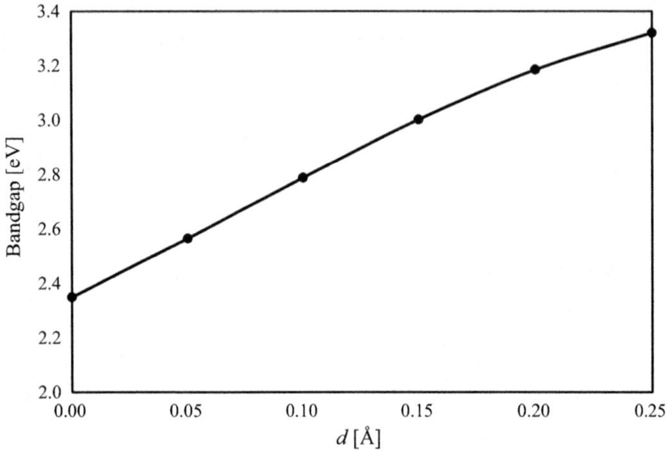

Fig. 12.11 Variation of corrected bandgap of $SrTi_8O_{11}N$ model, when displacing titanium towards nitrogen in Ti-N-Ti bond. *d* is the displacement distance from the cubic corner. (BHHLYP method)

Figure 12.12 depicts the selected molecular orbitals of $SrTi_8O_{11}N$ model at $d = 0.15$Å. The obtained wave-function of HOMO-2 is

$$\begin{aligned}\psi_{\mathrm{HOMO}-2} = {} & 0.12\phi_{\mathrm{O9(2px'')}} + 0.14\phi_{\mathrm{O9(2px''')}} \\ & - 0.11\phi_{\mathrm{O10(2px'')}} - 0.12\phi_{\mathrm{O10(2px''')}} - 0.09\phi_{\mathrm{O10(2pz'')}} - 0.10\phi_{\mathrm{O10(2pz''')}} \\ & - 0.11\phi_{\mathrm{O11(2py'')}} - 0.12\phi_{\mathrm{O11(2py''')}} + 0.09\phi_{\mathrm{O11(2pz'')}} + 0.10\phi_{\mathrm{O11(2pz''')}} \\ & + 0.12\phi_{\mathrm{O12(2py'')}} + 0.14\phi_{\mathrm{O12(2py''')}} \\ & - 0.10\phi_{\mathrm{O13(2px'')}} - 0.11\phi_{\mathrm{O13(2px''')}} \\ & + 0.10\phi_{\mathrm{O15(2px'')}} + 0.10\phi_{\mathrm{O15(2px''')}} + 0.10\phi_{\mathrm{O15(2py'')}} + 0.10\phi_{\mathrm{O15(2py''')}} \\ & - 0.10\phi_{\mathrm{O16(2py'')}} - 0.10\phi_{\mathrm{O16(2py''')}} \\ & + 0.12\phi_{\mathrm{O17(2px'')}} + 0.14\phi_{\mathrm{O17(2px''')}} \\ & - 0.11\phi_{\mathrm{O18(2px'')}} - 0.12\phi_{\mathrm{O18(2px''')}} + 0.09\phi_{\mathrm{O18(2pz'')}} + 0.10\phi_{\mathrm{O18(2pz''')}} \\ & - 0.11\phi_{\mathrm{O19(2py'')}} - 0.12\phi_{\mathrm{O19(2py''')}} - 0.09\phi_{\mathrm{O19(2pz'')}} - 0.10\phi_{\mathrm{O19(2pz''')}} \\ & + 0.12\phi_{\mathrm{O20(2py'')}} + 0.14\phi_{\mathrm{O20(2py''')}}\end{aligned} \tag{12.9}$$

HOMO-2 consists of oxygen 2p orbitals, corresponding to valence bond. The obtained wave-functions of HOMO-1 and HOMO are

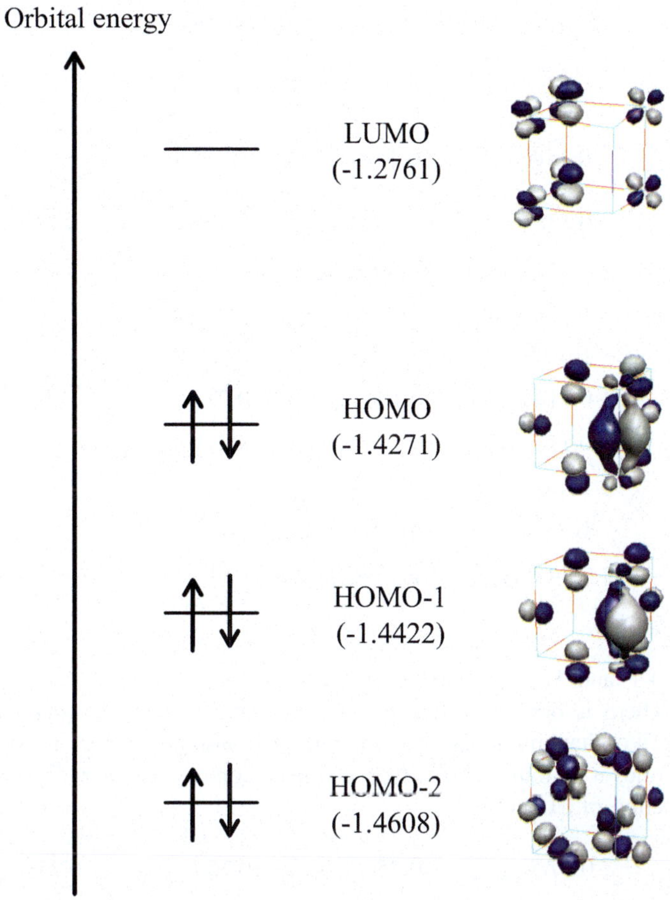

Fig. 12.12 Orbital energy diagram and selected molecular orbitals of $SrTi_8O_{11}N$ model at $d = 0.15$Å. The orbital energy is given in parenthesis. (BHHLYP method)

$$\begin{aligned}\psi_{\mathrm{HOMO-1}} = & -0.12\phi_{\mathrm{Ti2(3d}xz')} + 0.12\phi_{\mathrm{Ti2(3d}yz')} \\ & +0.12\phi_{\mathrm{Ti6(3d}xz')} - 0.12\phi_{\mathrm{Ti6(3d}yz')} \\ & -0.14\phi_{\mathrm{O9(2p}z'')} - 0.15\phi_{\mathrm{O9(2p}z''')} \\ & -0.14\phi_{\mathrm{O12(2p}z'')} - 0.15\phi_{\mathrm{O12(2p}z''')} \\ & +0.11\phi_{\mathrm{O13(2p}y'')} + 0.13\phi_{\mathrm{O13(2p}y''')} \\ & +0.12\phi_{\mathrm{N14(2p}x')} + 0.18\phi_{\mathrm{N14(2p}x'')} + 0.23\phi_{\mathrm{N14(2p}x''')}\end{aligned}$$

$$
\begin{aligned}
&- 0.12\phi_{\mathrm{N14(2py')}} - 0.18\phi_{\mathrm{N14(2py'')}} - 0.23\phi_{\mathrm{N14(2py''')}} \\
&- 0.11\phi_{\mathrm{O16(2px'')}} - 0.13\phi_{\mathrm{O16(2px''')}} \\
&+ 0.14\phi_{\mathrm{O17(2pz'')}} + 0.15\phi_{\mathrm{O17(2pz''')}} \\
&+ 0.14\phi_{\mathrm{O20(2pz'')}} + 0.15\phi_{\mathrm{O20(2pz''')}}
\end{aligned} \tag{12.10}
$$

$$
\begin{aligned}
\psi_{\mathrm{HOMO}} = &-0.13\phi_{\mathrm{Ti2(3dxz')}} - 0.13\phi_{\mathrm{Ti2(3dyz')}} + 0.13\phi_{\mathrm{Ti6(3dxz')}} + 0.13\phi_{\mathrm{Ti6(3dyz')}} \\
&+ 0.14\phi_{\mathrm{O9(2pz'')}} + 0.15\phi_{\mathrm{O9(2pz''')}} \\
&- 0.14\phi_{\mathrm{O12(2pz'')}} - 0.15\phi_{\mathrm{O12(2pz''')}} \\
&- 0.10\phi_{\mathrm{O13(2py'')}} - 0.11\phi_{\mathrm{O13(2py''')}} \\
&+ 0.11\phi_{\mathrm{N14(2px')}} + 0.18\phi_{\mathrm{N14(2px'')}} + 0.21\phi_{\mathrm{N14(2px''')}} \\
&+ 0.11\phi_{\mathrm{N14(2py')}} + 0.18\phi_{\mathrm{N14(2py'')}} + 0.21\phi_{\mathrm{N14(2py''')}} \\
&- 0.10\phi_{\mathrm{O16(2px'')}} - 0.11\phi_{\mathrm{O16(2px''')}} \\
&- 0.14\phi_{\mathrm{O17(2pz'')}} - 0.15\phi_{\mathrm{O17(2pz''')}} \\
&+ 0.14\phi_{\mathrm{O20(2pz'')}} + 0.15\phi_{\mathrm{O20(2pz''')}}
\end{aligned} \tag{12.11}
$$

HOMO-1 and HOMO consist of titanium 3d, nitrogen 2p and oxygen 2p orbitals. There is orbital overlap between titanium t_{2g}-type 3d and nitrogen 2p orbitals. Two titanium lobes interact with two nitrogen lobes. From chemical bonding rule, it is found that π-type covalent bonding is formed. The obtained wave-function of LUMO is

$$
\begin{aligned}
\psi_{\mathrm{LUMO}} = &-0.27\phi_{\mathrm{Ti1(3dxz')}} - 0.35\phi_{\mathrm{Ti3(3dxz')}} + 0.35\phi_{\mathrm{Ti3(3dyz')}} + 0.27\phi_{\mathrm{Ti4(3dyz')}} \\
&- 0.27\phi_{\mathrm{Ti5(3dxz')}} - 0.35\phi_{\mathrm{Ti7(3dxz')}} + 0.35\phi_{\mathrm{Ti7(3dyz')}} + 0.27\phi_{\mathrm{Ti8(3dyz')}}
\end{aligned} \tag{12.12}
$$

LUMO consists of titanium t_{2g}-type 3d orbitals, corresponding to conduction band. It is found that electron of HOMO is excited to LUMO by the smaller excitation energy.

Figure 12.13 depicts the molecular orbitals of oxygen-deficient $SrTi_8O_{10}N$ model at d = 0.15Å. The obtained wave-function of HOMO-1 is

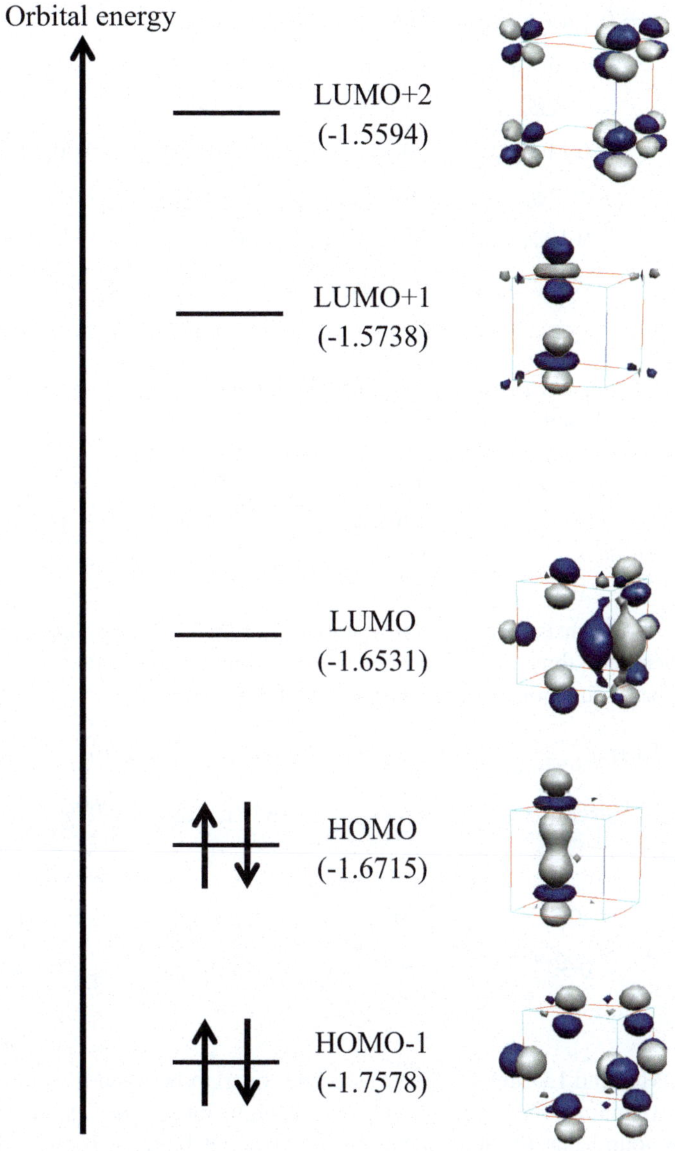

Fig. 12.13 Orbital energy diagram and selected molecular orbitals of $SrTi_8O_{10}N$ model at $d = 0.15$Å. The orbital energy is given in parenthesis. (BHHLYP method)

$$\begin{aligned}\psi_{\mathrm{HOMO-1}} = {} & 0.10\phi_{\mathrm{O9(2pz')}} + 0.15\phi_{\mathrm{O9(2pz'')}} + 0.16\phi_{\mathrm{O9(2pz''')}} \\ & + 0.10\phi_{\mathrm{O12(2pz')}} + 0.15\phi_{\mathrm{O12(2pz'')}} + 0.16\phi_{\mathrm{O12(2pz''')}} \\ & - 0.12\phi_{\mathrm{O13(2py')}} - 0.17\phi_{\mathrm{O13(2py'')}} - 0.19\phi_{\mathrm{O13(2py''')}} \\ & - 0.11\phi_{\mathrm{N14(2px'')}} - 0.14\phi_{\mathrm{N14(2px''')}} + 0.11\phi_{\mathrm{N14(2py'')}} + 0.14\phi_{\mathrm{N14(2py''')}} \\ & + 0.12\phi_{\mathrm{O16(2px')}} + 0.17\phi_{\mathrm{O16(2px'')}} + 0.19\phi_{\mathrm{O16(2px'')}} \\ & - 0.10\phi_{\mathrm{O17(2pz')}} - 0.15\phi_{\mathrm{O17(2pz'')}} - 0.16\phi_{\mathrm{O17(2pz''')}} \\ & - 0.10\phi_{\mathrm{O20(2pz')}} - 0.15\phi_{\mathrm{O20(2pz'')}} - 0.16\phi_{\mathrm{O20(2pz''')}}\end{aligned} \tag{12.13}$$

HOMO-1 consists of oxygen 2p orbitals, corresponding to valence band. The obtained wave-function of LUMO + 2 is

$$\begin{aligned}\psi_{\mathrm{LUMO+2}} = {} & 0.33\phi_{\mathrm{Ti1(3dyz')}} - 0.30\phi_{\mathrm{Ti2(3dxz')}} + 0.30\phi_{\mathrm{Ti2(3dyz')}} - 0.33\phi_{\mathrm{Ti4(3dxz')}} \\ & + 0.33\phi_{\mathrm{Ti5(3dyz')}} - 0.30\phi_{\mathrm{Ti6(3dxz')}} + 0.30\phi_{\mathrm{Ti6(3dyz')}} - 0.33\phi_{\mathrm{Ti8(3dxz')}}\end{aligned} \tag{12.14}$$

LUMO + 2 consists of titanium t_{2g}-type 3d orbitals, corresponding conduction band. There are three MOs between valence bond and conduction band. The obtained wave-functions of HOMO and LUMO + 1 are

$$\begin{aligned}\psi_{\mathrm{HOMO}} = {} & 0.16\phi_{\mathrm{Ti3}(3s)} - 0.25\phi_{\mathrm{Ti3}(3dx^{2'})} - 0.25\phi_{\mathrm{Ti3}(3dy^{2'})} + 0.60\phi_{\mathrm{Ti3}(3dz^{2'})} + 0.18\phi_{\mathrm{Ti3}(3dz^{2''})} \\ & + 0.16\phi_{\mathrm{Ti7}(3s)} - 0.25\phi_{\mathrm{Ti7}(3dx^{2'})} - 0.25\phi_{\mathrm{Ti7}(3dy^{2'})} + 0.60\phi_{\mathrm{Ti7}(3dz^{2'})} + 0.18\phi_{\mathrm{Ti7}(3dz^{2'}}\end{aligned} \tag{12.15}$$

$$\begin{aligned}\psi_{\mathrm{LUMO+1}} = {} & 0.30\phi_{\mathrm{Ti3}(3dx^{2'})} + 0.30\phi_{\mathrm{Ti3}(3dy^{2'})} - 0.63\phi_{\mathrm{Ti3}(3dz^{2'})} - 0.14\phi_{\mathrm{Ti3}(3dz^{2''})} \\ & - 0.30\phi_{\mathrm{Ti7}(3dx^{2'})} - 0.30\phi_{\mathrm{Ti7}(3dy^{2'})} + 0.63\phi_{\mathrm{Ti7}(3dz^{2'})} + 0.14\phi_{\mathrm{Ti7}(3dz^{2''})}\end{aligned} \tag{12.16}$$

In HOMO and LUMO + 1, there is orbital overlap between titanium $3d_z^2$ orbitals. One titanium lobe interacts with one titanium lobe. From chemical bonding rule, it is found that the long range σ-type covalent bonding is formed between titanium 3d orbitals. Note that LUMO + 1 is inversion σ-type covalent bonding to HOMO. The obtained wave-function of LUMO is

$$
\begin{aligned}
\psi_{\text{LUMO}} = {} & -0.10\phi_{\text{Ti2(3d}xz')} - 0.10\phi_{\text{Ti2(3d}yz')} + 0.10\phi_{\text{Ti6(3d}xz')} - 0.10\phi_{\text{Ti6(3d}yz')} \\
& + 0.15\phi_{\text{O9(2p}z'')} + 0.16\phi_{\text{O9(2p}z''')} \\
& - 0.15\phi_{\text{O12(2p}z'')} - 0.16\phi_{\text{O12(2p}z''')} \\
& - 0.11\phi_{\text{O13(2p}y'')} - 0.13\phi_{\text{O13(2p}y''')} \\
& + 0.11\phi_{N14(\text{2p}x')} + 0.17\phi_{N14(\text{2p}x'')} + 0.20\phi_{N14(\text{2p}x''')} \\
& + 0.11\phi_{N14(\text{2p}y')} + 0.17\phi_{N14(\text{2p}y'')} + 0.20\phi_{N14(\text{2p}y''')} \\
& - 0.11\phi_{\text{O16(2p}x'')} - 0.13\phi_{\text{O16(2p}x''')} \\
& - 0.15\phi_{\text{O17(2p}z'')} - 0.16\phi_{\text{O17(2p}z''')} \\
& + 0.15\phi_{\text{O20(2p}z'')} + 0.16\phi_{\text{O20(2p}z''')}
\end{aligned}
\tag{12.17}
$$

In LUMO, there is orbital overlap between titanium t_{2g}-type 3d and nitrogen 2p orbitals. Two titanium lobes interact with two nitrogen lobes. From chemical bonding rule, it is found that π-type covalent bonding is formed. The corrected bandgap is 0.37 eV. It is found that oxygen vacancy disturbs a visible light response, due to small bandgap.

Figure 12.14 depicts the schematic drawing of the effect of nitrogen-doping. A visible light response is enhanced in the perfect cubic unit. It is because the bandgap corresponds to the wave-length of visible light. However, in oxygen-deficient cubic unit, there is no visible light response, due to bandgap decrease.

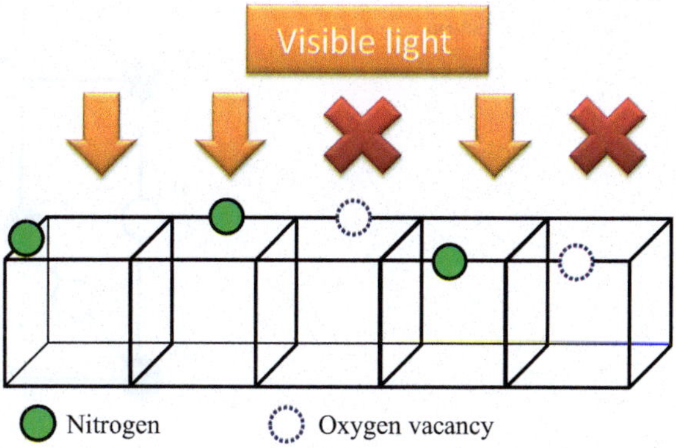

Fig. 12.14 Schematic drawing of the effect of nitrogen-doping in $SrTiO_3$ perovskite

12.3.3 Carbon-Doping

In nitrogen-doping, oxygen vacancy is combined, due to the difference of the formal charges. It was found that oxygen vacancy disturbs visible light response. As the alternative dopant, carbon is proposed. It is because the formal charge of carbon can be controllable. When C^{2-} is doped at oxygen site, the perfect crystal can be realized. Here, the effect of carbon-doping at oxygen site on photocatalytic activity is investigated.

Figure 12.15 depicts one carbon-doped $SrTi_8O_{11}C$ model. BHHLYP calculation is performed for $SrTi_8O_{11}C$ model. Basis sets for titanium, oxygen, nitrogen and strontium are MINI(5.3.3.3/5.3/5), 6-31G*, 6-31G* and MINI(4.3.3.3.3/4.3.3/4), respectively. The elongation of Ti-C-Ti bond is taken into account as partially structural relaxation. Titanium is displaced from the cubic corner towards carbon or oxygen in neighbouring $SrTi_8O_{11}C$ unit or $SrTi_8O_{12}$ unit.

Figure 12.16 shows the potential energy curve of $SrTi_8O_{11}C$ model, when displacing titanium. The local minimum is given around 0.10Å. It implies that Ti-C-Ti bond is longer than Ti-O-Ti bond. It is because Coulomb interaction between titanium and carbon is smaller than between titanium and oxygen, due to small Mulliken charge density of carbon (−0.248). Figure 12.17 shows the variation of corrected bandgap when displacing titanium. At the local minimum, the corrected bandgap is 2.41 eV, corresponding to 513 nm. It is found that the desirable bandgap is obtained in $SrTi_8O_{11}C$ model.

Figure 12.18 depicts the selected molecular orbitals of $SrTi_8O_{11}C$ model at d = 0.10Å. The obtained wave-functions of MO132 and MO142 are

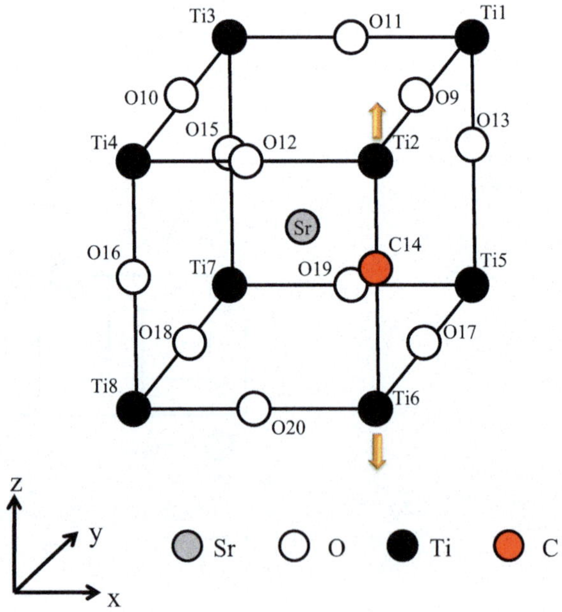

Fig. 12.15 $SrTi_8O_{11}C$ model of carbon-doped $SrTiO_3$ perovskite. The *arrows* depicts a titanium displacement direction

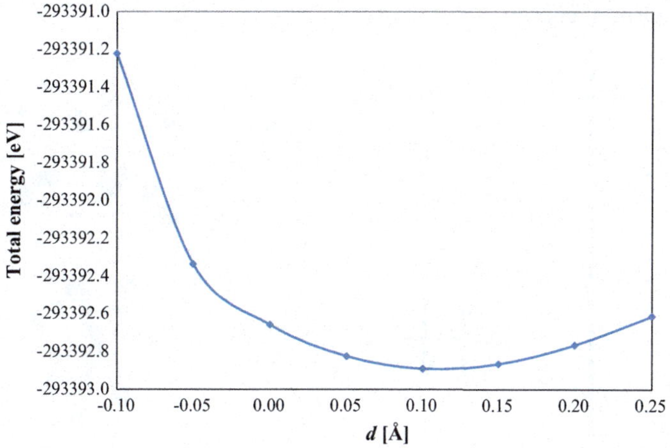

Fig. 12.16 Potential energy curve of $SrTi_8O_{11}C$ model, when displacing titanium towards carbon or oxygen in neighbouring $SrTi_8O_{11}C$ unit or $SrTi_8O_{12}$ unit, respectively. *d* is the displacement distance. (BHHLYP method)

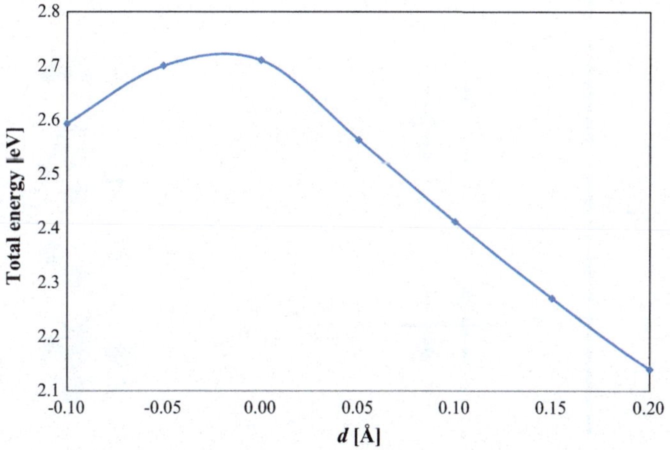

Fig. 12.17 Variation of corrected bandgap of $SrTi_8O_{11}C$ model, when displacing titanium towards carbon or oxygen in neighbouring $SrTi_8O_{11}C$ unit or $SrTi_8O_{12}$ unit, respectively. *d* is the displacement distance. (BHHLYP method)

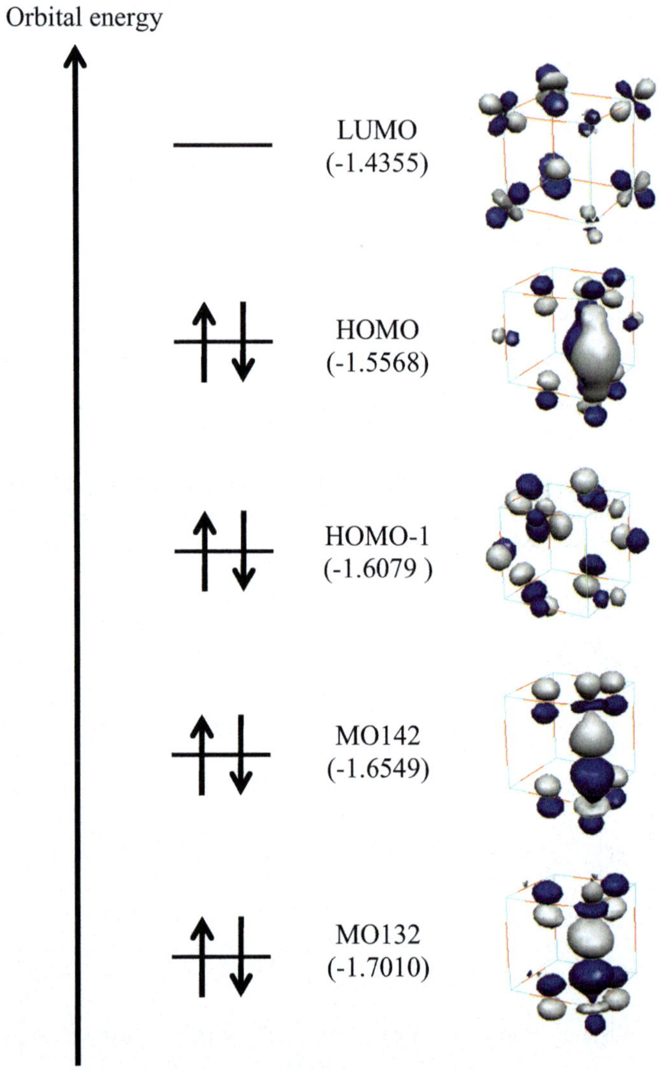

Fig. 12.18 Orbital energy diagram and molecular orbitals of $SrTi_8O_{11}C$ model at d = 0.10Å. The orbital energy is given in parenthesis. (BHHLYP method)

$$
\begin{aligned}
\psi_{\mathrm{MO132}} = {} & 0.11\phi_{\mathrm{Ti2(2pz'')}} - 0.11\phi_{\mathrm{Ti2(3dx^2)}} - 0.11\phi_{\mathrm{Ti2(3dy^2)}} + 0.24\phi_{\mathrm{Ti2(3dz^2)}} \\
& + 0.11\phi_{\mathrm{Ti6(2pz'')}} + 0.11\phi_{\mathrm{Ti6(3dx^2)}} + 0.11\phi_{\mathrm{Ti6(3dy^2)}} - 0.24\phi_{\mathrm{Ti6(3dz^2)}} \\
& + 0.30\phi_{\mathrm{C14(2pz')}} + 0.27\phi_{\mathrm{C14(2pz'')}} \\
& - 0.20\phi_{\mathrm{O9(2pz')}} - 0.14\phi_{\mathrm{O9(2pz'')}} - 0.20\phi_{\mathrm{O12(2pz')}} - 0.14\phi_{\mathrm{O12(2pz'')}} \\
& - 0.20\phi_{\mathrm{O17(2pz')}} - 0.14\phi_{\mathrm{O17(2pz'')}} - 0.20\phi_{\mathrm{O20(2pz')}} - 0.14\phi_{\mathrm{O20(2pz'')}}
\end{aligned} \tag{12.18}
$$

$$
\begin{aligned}
\psi_{\mathrm{MO142}} = {} & -0.10\phi_{\mathrm{Ti2(3dx^2)}} - 0.10\phi_{\mathrm{Ti2(3dy^2)}} + 0.35\phi_{\mathrm{Ti2(3dz^2)}} \\
& + 0.10\phi_{\mathrm{Ti6(3dx^2)}} + 0.10\phi_{\mathrm{Ti6(3dy^2)}} - 0.35\phi_{\mathrm{Ti6(3dz^2)}} \\
& + 0.36\phi_{\mathrm{C14(2pz')}} + 0.20\phi_{\mathrm{C14(2pz'')}} \\
& + 0.19\phi_{\mathrm{O9(2pz')}} + 0.14\phi_{\mathrm{O9(2pz'')}} + 0.19\phi_{\mathrm{O12(2pz')}} + 0.14\phi_{\mathrm{O12(2pz'')}} \\
& + 0.19\phi_{\mathrm{O17(2pz')}} + 0.14\phi_{\mathrm{O17(2pz'')}} + 0.19\phi_{\mathrm{O20(2pz')}} + 0.14\phi_{\mathrm{O20(2pz'')}}
\end{aligned} \tag{12.19}
$$

There is orbital overlap between titanium e_g-type 3d and carbon 2p orbitals. One titanium lobe interacts with one carbon lobe. From chemical bonding rule, it is found that σ-type covalent bonding is formed. The obtained wave-function of HOMO-1 is

$$
\begin{aligned}
\psi_{\mathrm{HOMO-1}} = {} & 0.13\phi_{\mathrm{O9(2px')}} + 0.10\phi_{\mathrm{O9(2px'')}} \\
& 0.15\phi_{\mathrm{O10(2px')}} - 0.11\phi_{\mathrm{O10(2px'')}} - 0.19\phi_{\mathrm{O10(2pz')}} - 0.14\phi_{\mathrm{O10(2pz'')}} \\
& - 0.15\phi_{\mathrm{O11(2py')}} - 0.11\phi_{\mathrm{O11(2py'')}} + 0.19\phi_{\mathrm{O11(2pz')}} + 0.14\phi_{\mathrm{O11(2pz'')}} \\
& + 0.13\phi_{\mathrm{O12(2py')}} + 0.10\phi_{\mathrm{O12(2py'')}} \\
& - 0.18\phi_{\mathrm{O13(2px')}} - 0.14\phi_{\mathrm{O13(2px'')}} \\
& + 0.20\phi_{\mathrm{O15(2px')}} + 0.15\phi_{\mathrm{O15(2px'')}} + 0.20\phi_{\mathrm{O15(2py')}} + 0.15\phi_{\mathrm{O15(2py'')}} \\
& - 0.18\phi_{\mathrm{O16(2py')}} - 0.14\phi_{\mathrm{O16(2py'')}} \\
& + 0.13\phi_{\mathrm{O17(2px')}} + 0.10\phi_{\mathrm{O17(2px'')}} \\
& - 0.15\phi_{\mathrm{O18(2px')}} - 0.11\phi_{\mathrm{O18(2px'')}} + 0.19\phi_{\mathrm{O18(2pz')}} + 0.14\phi_{\mathrm{O18(2pz'')}} \\
& - 0.15\phi_{\mathrm{O19(2py')}} - 0.11\phi_{\mathrm{O19(2py'')}} - 0.19\phi_{\mathrm{O19(2pz')}} - 0.14\phi_{\mathrm{O19(2pz'')}} \\
& + 0.13\phi_{\mathrm{O20(2py')}} + 0.10\phi_{\mathrm{O20(2py'')}}
\end{aligned} \tag{12.20}
$$

HOMO-1 consists only of oxygen 2p orbital, corresponding to valence bond. The obtained wave-function of LUMO is

$$\begin{aligned}\psi_{\mathrm{LUMO}} = {} & 0.14\phi_{\mathrm{Ti1(3dxy)}} + 0.30\phi_{\mathrm{Ti1(3dxz)}} + 0.15\phi_{\mathrm{Ti1(3dyz)}} \\ & + 0.10\phi_{\mathrm{Ti2(3dxy)}} - 0.14\phi_{\mathrm{Ti2(3dxz)}} + 0.14\phi_{\mathrm{Ti2(3dyz)}} \\ & + 0.15\phi_{\mathrm{Ti3(3dxy)}} + 0.30\phi_{\mathrm{Ti3(3dxz)}} - 0.30\phi_{\mathrm{Ti3(3dyz)}} \\ & + 0.14\phi_{\mathrm{Ti4(3dxy)}} - 0.15\phi_{\mathrm{Ti4(3dxz)}} - 0.30\phi_{\mathrm{Ti4(3dyz)}} \\ & - 0.14\phi_{\mathrm{Ti5(3dxy)}} + 0.30\phi_{\mathrm{Ti5(3dxz)}} + 0.15\phi_{\mathrm{Ti5(3dyz)}} \\ & - 0.10\phi_{\mathrm{Ti6(3dxy)}} - 0.14\phi_{\mathrm{Ti6(3dxz)}} + 0.14\phi_{\mathrm{Ti6(3dyz)}} \\ & - 0.15\phi_{\mathrm{Ti7(3dxy)}} + 0.30\phi_{\mathrm{Ti7(3dxz)}} - 0.30\phi_{\mathrm{Ti7(3dyz)}} \\ & - 0.14\phi_{\mathrm{Ti8(3dxy)}} - 0.15\phi_{\mathrm{Ti8(3dxz)}} - 0.30\phi_{\mathrm{Ti8(3dyz)}}\end{aligned} \tag{12.21}$$

LUMO consists only of titanium t_{2g}-type 3d orbital, corresponding to conduction bond. The obtained wave-function of HOMO is

$$\begin{aligned}\psi_{\mathrm{HOMO}} = {} & -0.23\phi_{\mathrm{Ti2(3dxz)}} + 0.23\phi_{\mathrm{Ti2(3dyz)}} + 0.23\phi_{\mathrm{Ti6(3dxz)}} - 0.23\phi_{\mathrm{Ti6(3dyz)}} \\ & + 0.30\phi_{\mathrm{C14(2px')}} + 0.21\phi_{\mathrm{C14(2px'')}} - 0.30\phi_{\mathrm{C14(2py')}} - 0.21\phi_{\mathrm{C14(2py'')}} \\ & - 0.16\phi_{\mathrm{O9(2pz')}} - 0.14\phi_{\mathrm{O9(2pz'')}} - 0.16\phi_{\mathrm{O12(2pz')}} - 0.14\phi_{\mathrm{O12(2pz'')}} \\ & + 0.11\phi_{\mathrm{O13(2py')}} + 0.10\phi_{\mathrm{O13(2py'')}} - 0.11\phi_{\mathrm{O16(2px')}} - 0.10\phi_{\mathrm{O16(2px'')}} \\ & + 0.16\phi_{\mathrm{O17(2pz')}} + 0.14\phi_{\mathrm{O17(2pz'')}} + 0.16\phi_{\mathrm{O20(2pz')}} + 0.14\phi_{\mathrm{O20(2pz'')}}\end{aligned} \tag{12.22}$$

There is orbital overlap between titanium t_{2g}-type 3d and oxygen 2p orbitals. Two titanium lobes interact with two oxygen lobes. From chemical bonding rule, it is found that π-type covalent bonding is formed. It is found that electron of HOMO is excited to LUMO with the smaller excitation energy.

Let us consider the effect of the structural relaxation in neighbouring $SrTi_8O_{11}C$ or $SrTi_8O_{12}$ units. Figure 12.19 shows the potential energy curve, when displacing titanium along Ti-O-Ti bond in $SrTi_8O_{12}$ model. When titanium is displaced towards to oxygen in Ti-O-Ti bond, the local minimum is given around 0.10Å. As shown in Fig. 12.20a, when neighbouring unit is $SrTi_8O_{12}$, the structural distortion disappears in total. Even if carbon is doped at neighbouring unit, the total structural distortion disappears by alternate stacking.

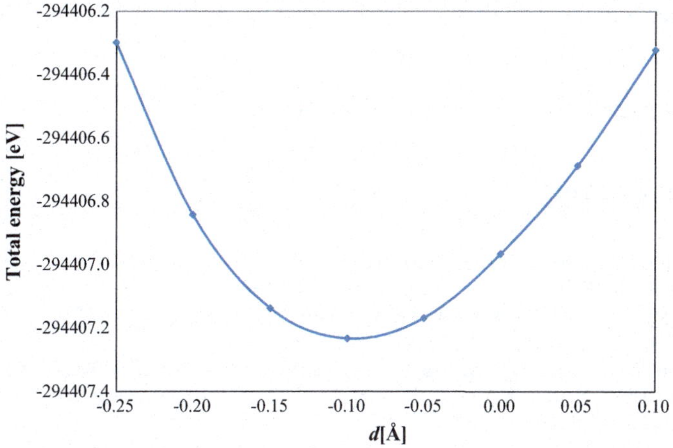

Fig. 12.19 Potential energy curve of $SrTi_8O_{12}$ model, when displacing titanium towards oxygen in Ti-O-Ti bond. *d* is the displacement distance from the cubic corner. (BHHLYP method)

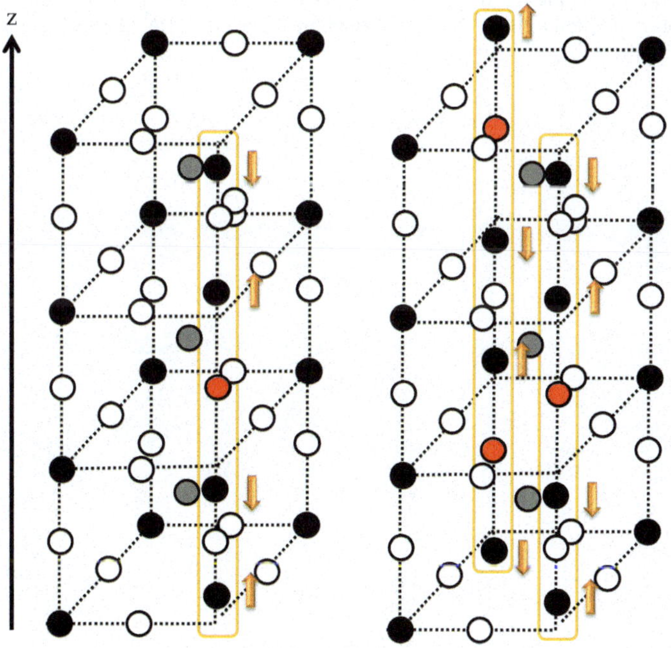

Fig. 12.20 Schematic drawing of Ti-C elongation and Ti-O shrink in carbon-doped $SrTiO_3$ perovskite: **a** stacking of $SrTiO_8O_{11}C$ and $SrTiO_8O_{12}$, **b** alternate stacking of $SrTiO_8O_{11}C$. The *arrows* depicts a titanium displacement direction

References

1. Onishi T (2012) Adv Quant Chem 64:70–78
2. Onishi T (2008) Int J Quant Chem 108:2856–2861

Further Readings

3. Onishi T (2010) Top Catal 53:566–570
4. Onishi T (2013) Prog. Theor Chem Phys 27:233–248
5. de Ligny D, Richet P (1996) Phys Rev B 53(6):3013–3022
6. Fujishima A, Honda K (1972) Nature 288:37–38
7. Schmidt MW, Baldridge KK, Boatz JA, Elbert ST, Gordon MS, Jensen JH, Koseki S, Matsunaga N, Nguyen KA, Su S, Windus TL, Dupuis M, Montgomery JA (1993) J Comput Chem 14:1347–1363
8. Varetto U < MOLEKEL 4.3. >; Swiss National Supercomputing Centre. Manno, Switzerland
9. Huzinaga S, Andzelm J, Radzio-Andzelm E, Sakai Y, Tatewaki H, Klobukowski M (1984) Gaussian basis sets for molecular calculations. Elsevier, Amsterdam
10. Hariharan PC, Pople JA (1973) Theoret Chim Acta 28:213–222
11. Francl MM, Pietro WJ, Hehre WJ, Binkley JS, Gordon MS, DeFrees DJ, Pople JA (1982) J Chem Phys 77(7):3654–3665
12. Rassolov VA, Pople JA, Ratner MA, Windus TL (1998) J Chem Phys 109(4):1223–1229
13. Rassolov VA, Ratner MA, Pople JA, Redfern PC, Curtiss LA (2001) J Compt Chem 22 (9):976–984

Chapter 13
Secondary Battery: Lithium Ion and Sodium Ion Conductions

Abstract Lithium ion battery has been widely in many electronic devices. Due to flammability of liquid organic electrolyte, solid electrolyte has been explored from the viewpoint of battery safety. To investigate the mechanism of lithium ion conduction in solid electrolyte of $La_{2/3-x}Li_{3x}TiO_3$ perovskite, hybrid-DFT calculation is performed. From the obtained potential energy curve, the activation energy for lithium ion conduction can be estimated. From chemical bonding rule, it is found that lithium ion forms ionic bonding during lithium ion conduction. Based on the knowledge, $K_xBa_{(1-x)/2}MnF_3$ perovskite was designed as thermally stable lithium ion conductor. Recently, sodium ion battery has attracted much interest, because of abundant sodium resource. However, as sodium ion has larger ionic radius, it is more difficult to design sodium ion conductor. In this chapter, our designed sodium ion conductors such as $CsMn(CN)_3$, $Al(CN)_3$ and $NaAlO(CN)_2$ are introduced. In $CsMn(CN)_3$, $Al(CN)_3$, sodium ion migrates through counter cation vacancy, as same as $La_{2/3-x}Li_{3x}TiO_3$ perovskite. In $NaAlO(CN)_2$, the anisotropic sodium ion conduction occurs. Sodium ion can migrates through only $Al_4(CN)_4$ bottleneck.

Keywords Secondary battery · Solid electrolyte · Lithium ion conduction · Sodium ion conduction · Materials design

13.1 Introduction of Secondary Battery

Secondary battery is an energy storage system using both chemical reactions and ion conductions. In general, lithium ion battery has advantages in larger gravimetric energy density (100–200 Wh kg^{-1}) and high voltage. In lithium ion battery, not neutral lithium but lithium ion migrates from one electrode to another through electrolyte, as shown in Fig. 13.1. On the other hand, in sodium ion battery, sodium ion migrates instead of lithium ion. In general, organic solvent has been widely utilized as electrolyte. It has a flammable problem during operation. The replacement of organic solvent by solid electrolyte has been much expected from the viewpoint of battery safety.

T. Onishi, *Quantum Computational Chemistry*,
DOI 10.1007/978-981-10-5933-9_13

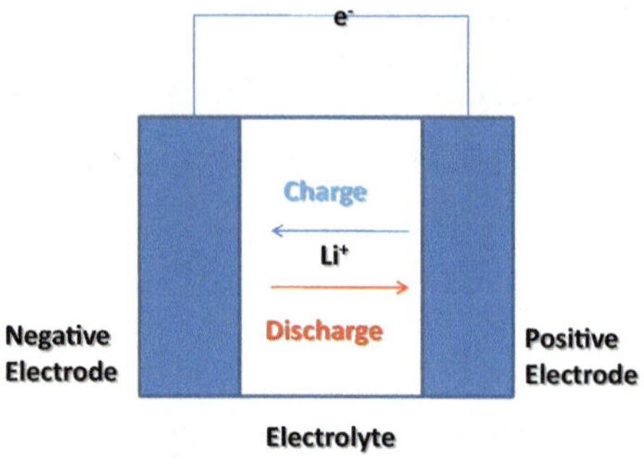

Fig. 13.1 Schematic drawing of lithium ion battery

Lithium ion is one of the best cations in secondary battery. It is due to light weight and small ionic radius. However, lithium resource is limited on earth. In addition, it is often reported that the reduction reaction of lithium ion causes unforeseen flammable accident. Recently, sodium ion battery has been explored as a substitute of lithium ion battery, due to the abundance of sodium resource. From the viewpoint of chemistry, as sodium ion has larger ionic radius compared with lithium ion, it is more difficult to explore sodium ion conductor. In this chapter, the ion conduction mechanism in solid state electrolyte is explained.

13.2 Lithium Ion Conductor

13.2.1 $La_{2/3-x}Li_{3x}TiO_3$ Perovskite

It was reported that $La_{2/3-x}Li_{3x}TiO_3$ perovskite exhibits high lithium ion conductivity at room temperature. Figure 13.2 depicts ATi_8O_{12} model in $La_{2/3-x}Li_{3x}TiO_3$ perovskite, where A denotes counter cation (La or Li). For the simplicity, the simple cubic structure with lattice constant 3.871 Å ($x = 0.116$) is considered. Due to the difference of formal charges of counter cations, vacancy is produced at counter cation site. Figure 13.3 depicts the schematic drawing of lithium ion conduction in $La_{2/3-x}Li_{3x}TiO_3$ perovskite. Though lanthanum cation is kept fixed due to the large ionic radius, lithium cation can migrates through vacancy. BHHLYP calculation is performed for $LiTi_8O_{12}$ model. Basis sets used for titanium, oxygen and lithium are MINI(5.3.3.3/5.3/5), 6-31G* and MINI(7.3), respectively.

Figure 13.4 shows the potential energy curve of $LiTi_8O_{12}$ model, when displacing lithium ion along x axis. Note that Ti_4O_4 square part is called bottleneck. Figure 13.5 depicts MO61, HOMO and LUMO of $LiTi_8O_{12}$ model, at the centre, local minimum and bottleneck. The obtained wave-functions of MO61 at the centre, local minimum and bottleneck are

$$\psi_{\mathrm{MO61(centre)}} = 0.99\phi_{\mathrm{Li(1s)}} \tag{13.1}$$

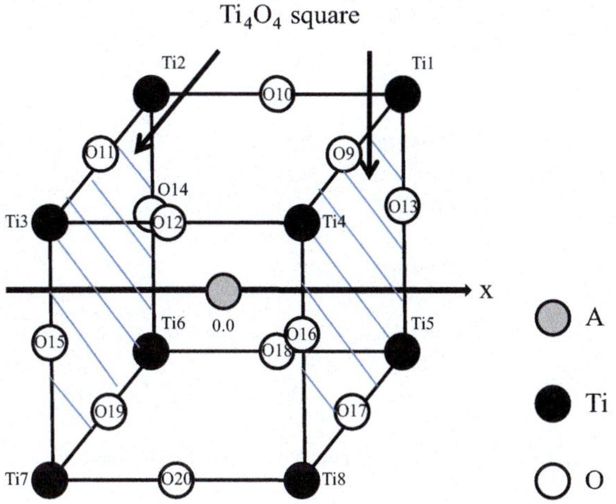

Fig. 13.2 ATi_8O_{12} model of $La_{2/3-x}Li_{3x}TiO_3$ perovskite. A denotes La or Li. The site numbers are shown for titanium and oxygen. Reference [1] by permission from Elsevier

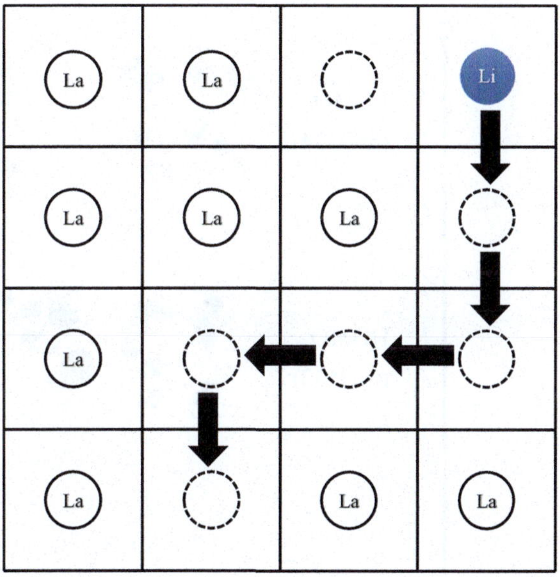

Fig. 13.3 Schematic drawing of lithium ion conduction in $La_{2/3-x}Li_{3x}TiO_3$ perovskite. Reference [1] by permission from Elsevier

$$\psi_{MO61(min)} = 0.99\phi_{Li(1s)} \tag{13.2}$$

$$\psi_{MO61(bottleneck)} = 0.99\phi_{Li(1s)} \tag{13.3}$$

MO61s consist of lithium 1s orbital. There is no orbital overlap between lithium ion and others. From chemical bonding rule, it is found that lithium ion forms ionic bonding during lithium ion conduction. The obtained wave-functions of HOMO at the centre, local minimum and bottleneck are

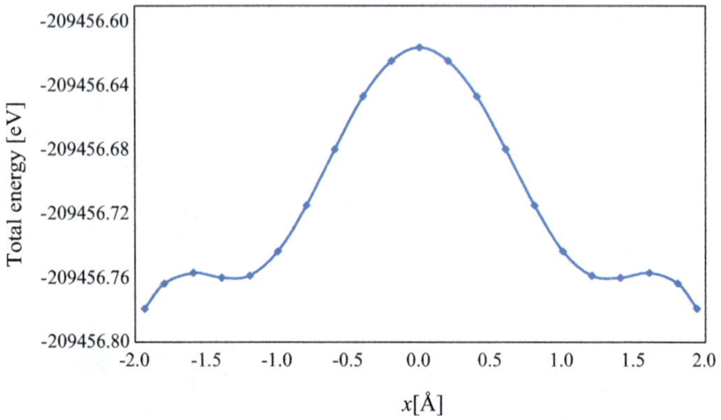

Fig. 13.4 Potential energy curve of $LiTi_8O_{12}$ model, when displacing lithium ion along x axis

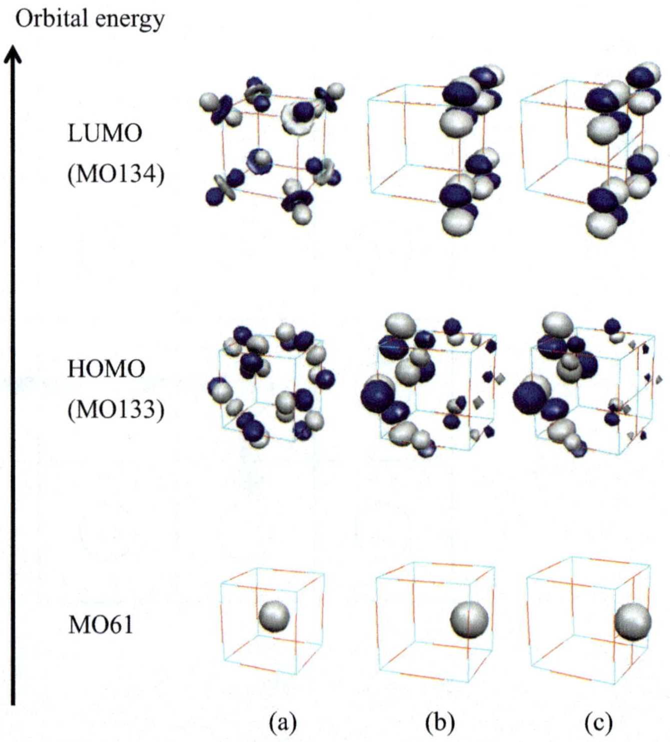

Fig. 13.5 Selected molecular orbitals of $LiTi_8O_{12}$ model at **a** the centre, **b** local minimum and **c** bottleneck

$$
\begin{aligned}
\psi_{\mathrm{HOMO(centre)}} = {} & 0.14\phi_{\mathrm{O9(2px')}} + 0.10\phi_{\mathrm{O9(2px'')}} - 0.14\phi_{\mathrm{O9(2pz')}} - 0.10\phi_{\mathrm{O9(2pz'')}} \\
& - 0.14\phi_{\mathrm{O10(2py')}} - 0.10\phi_{\mathrm{O10(2py'')}} + 0.14\phi_{\mathrm{O10(2pz')}} + 0.10\phi_{\mathrm{O10(2pz'')}} \\
& - 0.14\phi_{\mathrm{O11(2px')}} - 0.10\phi_{\mathrm{O11(2px'')}} - 0.14\phi_{\mathrm{O11(2pz')}} - 0.10\phi_{\mathrm{O11(2pz'')}} \\
& + 0.14\phi_{\mathrm{O12(2py')}} + 0.10\phi_{\mathrm{O12(2py'')}} + 0.14\phi_{\mathrm{O12(2pz')}} + 0.10\phi_{\mathrm{O12(2pz'')}} \\
& - 0.14\phi_{\mathrm{O13(2px')}} - 0.10\phi_{\mathrm{O13(2px'')}} + 0.14\phi_{\mathrm{O13(2py')}} + 0.10\phi_{\mathrm{O13(2py'')}} \\
& + 0.14\phi_{\mathrm{O14(2px')}} + 0.10\phi_{\mathrm{O14(2px'')}} + 0.14\phi_{\mathrm{O14(2py')}} + 0.10\phi_{\mathrm{O14(2py'')}} \\
& + 0.14\phi_{\mathrm{O15(2px')}} + 0.10\phi_{\mathrm{O15(2px'')}} - 0.14\phi_{\mathrm{O15(2py')}} - 0.10\phi_{\mathrm{O15(2py'')}} \\
& - 0.14\phi_{\mathrm{O16(2px')}} - 0.10\phi_{\mathrm{O16(2px'')}} - 0.14\phi_{\mathrm{O16(2py')}} - 0.10\phi_{\mathrm{O16(2py'')}} \\
& + 0.14\phi_{\mathrm{O17(2px')}} + 0.10\phi_{\mathrm{O17(2px'')}} + 0.14\phi_{\mathrm{O17(2pz')}} + 0.10\phi_{\mathrm{O17(2pz'')}} \\
& - 0.14\phi_{\mathrm{O18(2py')}} - 0.10\phi_{\mathrm{O18(2py'')}} - 0.14\phi_{\mathrm{O18(2pz')}} - 0.10\phi_{\mathrm{O18(2pz'')}} \\
& - 0.14\phi_{\mathrm{O19(2px')}} - 0.10\phi_{\mathrm{O19(2px'')}} + 0.14\phi_{\mathrm{O19(2pz')}} + 0.10\phi_{\mathrm{O19(2pz'')}} \\
& + 0.14\phi_{\mathrm{O20(2py')}} + 0.10\phi_{\mathrm{O20(2py'')}} - 0.14\phi_{\mathrm{O20(2pz')}} - 0.10\phi_{\mathrm{O20(2pz'')}}
\end{aligned}
\tag{13.4}
$$

$$
\begin{aligned}
\psi_{\mathrm{HOMO(min)}} = {} & 0.10\phi_{\mathrm{O10(2py')}} + 0.07\phi_{\mathrm{O10(2py'')}} - 0.10\phi_{\mathrm{O10(2pz')}} - 0.07\phi_{\mathrm{O10(2pz'')}} \\
& + 0.11\phi_{\mathrm{O11(2px')}} + 0.08\phi_{\mathrm{O11(2px'')}} + 0.27\phi_{\mathrm{O11(2pz')}} + 0.21\phi_{\mathrm{O11(2pz'')}} \\
& - 0.10\phi_{\mathrm{O12(2py')}} - 0.07\phi_{\mathrm{O12(2py'')}} - 0.10\phi_{\mathrm{O12(2pz')}} - 0.07\phi_{\mathrm{O12(2pz'')}} \\
& - 0.11\phi_{\mathrm{O14(2px')}} - 0.08\phi_{\mathrm{O14(2px'')}} - 0.27\phi_{\mathrm{O14(2py')}} - 0.21\phi_{\mathrm{O14(2py'')}} \\
& - 0.11\phi_{\mathrm{O15(2px')}} - 0.08\phi_{\mathrm{O15(2px'')}} + 0.27\phi_{\mathrm{O15(2py')}} + 0.21\phi_{\mathrm{O15(2py'')}} \\
& + 0.10\phi_{\mathrm{O18(2py')}} + 0.07\phi_{\mathrm{O18(2py'')}} + 0.10\phi_{\mathrm{O18(2pz')}} + 0.07\phi_{\mathrm{O18(2pz'')}} \\
& + 0.11\phi_{\mathrm{O19(2px')}} + 0.08\phi_{\mathrm{O19(2px'')}} - 0.27\phi_{\mathrm{O19(2pz')}} - 0.21\phi_{\mathrm{O19(2pz'')}} \\
& - 0.10\phi_{\mathrm{O20(2py')}} - 0.07\phi_{\mathrm{O20(2py'')}} + 0.10\phi_{\mathrm{O20(2pz')}} + 0.07\phi_{\mathrm{O20(2pz'')}}
\end{aligned}
\tag{13.5}
$$

$$
\begin{aligned}
\psi_{\mathrm{HOMO(bottleneck)}} = {} & 0.09\phi_{\mathrm{O10(2py')}} + 0.07\phi_{\mathrm{O10(2py'')}} - 0.09\phi_{\mathrm{O10(2pz')}} - 0.07\phi_{\mathrm{O10(2pz'')}} \\
& + 0.10\phi_{\mathrm{O11(2px')}} + 0.07\phi_{\mathrm{O11(2px'')}} + 0.29\phi_{\mathrm{O11(2pz')}} + 0.22\phi_{\mathrm{O11(2pz'')}} \\
& - 0.09\phi_{\mathrm{O12(2py')}} - 0.07\phi_{\mathrm{O12(2py'')}} - 0.09\phi_{\mathrm{O12(2pz')}} - 0.07\phi_{\mathrm{O12(2pz'')}} \\
& - 0.10\phi_{\mathrm{O14(2px')}} - 0.07\phi_{\mathrm{O14(2px'')}} - 0.29\phi_{\mathrm{O14(2py')}} - 0.22\phi_{\mathrm{O14(2py'')}} \\
& - 0.10\phi_{\mathrm{O15(2px')}} - 0.07\phi_{\mathrm{O15(2px'')}} + 0.29\phi_{\mathrm{O15(2py')}} + 0.22\phi_{\mathrm{O15(2py'')}} \\
& + 0.09\phi_{\mathrm{O18(2py')}} + 0.07\phi_{\mathrm{O18(2py'')}} + 0.09\phi_{\mathrm{O18(2pz')}} + 0.07\phi_{\mathrm{O18(2pz'')}} \\
& + 0.10\phi_{\mathrm{O19(2px')}} + 0.07\phi_{\mathrm{O19(2px'')}} - 0.29\phi_{\mathrm{O19(2pz')}} - 0.22\phi_{\mathrm{O19(2pz'')}} \\
& - 0.09\phi_{\mathrm{O20(2py')}} - 0.07\phi_{\mathrm{O20(2py'')}} + 0.09\phi_{\mathrm{O20(2pz')}} + 0.07\phi_{\mathrm{O20(2pz'')}}
\end{aligned}
\tag{13.6}
$$

HOMOs consist of oxygen 2p orbitals, corresponding to oxygen 2p valence band. The obtained wave-functions of LUMO at the centre, local minimum and bottleneck are

$$\begin{aligned}\psi_{\mathrm{LUMO(centre)}} = {} & 0.21\phi_{\mathrm{Ti1(3d}xy)} + 0.21\phi_{\mathrm{Ti1(3d}xz)} + 0.21\phi_{\mathrm{Ti1(3d}yz)} \\ & + 0.21\phi_{\mathrm{Ti2(3d}xy)} + 0.21\phi_{\mathrm{Ti2(3d}xz)} - 0.21\phi_{\mathrm{Ti2(3d}yz)} \\ & + 0.21\phi_{\mathrm{Ti3(3d}xy)} - 0.21\phi_{\mathrm{Ti3(3d}xz)} - 0.21\phi_{\mathrm{Ti3(3d}yz)} \\ & + 0.21\phi_{\mathrm{Ti4(3d}xy)} - 0.21\phi_{\mathrm{Ti4(3d}xz)} + 0.21\phi_{\mathrm{Ti4(3d}yz)} \\ & - 0.21\phi_{\mathrm{Ti5(3d}xy)} + 0.21\phi_{\mathrm{Ti5(3d}xz)} + 0.21\phi_{\mathrm{Ti5(3d}yz)} \\ & - 0.21\phi_{\mathrm{Ti6(3d}xy)} + 0.21\phi_{\mathrm{Ti6(3d}xz)} - 0.21\phi_{\mathrm{Ti6(3d}yz)} \\ & - 0.21\phi_{\mathrm{Ti7(3d}xy)} - 0.21\phi_{\mathrm{Ti7(3d}xz)} - 0.21\phi_{\mathrm{Ti7(3d}yz)} \\ & - 0.21\phi_{\mathrm{Ti8(3d}xy)} - 0.21\phi_{\mathrm{Ti8(3d}xz)} + 0.21\phi_{\mathrm{Ti8(3d}yz)}\end{aligned} \tag{13.7}$$

$$\psi_{\mathrm{LUMO(min)}} = 0.50\phi_{\mathrm{Ti1(3d}yz)} + 0.50\phi_{\mathrm{Ti4(3d}yz)} + 0.50\phi_{\mathrm{Ti5(3d}yz)} + 0.50\phi_{\mathrm{Ti8(3d}yz)} \tag{13.8}$$

$$\psi_{\mathrm{LUMO(bottleneck)}} = 0.51\phi_{\mathrm{Ti1(3d}yz)} + 0.51\phi_{\mathrm{Ti4(3d}yz)} + 0.51\phi_{\mathrm{Ti5(3d}yz)} + 0.51\phi_{\mathrm{Ti8(3d}yz)} \tag{13.9}$$

LUMOs consist of titanium t_{2g}-type 3d orbitals, corresponding to titanium 3d conduction band. However, the coefficients are changeable in HOMO and LUMO, during lithium ion conduction.

The local maximum is given at the centre. In general, counter cation has a role of stabilizing cubic structure, due to the large ionic radius. However, the ionic radius of lithium ion is smaller, in comparison with lanthanum ion (see Table 9.1). For example, in eight-coordination, the ionic radii of lithium and lanthanum ions are 1.06 and 1.30 Å, respectively. Lithium ion has no role in stabilizing the cubic structure, but just a role in neutralization of solid. Hence, the higher total energy is given at the centre.

The local minimum is given near the bottleneck. It is responsible for Coulomb interaction between positively charged lithium ion and negatively charged oxygen anions of bottleneck. The activation energy can be estimated from the total energy difference between the centre and bottleneck (0.167 eV). It corresponds to the experimental values (0.15–0.40 eV).

13.2.2 $K_xBa_{(1-x)/2}MnF_3$ Perovskite

$KMnF_3$ perovskite has a cubic structure at room temperature, and displays cubic-tetragonal structural distortion at low temperature. At operation temperature of lithium ion battery, it keeps a cubic structure (see Fig. 13.6). It is expected that barium-doping causes no structural distortion, since the ionic radii of potassium and barium are 1.65 and 1.56, respectively. In addition, when barium is doped at counter cation site, one vacancy is produced per one barium-doping, due to the difference of formal charges. Note that the formal charges of potassium, barium and lithium are +1, +2 and +1, respectively. BHHLYP is performed for $LiMn_8F_{12}$ model. Basis sets used for manganese, fluorine and lithium are MINI(5.3.3.3/5.3/5), 6-31G* and MINI(7.3), respectively.

Figure 13.7 shows the potential energy curve of $LiMn_8F_{12}$ model, when displacing lithium ion along x axis. The local minimum is given near the bottleneck. The local maximum is given at the centre, though the highest total energy is given at the bottleneck. The activation energy can be estimated from the total energy difference between the local minimum and bottleneck. The value (0.27 eV) is enough small for lithium ion conduction. Figure 13.8 depicts the selected molecular orbital related to conductive lithium ion (MO73) at the centre, local minimum and bottleneck. The obtained wave-functions of MO73 at the centre, local minimum and bottleneck are

$$\psi_{MO73(centre)} = 1.00\phi_{Li(1s)} \tag{13.10}$$

$$\psi_{MO73(min)} = 1.00\phi_{Li(1s)} \tag{13.11}$$

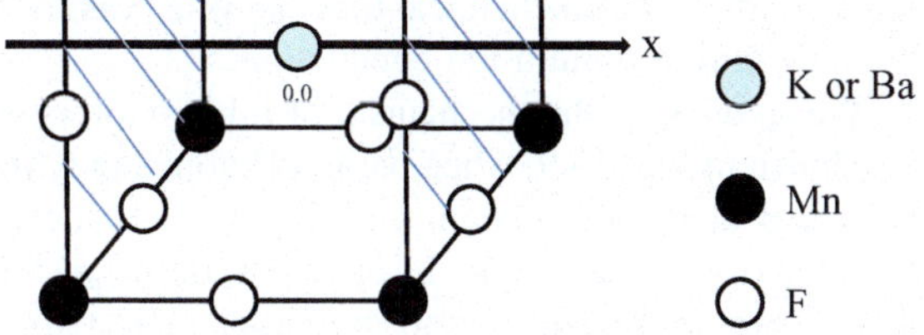

Fig. 13.6 Crystal structure of $K_xBa_{(1-x)/2}MnF_3$ perovskite

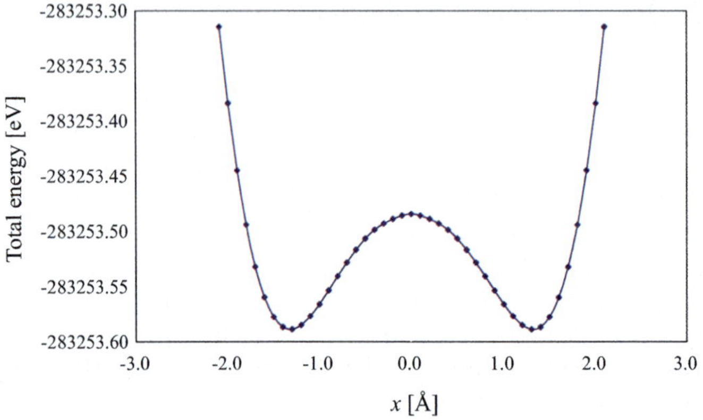

Fig. 13.7 Potential energy curve of $LiMn_8F_{12}$ model, when displacing lithium ion along x axis. Reference [2] by permission from Wiley

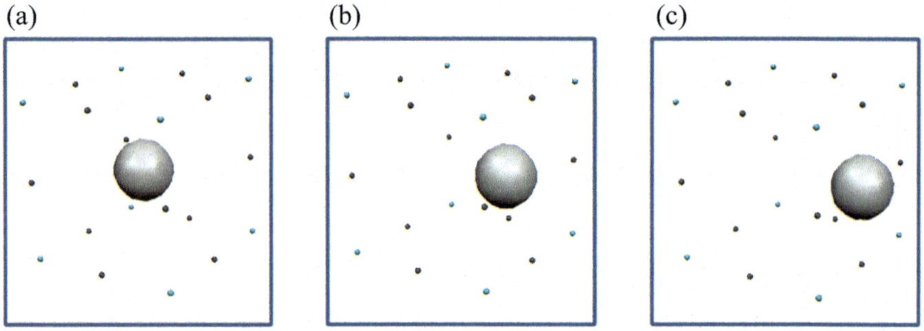

Fig. 13.8 Selected molecular orbital related to lithium ion (MO73) in $LiMn_8F_{12}$ model: **a** centre, **b** local minimum, **c** bottleneck

$$\psi_{\text{MO73(bottleneck)}} = 1.00\phi_{\text{Li(1s)}} \tag{13.12}$$

MO73s consist of lithium 1s orbital. There is no orbital overlap between lithium ion and others. From chemical bonding rule, it is found that lithium ion forms ionic bonding during lithium ion conduction.

We summarize the mechanism of lithium ion conduction in $KMnF_3$ perovskite. As shown in Fig. 13.9, when barium is doped in $KMnF_3$ perovskite, one vacancy is produced at counter cation site. Figure 13.10 depicts the schematic drawing of lithium ion conduction in Li-doped $K_xBa_{(1-x)/2}MnF_3$ perovskite. Potassium and barium are kept fixed, due to the larger ionic radii. Instead, lithium ion migrates through vacancy.

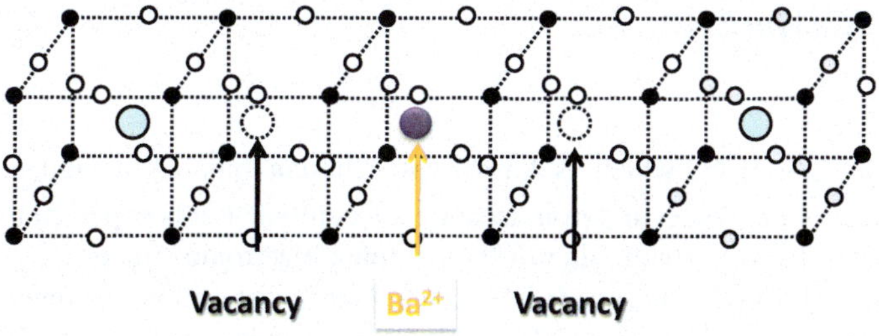

Fig. 13.9 Vacancy of counter cation site in $K_xBa_{(1-x)/2}MnF_3$ perovskite

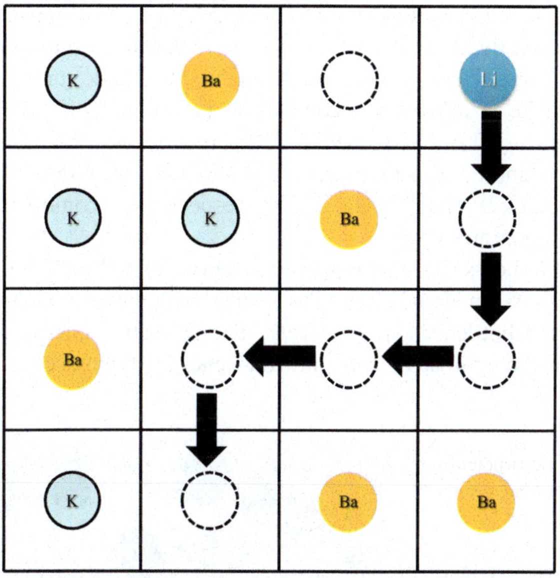

Fig. 13.10 Schematic drawing of lithium ion conduction in Li-doped $K_xBa_{(1-x)/2}MnF_3$ perovskite

13.3 Sodium Ion Conductor

Sodium ion has larger ionic radius, in comparison with lithium ion (see Table 9.1). It is considered that sodium ion conduction is blocked in the same material, due to the larger ionic radius. To overcome the problem, the larger cubic structure is favourable. One of candidate materials is transition metal cyanide. In general, M-CN-M distance is larger than M-O-M and M-F-M distances (M = transition metal). Here, our designed sodium ion conductive $CsMn(CN)_3$, $Al(CN)_3$ and $NaAlO(CN)_2$ are introduced.

13.3.1 $CsMn(CN)_3$

$Fe_4\left[Fe(CN)_6\right]_3 \cdot xH_2O$, which is known as Prussian blue, is candidate material. However, as water defect and iron vacancy are combined, it is expected that sodium ion conductivity is unstable, due to the complex electronic structure.

$CsMn(CN)_3$ was proposed. It is because there exist less water defect and less manganese vacancy. Note that $CsMn(CN)_3$ is often expressed as $Cs_2Mn[Mn(CN)_6]$. The formal charge of manganese is +2. The spin state of Mn^{2+} is sextet ($t_{2g}^3e_g^2$). Figure 13.11 shows the crystal structure of $CsMn(CN)_3$. There are two coordination patterns. One manganese is surrounded by six nitrogen atoms, and the other is surrounded by six carbon atoms. Mn–C, Mn–N and C–N distances are 1.93, 2.19 and 1.18 Å, respectively. Caesium ion can be replaced by sodium ion, due to the same formal charge. Here, in $Cs_{1-x}Na_xMn(CN)_3$, the same ion conduction mechanism is assumed as lithium ion conduction (see Fig. 13.12). BHHLYP is performed for $NaMn_8(CN)_{12}$ and $CsMn_8(CN)_{12}$ models (see Fig. 13.13). Basis sets used for manganese, caesium and sodium are MINI(5.3.3.3/5.3/5), MINI (4.3.2.2.2.2/4.2.2/4.2) and MINI(5.3.3/5), respectively, combined with 6-31G* basis set for carbon and nitrogen.

Figure 13.14 shows the potential energy curve of $CsMn_8(CN)_{12}$ model, when displacing caesium ion along x axis. The lowest and highest total energies are given at the centre and bottleneck, respectively. The activation energy for caesium ion conduction can be estimated from the total energy difference between the local

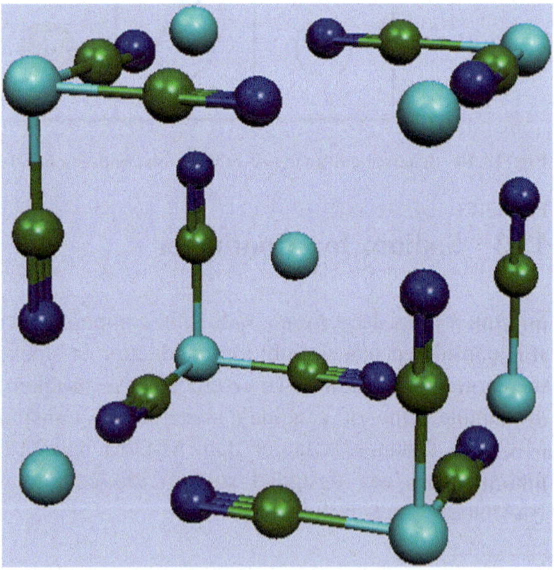

Fig. 13.11 Crystal structure of $CsMn(CN)_3$. A *blue*, *green*, *dark blue* and *centred blue dots* denote manganese, carbon, nitrogen and caesium, respectively

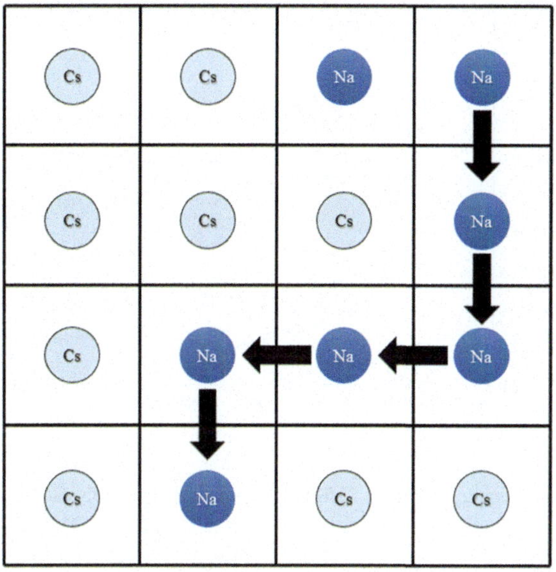

Fig. 13.12 Schematic drawing of sodium ion conduction in $Cs_{1-x}Na_xMn(CN)_3$

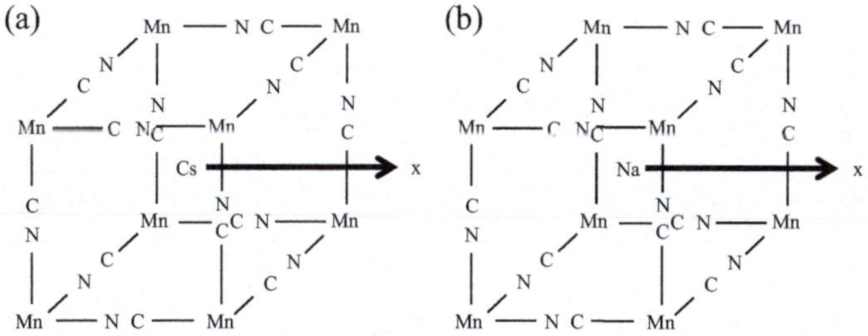

Fig. 13.13 **a** $CsMn_8(CN)_{12}$ and **b** $NaMn_8(CN)_{12}$ models of $Cs_{1-x}Na_xMn(CN)_3$. The origin of *x* axis is the cubic centre

minimum and local maximum. It becomes 4.14 eV. As the value is too large for ion conduction, caesium ion is kept fixed at the centre.

Figure 13.15 shows the potential energy curve of $NaMn_8(CN)_{12}$ model, when displacing sodium ion along *x* axis. The highest total energy is given at the centre. It is because the ionic radius of sodium ion is enough small for the cube, as same as lithium ion in $La_{2/3-x}Li_{3x}TiO_3$ perovskite. The local minima are given around $x = 1.8$ and -1.8 Å. It comes from the Coulomb interaction between positively charged sodium ion and negatively charged cyano ligand. However, the effect of the steric repulsion between sodium ion and other atoms is negligible at the bottleneck.

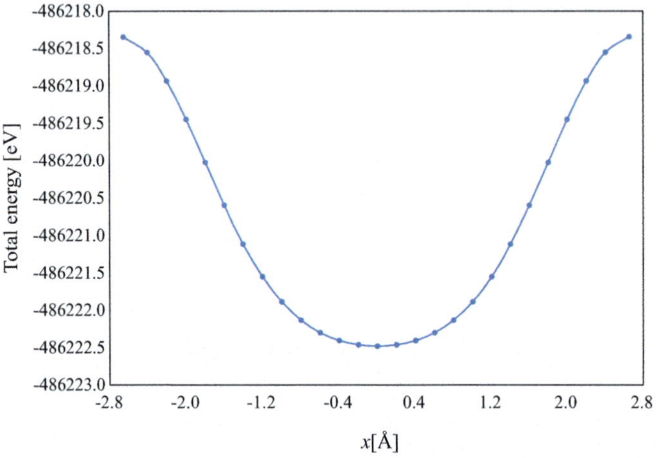

Fig. 13.14 Potential energy curve of $CsMn_8(CN)_{12}$ model, when displacing caesium ion along *x* axis

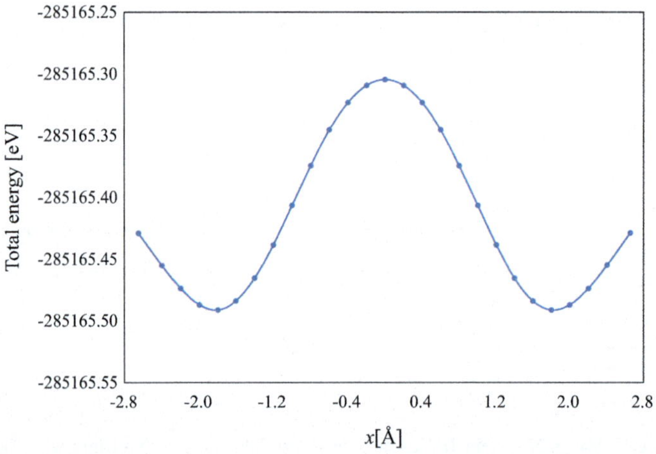

Fig. 13.15 Potential energy curve of $NaMn_8(CN)_{12}$ model, when displacing sodium ion along *x* axis

The activation energy for sodium ion conduction is 0.19 eV. The value is enough small for sodium ion conduction.

Figure 13.16 depicts the molecular orbitals related to outer shell electrons of sodium ion (sodium 2s and 2p electrons) in $NaMn_8(CN)_{12}$ model. Note that not only 2s but also 2p electrons work as outer shell electron in sodium ion. At the cubic centre, the obtained wave-functions of MOs related to outer shell electrons are

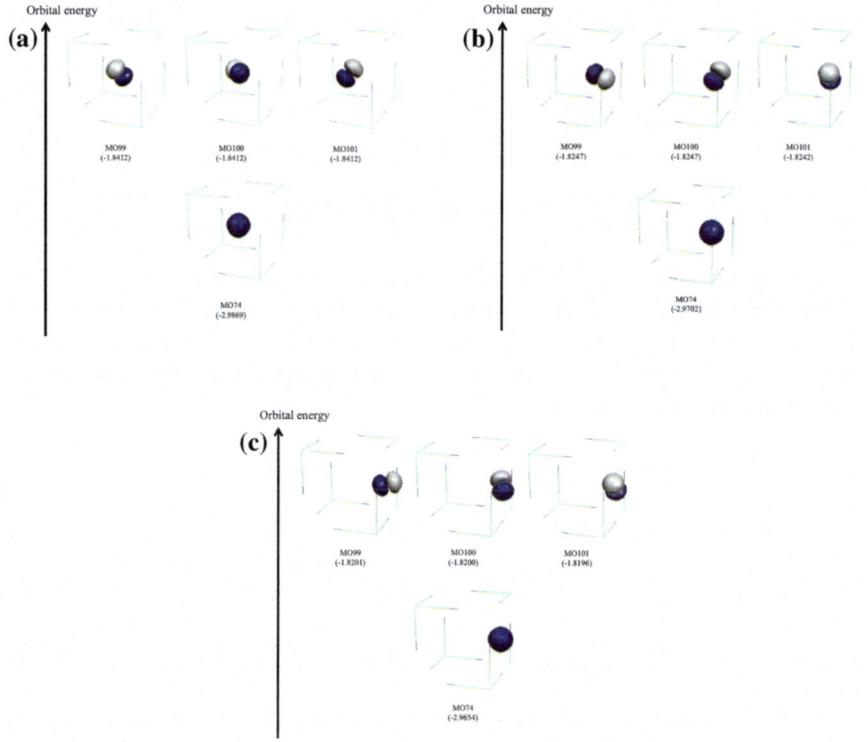

Fig. 13.16 Selected molecular orbitals related to outer shell electrons of sodium ion (sodium 2s and 2p electrons) in $NaMn_8(CN)_{12}$ model: **a** centre, **b** local minima, **c** bottleneck

$$\psi_{\mathrm{MO74(centre)}} = -0.27\phi_{\mathrm{Na(1s)}} + 1.03\phi_{\mathrm{Na(2s)}} \tag{13.13}$$

$$\psi_{\mathrm{MO99(centre)}} = -0.50\phi_{\mathrm{Na(2p}x)} - 0.50\phi_{\mathrm{Na(2p}y)} + 0.71\phi_{\mathrm{Na(2p}z)} \tag{13.14}$$

$$\psi_{\mathrm{MO100(centre)}} = -0.71\phi_{\mathrm{Na(2p}x)} + 0.71\phi_{\mathrm{Na(2p}y)} \tag{13.15}$$

$$\psi_{\mathrm{MO101(centre)}} = 0.50\phi_{\mathrm{Na(2p}x)} + 0.50\phi_{\mathrm{Na(2p}y)} + 0.71\phi_{\mathrm{Na(2p}z)} \tag{13.16}$$

In MO74, sodium 2s orbital has no orbital overlap with other atoms. In MO99, MO100 and MO101, sodium 2p orbital has no orbital overlap with other atoms. Note that sodium 2p orbital is rotated from the standard direction in MO99, MO100 and MO101. From chemical bonding rule, it is found that sodium forms ionic bonding with other atoms at the centre. At the local minima, the obtained wave-functions of MOs related to outer shell electrons are

$$\psi_{\mathrm{MO74(min)}} = -0.27\phi_{\mathrm{Na(1s)}} + 1.03\phi_{\mathrm{Na(2s)}} \tag{13.17}$$

$$\psi_{\mathrm{MO99(min)}} = 0.94\phi_{\mathrm{Na(2px)}} - 0.20\phi_{\mathrm{Na(2py)}} - 0.26\phi_{\mathrm{Na(2pz)}} \tag{13.18}$$

$$\psi_{\mathrm{MO100(min)}} = 0.33\phi_{\mathrm{Na(2px)}} + 0.69\phi_{\mathrm{Na(2py)}} + 0.64\phi_{\mathrm{Na(2pz)}} \tag{13.19}$$

$$\psi_{\mathrm{MO101(min)}} = -0.69\phi_{\mathrm{Na(2py)}} + 0.72\phi_{\mathrm{Na(2pz)}} \tag{13.20}$$

In MO74, sodium 2s orbital has no orbital overlap with other atoms. In MO99, MO100 and MO101, sodium 2p orbital has no orbital overlap with other atoms. From chemical bonding rule, it is found that sodium forms ionic bonding with other atoms at the local minima. At the bottleneck, the obtained wave-functions of MOs related to outer shell electrons are

$$\psi_{\mathrm{MO74(bottleneck)}} = -0.27\phi_{\mathrm{Na(1s)}} + 1.03\phi_{\mathrm{Na(2s)}} \tag{13.21}$$

$$\psi_{\mathrm{MO99(bottleneck)}} = 0.93\phi_{\mathrm{Na(2px)}} + 0.24\phi_{\mathrm{Na(2py)}} + 0.27\phi_{\mathrm{Na(2pz)}} \tag{13.22}$$

$$\psi_{\mathrm{MO100(bottleneck)}} = -0.36\phi_{\mathrm{Na(2px)}} + 0.68\phi_{\mathrm{Na(2py)}} + 0.63\phi_{\mathrm{Na(2pz)}} \tag{13.23}$$

$$\psi_{\mathrm{MO101(bottleneck)}} = -0.69\phi_{\mathrm{Na(2py)}} + 0.72\phi_{\mathrm{Na(2pz)}} \tag{13.24}$$

In MO74, sodium 2s orbital has no orbital overlap with other atoms. Though the orbital energies of MO99, MO100 and MO101 are slightly different, sodium 2p orbital has no orbital overlap with other atoms. From chemical bonding rule, it is found that sodium forms ionic bonding with other atoms at the bottleneck.

It is found that sodium-doped $CsMn(CN)_3$ can be applicable as sodium ion conductor. The sodium ion conduction comes from vacancy at counter cation site. Sodium ion forms ionic bonding during sodium ion conduction.

13.3.2 $Al(CN)_3$

Let us consider another cyanide $Al(CN)_3$. As there exists no 3d electron in aluminium, the simple chemical bonding is formed in Al–CN–Al bond, compared with Mn–CN–Mn bond in $CsMn(CN)_3$. In order to compare the difference of lattice distance, we also consider conventional $LaAlO_3$ perovskite. Figure 13.17 depicts $NaAl_8O_{12}$ model of $LaAlO_3$ perovskite and $NalAl_8(CN)_{12}$ model of Na-doped Al $(CN)_3$. BHHLYP is performed for $NaAl_8O_{12}$ and $NaAl_8(CN)_{12}$ models. Basis sets used for aluminium, carbon, nitrogen and sodium are 6-31G* basis set.

Figure 13.18 shows the potential energy curve of $NaAl_8O_{12}$ model, when displacing sodium ion along x axis. The lowest and highest total energies are given at the centre and bottleneck, respectively. The activation energy for sodium ion

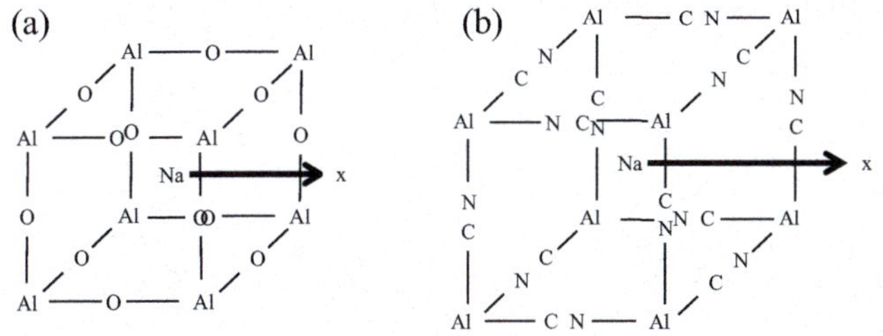

Fig. 13.17 **a** $NaAl_8O_{12}$ model of $LaAlO_3$ perovskite and **b** $NaAl_8(CN)_{12}$ model of $Al(CN)_3$

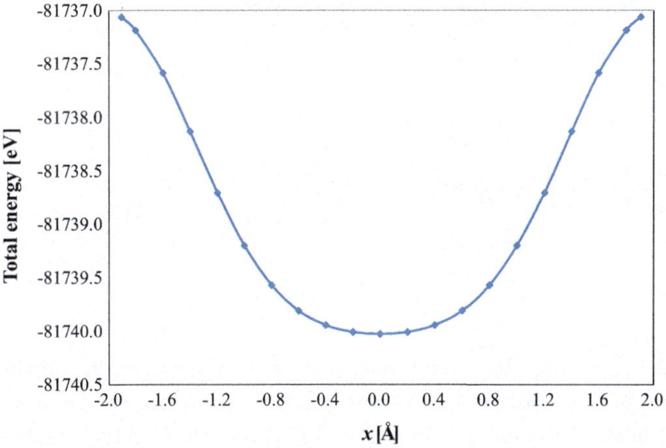

Fig. 13.18 Potential energy curve of $NaAl_8O_{12}$ model, when displacing sodium ion along x axis

conduction becomes 2.96 eV. It is concluded that sodium is kept fixed at counter cation site, due to the small lattice distance.

Figure 13.19 shows the potential energy curve of $NalAl_8(CN)_{12}$ model, when displacing sodium ion along x axis. The lowest and highest total energies are given at the bottleneck and centre, respectively. The activation energy for sodium ion conduction is 0.71 eV. In comparison with $CsMn(CN)_3$, no local minimum is given. It is because the steric repulsion between sodium ion and other atoms is suppressed at the bottleneck. It is concluded that sodium ion conduction occurs in sodium-doped $Al(CN)_3$.

Figure 13.20 depicts the selected molecular orbitals related to outer shell electrons of sodium ion (sodium 2s and 2p electrons) in $NalAl_8(CN)_{12}$ model. At the cubic centre, the obtained wave-function related to outer shell electrons are

$$\psi_{\text{MO66(centre)}} = -0.25\phi_{\text{Na(1s)}} + 1.02\phi_{\text{Na(2s)}} \tag{13.25}$$

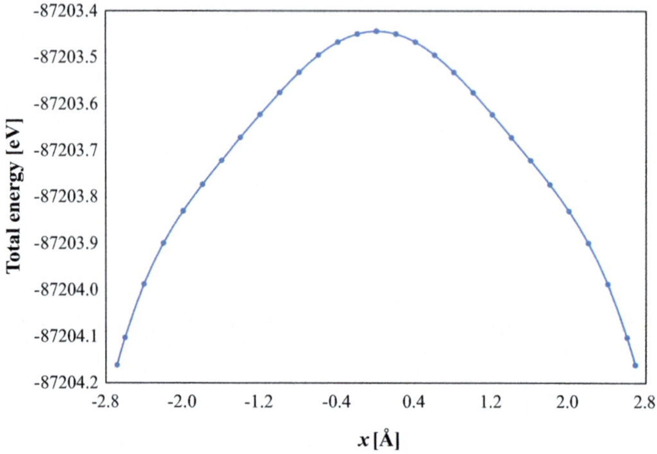

Fig. 13.19 Potential energy curve of $NaIAl_8(CN)_{12}$ model, when displacing sodium ion along x axis

$$\psi_{\mathrm{MO67(centre)}} = 0.65\phi_{\mathrm{Na(2p}x)} + 0.75\phi_{\mathrm{Na(2pz)}} \tag{13.26}$$

$$\psi_{\mathrm{MO68(centre)}} = 0.75\phi_{\mathrm{Na(2p}x)} - 0.65\phi_{\mathrm{Na(2pz)}} \tag{13.27}$$

$$\psi_{\mathrm{MO69(centre)}} = 1.00\phi_{\mathrm{Na(2p}y)} \tag{13.28}$$

In MO66, sodium 2s orbital has no orbital overlap with other atoms. In degenerated MO67, MO68 and MO69, sodium 2p orbital has no orbital overlap with other atoms. Note that sodium 2p orbital is rotated from the standard direction in MO67, MO68 and MO69. It is because sodium $2p_x$, $2p_y$ and $2p_z$ orbitals are hybridized. From chemical bonding rule, it is found that sodium forms ionic bonding with other atoms at the centre. At the bottleneck, the obtained wave-function related to outer shell electrons are

$$\psi_{\mathrm{MO66(bottleneck)}} = -0.25\phi_{\mathrm{Na(1s)}} + 1.02\phi_{\mathrm{Na(2s)}} \tag{13.29}$$

$$\psi_{\mathrm{MO67(bottleneck)}} = 1.00\phi_{\mathrm{Na(2pz)}} \tag{13.30}$$

$$\psi_{\mathrm{MO68(bottleneck)}} = 1.00\phi_{\mathrm{Na(2p}y)} \tag{13.31}$$

$$\psi_{\mathrm{MO69(bottleneck)}} = 1.00\phi_{\mathrm{Na(2p}x)} \tag{13.32}$$

In MO66, sodium 2s orbital has no orbital overlap with other atoms. In MO67, MO68 and MO69, sodium 2p orbital has no orbital overlap with other atoms, though the orbital energy of MO69 is slightly larger than MO67 and MO68. From chemical bonding rule, it is found that sodium forms ionic bonding with other atoms at the bottleneck.

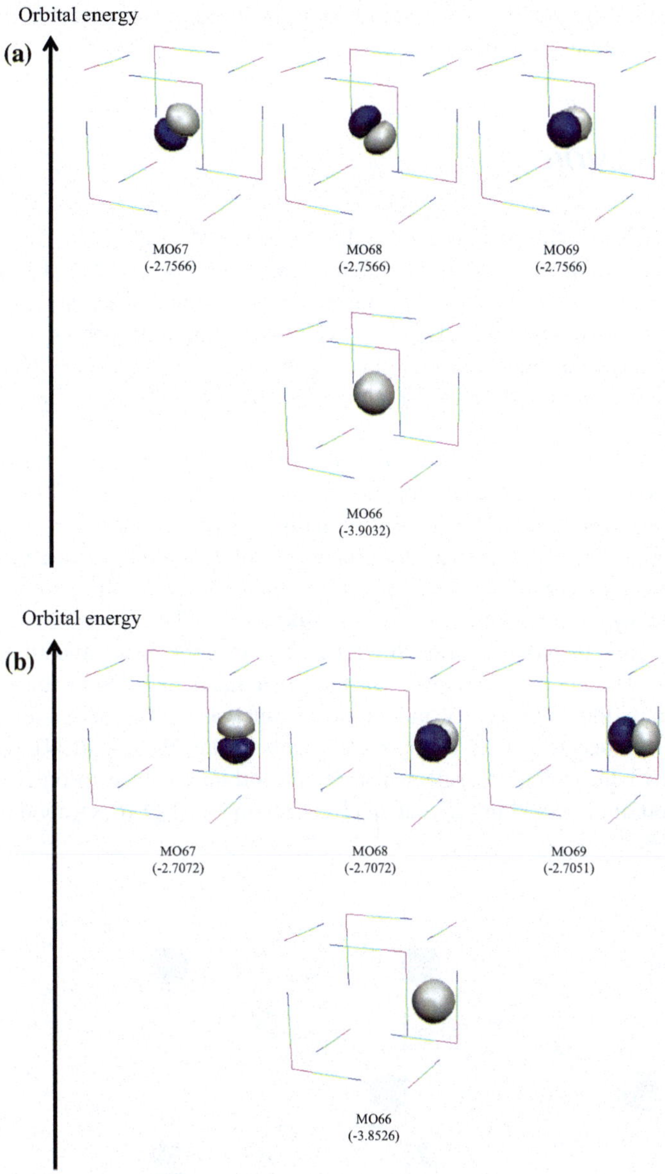

Fig. 13.20 Selected molecular orbitals related to outer shell electrons of sodium ion (sodium 2s and 2p electrons) in $NaAl_8(CN)_{12}$ model: **a** centre, **b** bottleneck. Reference [3] by permission from Wiley

It is found that sodium-doped $Al(CN)_3$ can be applicable as sodium ion conductor. The sodium ion conduction comes from vacancy at counter cation site. Sodium ion forms ionic bonding during sodium ion conduction. However, as no counter cation exists, trivalent aluminium must be substituted by divalent or monovalent cation, to introduce sodium ion.

13.3.3 $NaAlO(CN)_2$

In $NaAlO(CN)_2$, sodium ion is allocated at the centre of $Ti_8O_4(CN)_8$ cuboid without the replacement of aluminium by different-valent cation (see Fig. 13.21). As the long and short lattice distances are mixed in the cuboid, there are two types of bottlenecks: $Al_4(CN)_4$ and $Al_4O_2(CN)_2$. The two different directions of sodium ion conductions are considered. BHHLYP is performed for $NaAl_8O_4(CN)_8$ model of $NaAlO(CN)_2$. Basis sets used for aluminium, carbon, nitrogen and sodium are 6-31G* basis set.

Figures 13.22 and 13.23 show the potential energy curves of $NaAl_8O_4(CN)_8$ model, when displacing sodium ion along z and x axes, respectively. In sodium ion conduction along z axis, the local minimum is given, though it is not given in $NaAl_8(CN)_{12}$ model. It is because the effect of Coulomb interaction between sodium ion and oxygen anion is larger at the centre. Note that the formal charges of oxygen and cyano ligand are −2 and −1, respectively. The activation energy can be estimated from the total energy difference between the local maximum and the bottleneck. The value is 0.06 eV. On the other hand, in sodium ion conduction along x axis, the total energy monotonously increases. The activation energy is 5.55 eV. It is because $Al_4O_2(CN)_2$ rectangle is smaller than $Al_4(CN)_4$ square.

Figure 13.24 depicts the selected molecular orbitals related to outer shell electrons of sodium ion (sodium 2s and 2p electrons) in $NaAl_8O_4(CN)_8$ model at cuboid

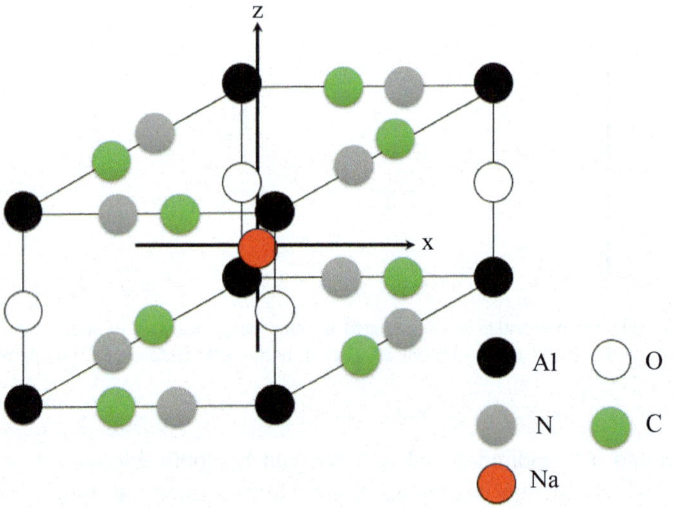

Fig. 13.21 $NaAl_8O_4(CN)_8$ model of $NaAlO(CN)_2$

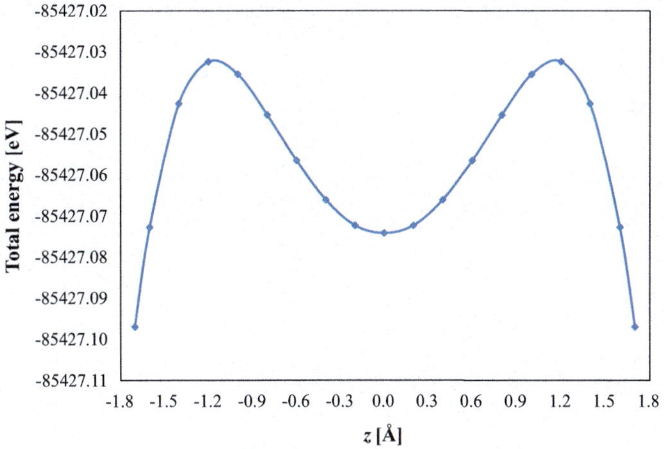

Fig. 13.22 Potential energy curves of $NaAl_8O_4(CN)_8$ model, when displacing sodium ion along z axis

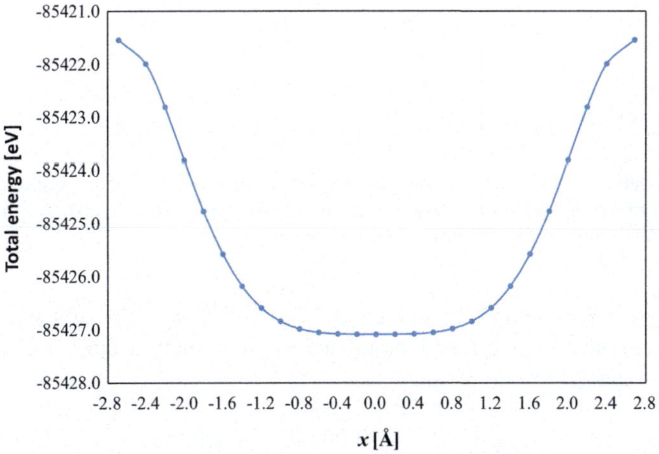

Fig. 13.23 Potential energy curves of $NaAl_8O_4(CN)_8$ model, when displacing sodium ion along x axis

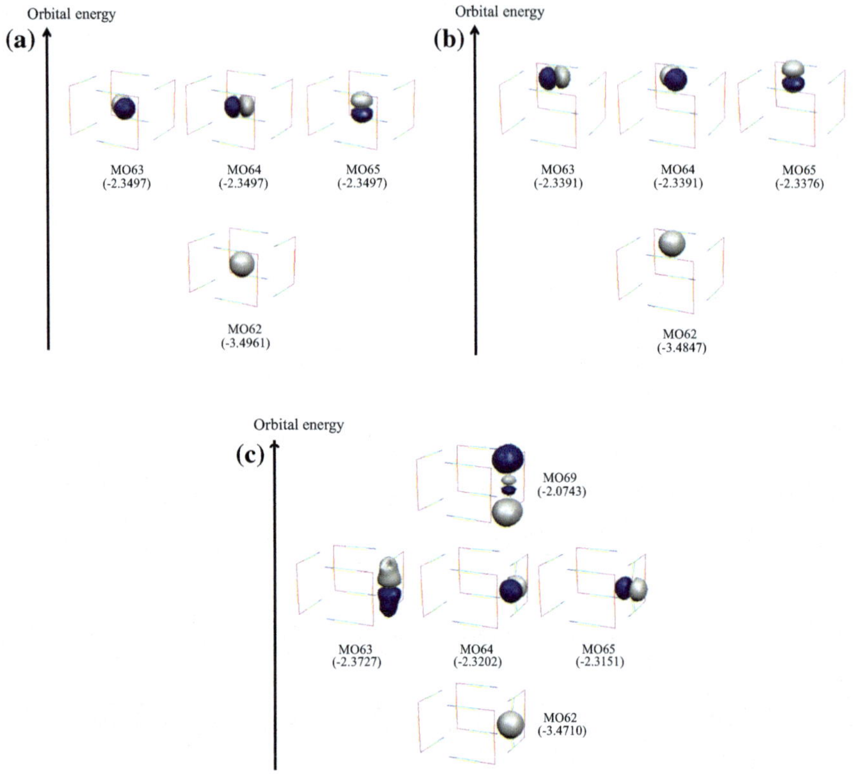

Fig. 13.24 Selected molecular orbitals related to outer shell electrons of sodium ion (sodium 2s and 2p electrons) in $NaAl_8O_4(CN)_8$ model: **a** cuboid centre ($x = z = 0.0$ Å), **b** $z = 1.7$ Å, **c** $x = 2.69$ Å. Reference [3] by permission from Wiley

centre ($x = z = 0.0$ Å) and the bottlenecks ($z = 1.7$ Å and $x = 2.69$ Å). At cuboid centre, the obtained wave-functions related to outer shell electrons of sodium ion (sodium 2s and 2p electrons) are

$$\psi_{\mathrm{MO62(centre)}} = -0.25\phi_{\mathrm{Na(1s)}} + 1.02\phi_{\mathrm{Na(2s)}} \tag{13.33}$$

$$\psi_{\mathrm{MO63(centre)}} = -0.69\phi_{\mathrm{Na(2px)}} + 0.71\phi_{\mathrm{Na(2py)}} \tag{13.34}$$

$$\psi_{\mathrm{MO64(centre)}} = 0.71\phi_{\mathrm{Na(2px)}} + 0.69\phi_{\mathrm{Na(2py)}} \tag{13.35}$$

$$\psi_{\mathrm{MO65(centre)}} = 1.00\phi_{\mathrm{Na(2pz)}} \tag{13.36}$$

In MO62, sodium 2s orbital has no orbital overlap with other atoms. In degenerated MO63, MO64 and MO65, sodium 2p orbital has no orbital overlap with other atoms. From chemical bonding rule, it is found that sodium ion forms ionic bonding at cuboid centre. At $Al_4(CN)_4$ bottleneck (z = 1.7 Å), the obtained wave-functions related to outer shell electrons of sodium ion are

$$\psi_{MO62(z=1.7\,\text{Å})} = -0.25\phi_{Na(1s)} + 1.02\phi_{Na(2s)} \tag{13.37}$$

$$\psi_{MO63(z=1.7\,\text{Å})} = 0.71\phi_{Na(2px)} + 0.70\phi_{Na(2py)} \tag{13.38}$$

$$\psi_{MO64(z=1.7\,\text{Å})} = -0.70\phi_{Na(2px)} + 0.71\phi_{Na(2py)} \tag{13.39}$$

$$\psi_{MO65(z=1.7\,\text{Å})} = 1.00\phi_{Na(2px)} \tag{13.40}$$

In MO62, sodium 2s orbital has no orbital overlap with other atoms. In MO63, MO64 and MO65, sodium 2p orbital has no orbital overlap with other atoms, though the orbital energy of MO65 is slightly larger than MO63 and MO64. From chemical bonding rule, it is found that sodium ion forms ionic bonding with other atoms at the bottleneck. At $Al_4O_2(CN)_2$ bottleneck (x = 2.69 Å), the obtained wave-functions related to outer shell electrons of sodium ion are

$$\psi_{MO62(x=2.69\,\text{Å})} = -0.24\phi_{Na(1s)} + 1.02\phi_{Na(2s)} \tag{13.41}$$

$$\begin{aligned}\psi_{MO63(x=2.69\,\text{Å})} &= 0.91\phi_{Na(2pz)} \\ &\quad + 0.07\phi_{N10(2s')} + 0.07\phi_{N10(2s'')} - 0.07\phi_{N18(2s')} - 0.07\phi_{N18(2s'')}\end{aligned} \tag{13.42}$$

$$\psi_{MO64(x=2.69\,\text{Å})} = 1.00\phi_{Na(2py)} \tag{13.43}$$

$$\psi_{MO65x=2.69\,\text{Å})} = 1.00\phi_{Na(2px)} \tag{13.44}$$

$$\begin{aligned}\psi_{MO69(x=2.69\,\text{Å})} &= 0.38\phi_{Na(2pz)} \\ &\quad + 0.13\phi_{N10(1s)} - 0.26\phi_{N10(2s')} - 0.28\phi_{N10(2s'')} - 0.12\phi_{N10(2py')} \\ &\quad - 0.13\phi_{N18(1s)} + 0.26\phi_{N18(2s')} + 0.28\phi_{N18(2s'')} + 0.12\phi_{N18(2py')}\end{aligned} \tag{13.45}$$

In MO62, sodium 2s orbital has no orbital overlap with other atoms. Though MO64 and MO65 consist of sodium 2p orbital, there is orbital overlap between sodium 2p orbital and nitrogen 2s orbitals in MO63. From chemical bonding rule, it is found that covalent bonding is formed. MO69 is inversion covalent bonding to MO63.

In $NaAlO_2(CN)_2$, the anisotropic sodium conduction occurs. Sodium ion can migrate through $Al_4(CN)_4$ bottleneck. As the activation energy along z axis is enough small, it is expected as one-dimensional sodium ion conductor.

13.3.4 Materials Design of Sodium Ion Conductor

When designing sodium ion conductor, the two factors must be taken into consideration at least.

(1) **Large and inflexible bottleneck**

It is difficult to realize a large bottleneck in M_4X_4-type square. In addition, inflexible bottleneck is desirable. In flexible bottleneck, sodium ion may be strongly bonded to bottleneck.

(2) **Coulomb interaction**

Coulomb interaction between sodium ion and bottleneck affects the magnitude of activation energy. In $NaAlO(CN)_2$, there is energetic advantage, due to Coulomb interaction between sodium cation and oxygen anion.

References

1. Onishi T (2012) Adv Quant Chem 64:47–59
2. Onishi T (2009) Int J Quant Chem 109:3659–3665
3. Onishi T (2012) Int J Quant Chem 112:3777–3781

Further Readings

4. Onishi T (2009) Solid State Ionics 180:592–597
5. Ingaguma Y, Liquan C, Itoh M, Nakamura T, Uchida T, Ikuta H, Wakihara M (1993) Solid State Commun 86(10):689–693
6. Inaguma Y, Jianding Y, Shan YJ, Itoh M, Nakamura T (1995) J Electrochem Soc 142(1): L8–L11
7. Nakayama M, Usui T, Uchimoto Y, Wakihara M, Yamamoto M (2005) J Phys Chem B 109:4135–4143
8. Inaguma Y (2006) J Ceramic Soc Jpn 114(12):1103–1110
9. Onishi T (2009) Polyhedron 28:1792–1795
10. Onishi T (2015) J Compt Chem Jpn 14(2):36–42
11. Schmidt MW, Baldridge KK, Boatz JA, Elbert ST, Gordon MS, Jensen JH, Koseki S, Matsunaga N, Nguyen KA, Su S, Windus TL, Dupuis M, Montgomery JA (1993) J Comput Chem 14:1347–1363
12. Varetto U <MOLEKEL 4.3.>; Swiss National Supercomputing Centre. Manno, Switzerland
13. Huzinaga S, Andzelm J, Radzio-Andzelm E, Sakai Y, Tatewaki H, Klobukowski M (1984) Gaussian basis sets for molecular calculations. Elsevier, Amsterdam

14. Hariharan PC, Pople JA (1973) Theoret Chim Acta 28:213–222
15. Francl MM, Pietro WJ, Hehre WJ, Binkley JS, Gordon MS, DeFrees DJ, Pople JA (1982) J Chem Phys 77(7):3654–3665
16. Rassolov VA, Pople JA, Ratner MA, Windus TL (1998) J Chem Phys 109(4):1223–1229
17. Rassolov VA, Ratner MA, Pople JA, Redfern PC, Curtiss LA (2001) J Compt Chem 22 (9):976–984

Chapter 14
Solid Oxide Fuel Cell: Oxide Ion and Proton Conductions

Abstract "Hydrogen" (hydrogen molecule) has attracted much industrial interest as future energy resource. Fuel cell is the efficient system that produces the electric energy from hydrogen molecule. Solid oxide fuel cell has been much expected, due to high efficiency of power generation. Solid oxide fuel cell is classified into oxide ion conducing type and proton conducting type. In oxide ion conducting type, oxide ion migrates through oxygen vacancy. Oxide ion forms covalent bonding with counter cations. Oxide ion conductivity can be controlled by changing dopant. In proton conducting type, proton forms covalent bonding with oxygen atoms. In diagonal path, OH and OHO bonds are alternately formed. During proton conduction, the proton pumping is combined. It implies that proton is pumped towards the square centre through OH conduction. The conflict with oxide ion conduction during proton conduction is also discussed. Finally, the mismatch of the calculated activation energy with AC impedance measurement is mentioned.

Keywords Solid oxide fuel cell · Oxide ion conduction · Proton conduction · Proton pumping effect · Covalent bonding

14.1 Introduction of Solid Oxide Fuel Cell

Oil production will end in the future, though the date cannot be correctly predicted. "Hydrogen" has been much expected as next-generation energy resource, and will be replaced by present oil resource. Note that "Hydrogen" denotes hydrogen molecule. Hydrogen molecule is produced through many methods. For example, in steam reforming of methane, hydrogen molecule is produced. Methane is the main ingredient of natural gas.

$$CH_4 + 2H_2O \rightarrow 4H_2 + CO_2 \tag{14.1}$$

Recently, methane can be produced from fermentation of raw garbage. Hydrogen molecule is produced as by-product in steel plant. The direct synthesis of

T. Onishi, *Quantum Computational Chemistry*,
DOI 10.1007/978-981-10-5933-9_14

hydrogen molecule has been also developed. Though electrolysis using photocatalyst is well known, it does not reach practical use.

Fuel cell is the efficient system that produces electricity from hydrogen molecule. In solid oxide fuel cell (SOFC) and polymer electrolyte fuel cell (PEFC), solid oxides and polymer are used as solid electrolyte, respectively. SOFC operates between 500 and 1000°. PEFC can operate below 100°. SOFC has been expected in home fuel cell system. It is because SOFC exhibits high efficiency of power generation, in comparison with PEFC. In addition, heat waste can be utilized in practical use. Perovskite-type compounds are widely used in electrolyte and electrode of SOFC. Figure 14.1 depicts the schematic drawing of SOFC. SOFC is classified into two types, according to the difference of ion conduction type. One is oxide ion conducting type. The other is proton conducting type. In both cases, water molecule is finally produced through chemical reactions and ion conduction.

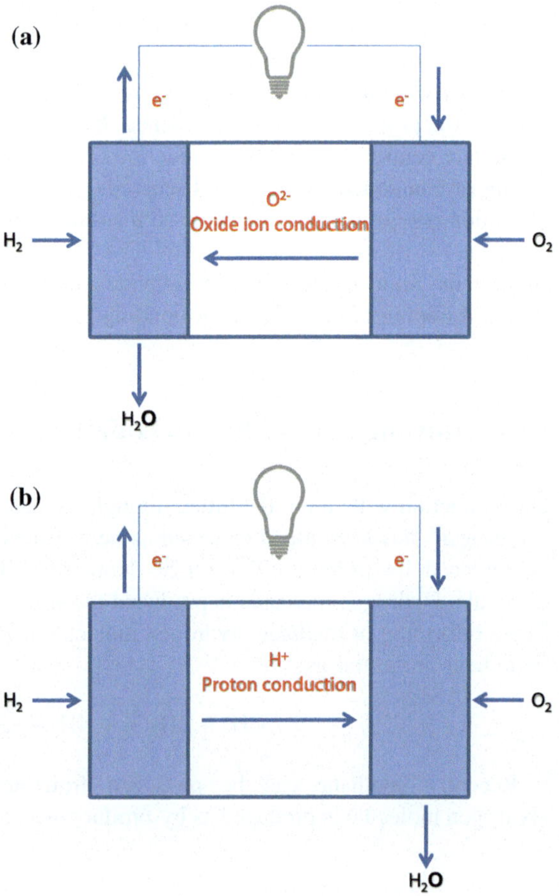

Fig. 14.1 Schematic drawing of solid oxide fuel cell: **a** oxide ion conducting type, **b** proton conducting type

$$H_2 + \frac{1}{2}O_2 \rightarrow H_2O \quad (14.2)$$

In this chapter, the mechanism of oxide ion conduction and proton conduction in perovskites is explained, from the viewpoint of energetics and bonding.

14.2 Oxide Ion Conduction in $LaAlO_3$ Perovskite

14.2.1 Introduction of Oxide Ion Conduction

Oxide ion conduction is observed in AMO_3 perovskite, where A and M denote counter cation and transition metal, respectively. As shown in Fig. 14.2, oxide ion migrates through oxygen vacancy. To incorporate oxygen vacancy in AMO_3 perovskite, A is replaced by different-valent counter cation. For example, in $LaAlO_3$ perovskite, trivalent lanthanum (La^{3+}) is replaced by divalent strontium (Sr^{2+}). When two lanthanum cations are replaced by two divalent counter cations, one oxygen vacancy is produced. Note that the formal charge of oxygen anion is −2, and the ionic radius of replaced counter cation should be close to La^{3+} to avoid the structural distortion. The local structural distortion may be caused near surface. Oxide ion conduction will be suppressed at locally distorted structure, due to strong chemical bonding formation between oxide ion and others. The effect is negligible, when considering oxide ion conduction in boundary solid structure.

$LaGaO_3$ perovskite was utilized as oxide ion conductor, due to lower operation temperature and lower activation energy. However, the alternative oxide ion conductor was expected, due to high cost of gallium. $LaAlO_3$ perovskite was considered as the candidate, due to low cost and light weight.

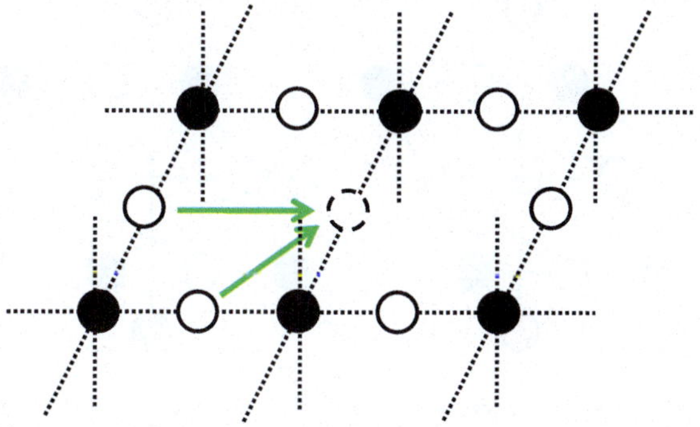

Fig. 14.2 Schematic drawing of oxide ion conduction on MO_2 layer in AMO_3 perovskite. *Black*, *white* and *dotted line circles* denote *M*, oxygen and oxygen vacancy, respectively. The *arrows* depicts oxide ion conduction path

14.2.2 Oxide Ion Conduction Mechanism

$LaAlO_3$ perovskite has a simple cubic structure at operation temperature. The lattice distance (Al–O–Al) is 3.81 Å. During oxide ion conduction, Al–O–Al bond is alternately broken and formed. To incorporate the effect of chemical bonding formation between oxide ion and up-and-down counter cations of AlO_2 layer, three $La_2Al_4O_3$, $LaSrAl_4O_3$ and $Sr_2Al_4O_3$ models are constructed, as shown in Fig. 14.3. The arrows depict two possible oxide ion conduction paths. One is diagonal path, where oxide ion migrates towards oxygen vacancy at nearest neighbouring oxygen site. The other is parallel path, where oxide ion migrates towards oxygen vacancy at next-nearest neighbouring oxygen site. BHHLYP calculation is performed for three models. Basis sets used for aluminium, oxygen, lanthanum and strontium are 6-31G*, 6-31G*, MINI(3.3.3.3.3.3/3.3.3.3/3.3/4) and MINI(4.3.3.3/4.3.3/4), respectively.

Figures 14.4 and 14.5 show the potential energy curves of $La_2Al_4O_3$ model, when displacing oxide ion along diagonal and parallel paths, respectively. In diagonal path, the local maximum is given at the middle of diagonal line, and the local minima are given around 0.6 and 2.1 Å. The highest energy is given at the lattice positions. The activation energy for oxide ion conduction can be estimated from the total energy difference between the local minimum and lattice position. The value is 2.73 eV. On the other hand, in parallel path, the local maxima and minima are given, as same as diagonal path. However, the highest total energy is given at the middle. Parallel path is impossible, due to too large activation energy (11.3 eV).

Figure 14.6 depicts the selected molecular orbitals related to outer shell electrons of oxide ion in $La_2Al_4O_3$ model, at the local minimum in diagonal path. The obtained wave-functions of MO72, MO81, MO83 and MO84 are

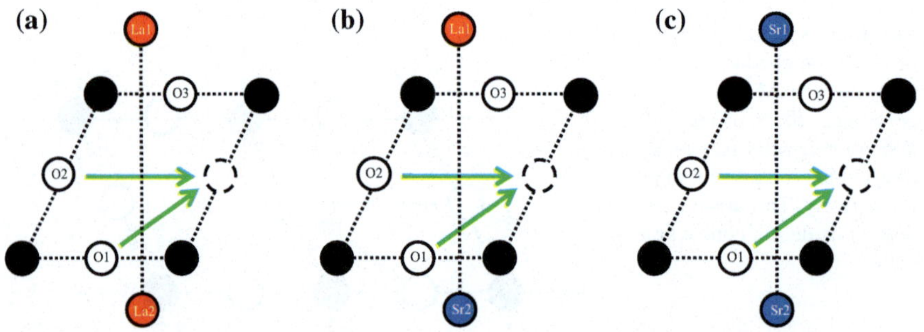

Fig. 14.3 **a** $La_2Al_4O_3$, **b** $LaSrAl_4O_3$ and **c** $Sr_2Al_4O_3$ models in Sr-doped $LaAlO_3$ perovskite. The *arrows* depict two possible oxide ion conduction paths. *Black*, *white*, *dotted line*, *red* and *blue circles* denote aluminium, oxygen, oxygen vacancy, lanthanum and strontium, respectively

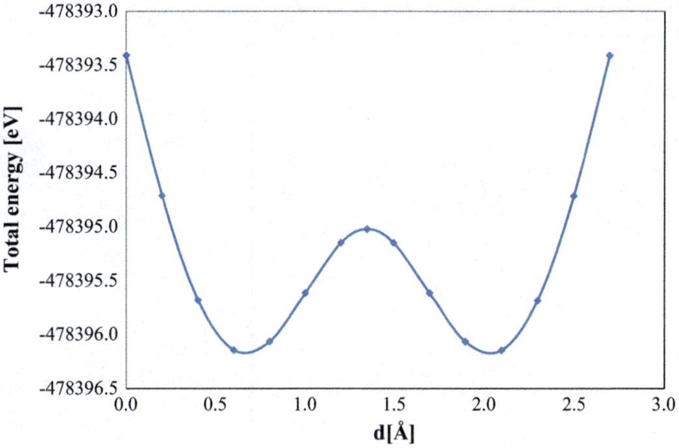

Fig. 14.4 Potential energy curve of $La_2Al_4O_3$ model, when displacing oxide ion along diagonal path. d is oxide ion conduction distance. Reference [1] by permission from Wiley

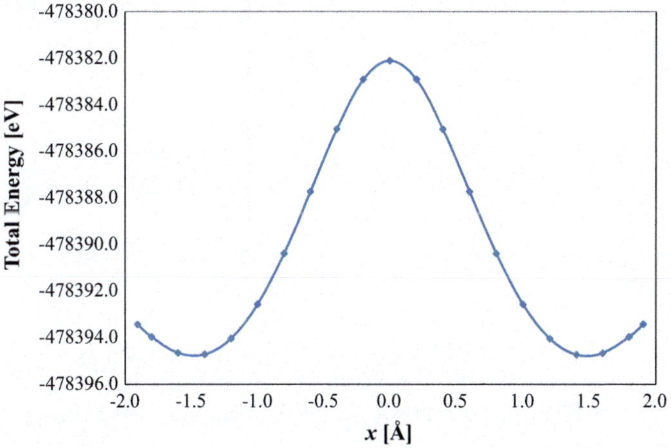

Fig. 14.5 Potential energy curve of $La_2Al_4O_3$ model, when displacing oxide ion along parallel path. x is oxide ion conduction distance

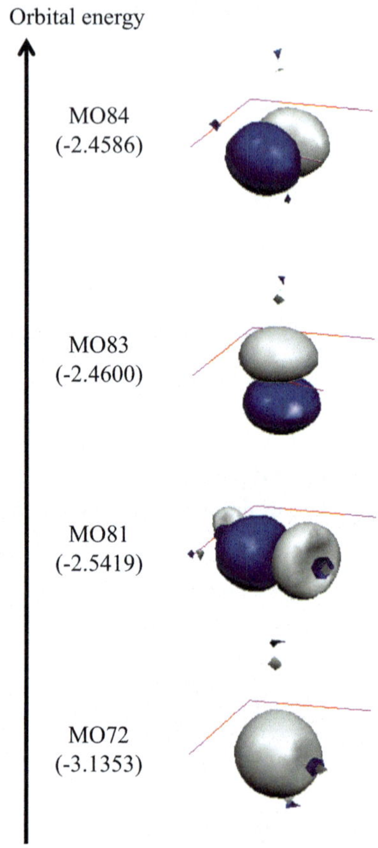

Fig. 14.6 Selected molecular orbitals related to outer shell electrons of oxide ion in $La_2Al_4O_3$ model, at the local minimum in diagonal path

$$\psi_{MO72} = -0.22\phi_{O1(1s)} + 0.44\phi_{O1(2s')} + 0.57\phi_{O1(2s'')} \tag{14.3}$$

$$\psi_{MO81} = -0.17\phi_{O1(2s'')} + 0.50\phi_{O1(2px')} + 0.43\phi_{O1(2px'')} + 0.11\phi_{O2(2py')} \tag{14.4}$$

$$\psi_{MO83} = 0.58\phi_{O1(2pz')} + 0.54\phi_{O1(2pz'')} \tag{14.5}$$

$$\psi_{MO84} = 0.57\phi_{O1(2py')} + 0.51\phi_{O1(2py'')} \tag{14.6}$$

2s and 2p orbitals of oxide ion have no orbital overlap with other atoms, though there is hybridization between O1 $2p_x$ and O2 $2p_y$ orbitals in MO81. From chemical bonding rule, it is found that oxide ion forms ionic bonding with other atoms at the local minimum.

Fig. 14.7 Selected molecular orbitals related to outer shell electrons of oxide ion in $La_2Al_4O_3$ model, at the middle in diagonal path

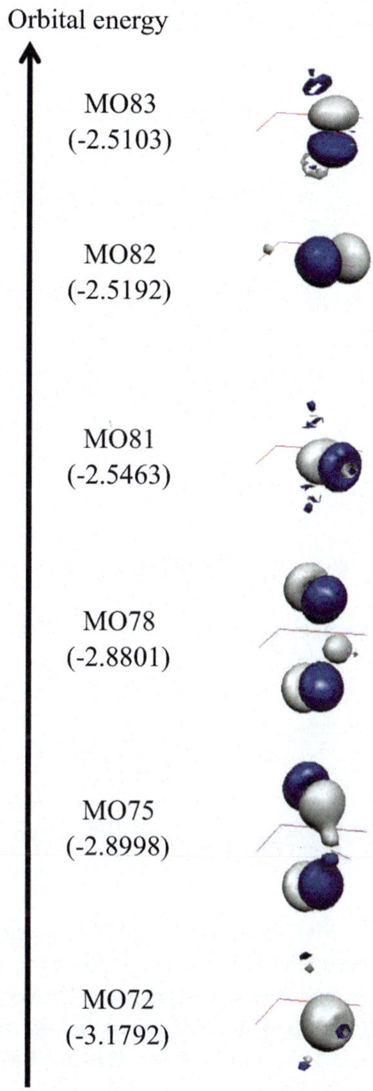

Figure 14.7 depicts the selected molecular orbitals related to outer shell electrons of oxide ion in $La_2Al_4O_3$ model, at the middle in diagonal path. The obtained wave-functions of MO72, MO75, MO78, MO81, MO82 and MO83 are

$$\psi_{MO72} = -0.22\phi_{O1(1s)} + 0.45\phi_{O1(2s')} + 0.58\phi_{O1(2s'')} \tag{14.7}$$

$$\begin{aligned}\psi_{\mathrm{MO75}} = {} & 0.09\phi_{\mathrm{O1(2pz')}} + 0.08\phi_{\mathrm{O1(2pz'')}} \\ & + 0.14\phi_{\mathrm{La1(3px)}} - 0.14\phi_{\mathrm{La1(3py)}} - 0.40\phi_{\mathrm{La1(4px)}} + 0.40\phi_{\mathrm{La1(4py)}} + 0.20\phi_{\mathrm{La1(4pz)}} \\ & - 0.42\phi_{\mathrm{La1(5px)}} + 0.42\phi_{\mathrm{La1(5py)}} + 0.20\phi_{\mathrm{La1(5pz)}} \\ & - 0.14\phi_{\mathrm{La2(3px)}} + 0.14\phi_{\mathrm{La2(3py)}} + 0.40\phi_{\mathrm{La2(4px)}} - 0.40\phi_{\mathrm{La2(4py)}} + 0.20\phi_{\mathrm{La2(4pz)}} \\ & + 0.42\phi_{\mathrm{La2(5px)}} - 0.42\phi_{\mathrm{La2(5py)}} + 0.20\phi_{\mathrm{La2(5pz)}}\end{aligned} \tag{14.8}$$

$$\begin{aligned}\psi_{\mathrm{MO78}} = {} & 0.10\phi_{\mathrm{O1(2s')}} + 0.13\phi_{\mathrm{O1(2s'')}} \\ & - 0.15\phi_{\mathrm{La1(3px)}} + 0.15\phi_{\mathrm{La1(3py)}} + 0.41\phi_{\mathrm{La1(4px)}} - 0.41\phi_{\mathrm{La1(4py)}} - 0.16\phi_{\mathrm{La1(4pz)}} \\ & + 0.45\phi_{\mathrm{La1(5px)}} - 0.45\phi_{\mathrm{La1(5py)}} - 0.16\phi_{\mathrm{La1(5pz)}} \\ & - 0.15\phi_{\mathrm{La2(3px)}} + 0.15\phi_{\mathrm{La2(3py)}} + 0.41\phi_{\mathrm{La2(4px)}} - 0.41\phi_{\mathrm{La2(4py)}} + 0.16\phi_{\mathrm{La2(4pz)}} \\ & + 0.45\phi_{\mathrm{La2(5px)}} - 0.45\phi_{\mathrm{La2(5py)}} + 0.16\phi_{\mathrm{La2(5pz)}}\end{aligned} \tag{14.9}$$

$$\psi_{\mathrm{MO81}} = -0.41\phi_{\mathrm{O1(2px')}} - 0.38\phi_{\mathrm{O1(2px'')}} + 0.41\phi_{\mathrm{O1(2py')}} + 0.38\phi_{\mathrm{O1(2py'')}} \tag{14.10}$$

$$\psi_{\mathrm{MO82}} = 0.39\phi_{\mathrm{O1(2px')}} + 0.35\phi_{\mathrm{O1(2px'')}} + 0.39\phi_{\mathrm{O1(2py')}} + 0.35\phi_{\mathrm{O1(2py'')}} \tag{14.11}$$

$$\begin{aligned}\psi_{\mathrm{MO83}} = {} & 0.57\phi_{\mathrm{O1(2pz')}} + 0.55\phi_{\mathrm{O1(2pz'')}} \\ & + 0.10\phi_{\mathrm{La1(4px)}} - 0.10\phi_{\mathrm{La1(4py)}} - 0.13\phi_{\mathrm{La1(4pz)}} \\ & + 0.13\phi_{\mathrm{La1(5px)}} - 0.13\phi_{\mathrm{La1(5py)}} - 0.19\phi_{\mathrm{La1(5pz)}} \\ & - 0.10\phi_{\mathrm{La2(4px)}} + 0.10\phi_{\mathrm{La2(4py)}} - 0.13\phi_{\mathrm{La2(4pz)}} \\ & - 0.13\phi_{\mathrm{La2(5px)}} + 0.13\phi_{\mathrm{La2(5py)}} - 0.19\phi_{\mathrm{La2(5pz)}}\end{aligned} \tag{14.12}$$

MO72 consists mainly of oxygen 2s orbital, though there is slight orbital overlap between oxide ion and lanthanum. In MO78, there are orbital overlaps between O1 2s and La1 p orbitals, and between O1 2s and La2 p orbitals. One oxygen lobe interacts with one lanthanum lobe of La1 and La2. There are two nodes between O1 and La1, and between O1 and La2. From chemical bonding rule, it is found that inversion σ-type covalent bonding is formed between oxide ion and two lanthanum cations. MO78 is inversion molecular orbital to MO72. In MO75 and MO83, there are orbital overlaps between O1 $2p_z$ and La1 p orbitals, and between O1 $2p_z$ and La2 p orbitals. One oxygen lobe interacts with one lanthanum lobe. From chemical bonding rule, it is found that σ-type covalent bonding is formed between oxide ion and two lanthanum cations in MO75. In MO83, there are two nodes between O1 and La1, and between O1 and La2. MO83 is inversion σ-type covalent bonding to MO75. MO81 and MO82 consist mainly of 2p orbitals of oxide ion, though the slight orbital overlap with lanthanum in MO81. It is concluded that oxide ion alternately forms and breaks σ-type covalent bonding with lanthanum, during oxide ion conduction.

To investigate doping effect on oxide ion conduction, $LaSrAl_4O_3$ and $Sr_2Al_4O_3$ models are considered. Note that $LaSrAl_4O_3$ and $Sr_2Al_4O_3$ units are locally realized in a part of solid. Figures 14.8 and 14.9 show the potential energy curves of $LaSrAl_4O_3$ and $Sr_2Al_4O_3$ models, when displacing oxide ion along diagonal path, respectively. In both models, the local maximum is given at the middle of diagonal line, and the local minima are given around 0.6 and 2.1 Å. The highest energy is given at the lattice positions. The activation energies of $LaSrAl_4O_3$ and $Sr_2Al_4O_3$

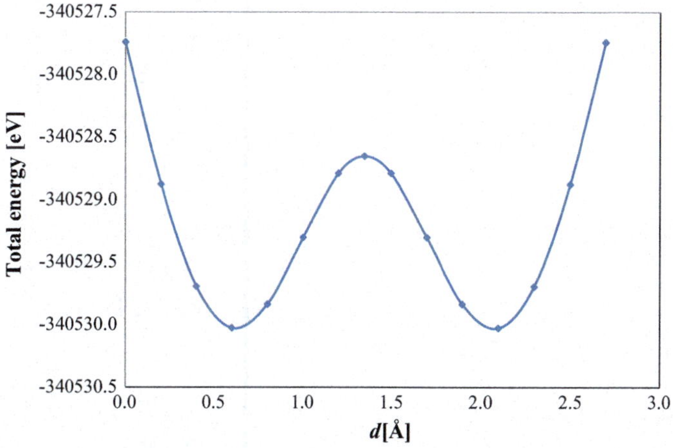

Fig. 14.8 Potential energy curve of $LaSrAl_4O_3$ model, when displacing oxide ion along diagonal path. d is oxide ion conduction distance. Reference [1] by permission from Wiley

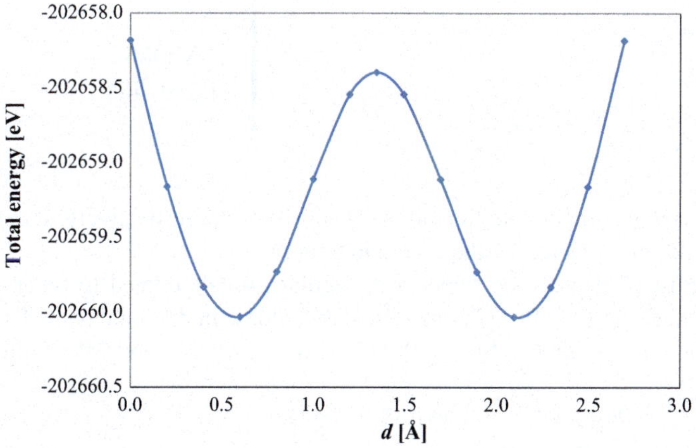

Fig. 14.9 Potential energy curves of $Sr_2Al_4O_3$ model, when displacing oxide ion along diagonal path. d is oxide ion conduction distance. Reference [1] by permission from Wiley

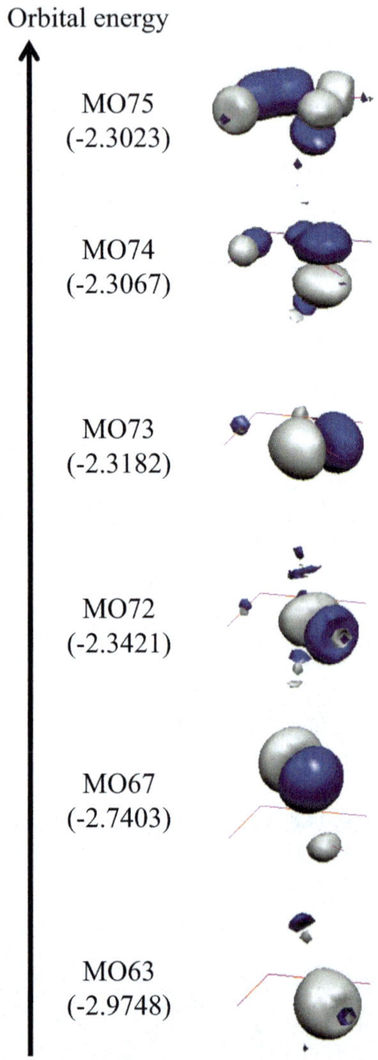

Fig. 14.10 Selected molecular orbitals related to outer shell electrons of oxide ion in $LaSrAl_4O_3$ model, at the middle in diagonal path

models for oxide ion conduction are 2.29 and 1.85 eV, respectively. It is found that strontium doping decreases the activation energy.

Figure 14.10 depicts the selected molecular orbitals related to outer shell electrons of oxide ion in $LaSrAl_4O_3$ model, at the middle in diagonal path. The obtained wave-functions of MO63, MO67, MO72, MO73, MO74, and MO75 are

$$\psi_{\mathrm{MO63}} = -0.22\phi_{\mathrm{O1(1s)}} + 0.45\phi_{\mathrm{O1(2s')}} + 0.57\phi_{\mathrm{O1(2s'')}} \tag{14.13}$$

$$\psi_{MO67} = 0.08\phi_{O1(2ps')} + 0.11\phi_{O1(2ps'')} - 0.21\phi_{La1(3px)} + 0.21\phi_{La1(3py)} + 0.58\phi_{La1(4px)} - 0.58\phi_{La1(4py)} + 0.21\phi_{La1(4pz)} + 0.63\phi_{La1(5px)} - 0.63\phi_{La1(5py)} + 0.21\phi_{La1(5pz)} \tag{14.14}$$

$$\psi_{MO72} = 0.21\phi_{O1(2s'')} - 0.40\phi_{O1(2px')} - 0.36\phi_{O1(2px'')} + 0.40\phi_{O1(2py')} + 0.36\phi_{O1(2py'')} \tag{14.15}$$

$$\psi_{MO73} = 0.38\phi_{O1(2px')} + 0.34\phi_{O1(2px'')} + 0.38\phi_{O1(2py')} + 0.34\phi_{O1(2py'')} \tag{14.16}$$

$$\psi_{MO74} = 0.50\phi_{O1(2pz')} + 0.47\phi_{O1(2pz'')} - 0.16\phi_{O2(2py')} - 0.12\phi_{O2(2py'')} + 0.16\phi_{O3(2px')} + 0.12\phi_{O3(2px'')} - 0.12\phi_{Sr2(4px)} + 0.12\phi_{Sr2(4py)} + 0.16\phi_{Sr2(4pz)} \tag{14.17}$$

$$\psi_{MO75} = 0.11\phi_{O1(2px')} - 0.11\phi_{O1(2py')} - 0.27\phi_{O1(2pz')} - 0.25\phi_{O1(2pz'')} - 0.28\phi_{O2(2py')} - 0.22\phi_{O2(2py'')} + 0.28\phi_{O3(2px')} + 0.22\phi_{O3(2px'')} - 0.14\phi_{Sr2(4pz)} \tag{14.18}$$

MO63 consists mainly of oxygen 2s orbital, though there is slight orbital overlap between oxide ion and lanthanum. In MO67, there is orbital overlap between O1 2s and La1 p orbitals. One oxygen lobe interacts with one lanthanum lobe. There is node between O1 and La1. From chemical bonding rule, it is found that inversion σ-type covalent bonding is formed between oxide ion and lanthanum cation. MO67 is inversion molecular orbital to MO63. MO72 consists mainly of 2p orbitals of oxide ion, though the slight orbital overlap between oxide ion and lanthanum. MO73 consists mainly of 2p orbitals of oxide ion, though the slight orbital overlap between oxide ion and other oxygen atoms. In MO74 and MO75, not only O1 but also O2 and O3 2p orbitals also participate.

The slight orbital overlap between oxide ion and strontium is observed. It is found that the total energy at the middle is much destabilized, due to the weak covalent bonding between oxide ion and strontium. As the result, the activation energy becomes smaller than $La_2Al_4O_3$ model.

Figure 14.11 depicts the molecular orbitals related to outer shell electrons of oxide ion in $Sr_2Al_4O_3$ model, at the middle in diagonal path. The obtained wave-functions of MO54, MO58, MO60, MO63, MO64, MO65 and MO66 are

$$\psi_{MO54} = -0.22\phi_{O1(1s)} + 0.46\phi_{O1(2s')} + 0.57\phi_{O1(2s'')} \tag{14.19}$$

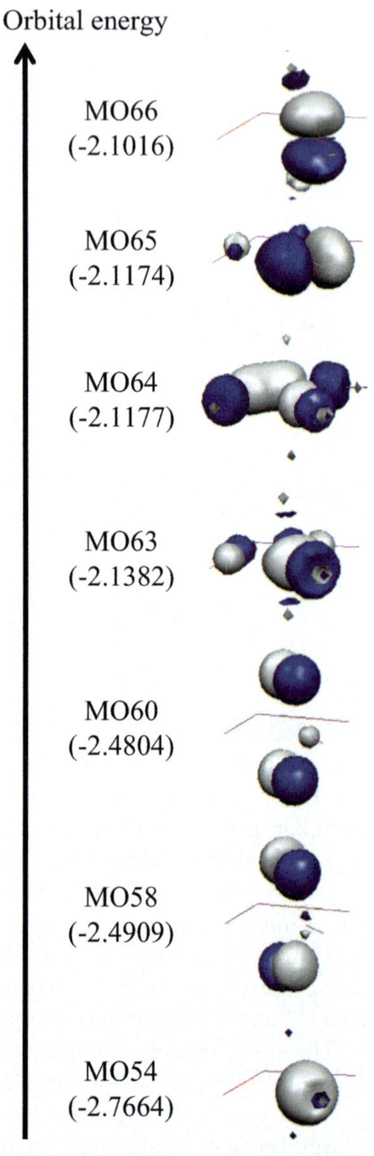

Fig. 14.11 Selected molecular orbitals related to outer shell electrons of oxide ion in $Sr_2Al_4O_3$ model, at the middle in diagonal path

$$\begin{aligned}\psi_{\mathrm{MO58}} = & -0.06\phi_{\mathrm{O1(2pz')}} - 0.06\phi_{\mathrm{O1(2pz'')}} \\ & +0.22\phi_{\mathrm{Sr1(3px)}} - 0.22\phi_{\mathrm{Sr1(3py)}} - 0.52\phi_{\mathrm{Sr1(4px)}} + 0.52\phi_{\mathrm{Sr1(4py)}} \\ & -0.22\phi_{\mathrm{Sr2(3px)}} + 0.22\phi_{\mathrm{Sr2(3py)}} + 0.52\phi_{\mathrm{Sr2(4px)}} - 0.52\phi_{\mathrm{Sr2(4py)}}\end{aligned} \tag{14.20}$$

$$\psi_{MO60} = 0.06\phi_{O1(2s')} + 0.07\phi_{O1(2s'')} + 0.23\phi_{Sr1(3px)} - 0.23\phi_{Sr1(3py)} - 0.54\phi_{Sr1(4px)} + 0.54\phi_{Sr1(4py)} + 0.23\phi_{Sr2(3px)} - 0.23\phi_{Sr2(3py)} - 0.54\phi_{Sr2(4px)} + 0.54\phi_{Sr2(4py)} \quad (14.21)$$

$$\psi_{MO63} = 0.19\phi_{O1(2s'')} - 0.35\phi_{O1(2px')} - 0.30\phi_{O1(2px'')} + 0.35\phi_{O1(2py')} + 0.30\phi_{O1(2py'')} - 0.16\phi_{O2(2py')} - 0.12\phi_{O2(2py'')} + 0.16\phi_{O3(2px')} + 0.12\phi_{O3(2px'')} \quad (14.22)$$

$$\psi_{MO64} = 0.12\phi_{O1(2s'')} - 0.21\phi_{O1(2px')} - 0.18\phi_{O1(2px'')} + 0.21\phi_{O1(2py')} + 0.18\phi_{O1(2py'')} + 0.29\phi_{O2(2py')} + 0.22\phi_{O2(2py'')} - 0.29\phi_{O3(2px')} - 0.22\phi_{O3(2px'')} \quad (14.23)$$

$$\psi_{MO65} = 0.36\phi_{O1(2px')} + 0.32\phi_{O1(2px'')} + 0.36\phi_{O1(2py')} + 0.32\phi_{O1(2py'')} + 0.11\phi_{O2(2py')} + 0.11\phi_{O3(2px')} \quad (14.24)$$

$$\psi_{MO66} = 0.58\phi_{O1(2pz')} + 0.52\phi_{O1(2pz'')} - 0.11\phi_{Sr1(4px)} + 0.11\phi_{Sr1(4py)} + 0.16\phi_{Sr1(4pz)} + 0.11\phi_{Sr2(4px)} - 0.11\phi_{Sr2(4py)} + 0.16\phi_{Sr2(4pz)} \quad (14.25)$$

MO54 consists mainly of oxygen 2s orbital. In MO60, there are orbital overlaps between O1 2s and Sr1 p orbitals, and between O1 2s and Sr2 p orbitals. One oxygen lobe interacts with one strontium lobe. There are nodes between O1 and Sr1, and between O1 and Sr2. From chemical bonding rule, it is found that inversion σ-type covalent bonding is formed between oxide ion and strontium cations. MO60 is inversion molecular orbital to MO54. In MO58 and MO66, there are orbital overlaps between O1 2p and Sr1 p orbitals, and between O1 2p and Sr2 p orbitals. One oxygen lobe interacts with one strontium lobe. From chemical bonding rule, it is found that σ-type covalent bonding is formed between oxide ion and strontium cations in MO58. In MO66, there are nodes between O1 and Sr1, and between O1 and Sr2. MO66 is inversion σ-type covalent bonding to MO58. In MO63, MO64 and MO65, not only O1 but also O2 and O3 2p orbitals also participate.

The orbital overlap between oxide ion and strontium is smaller. It is found that the total energy at the middle is much destabilized, due to the weak covalent bonding between oxide ion and strontium. As the result, the activation energy becomes smaller than $La_2Al_4O_3$ model.

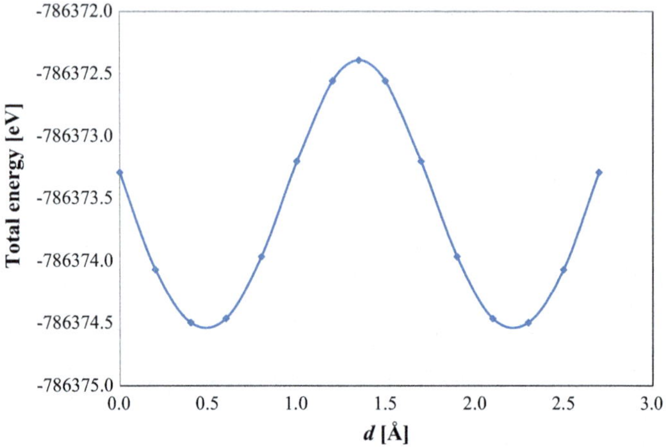

Fig. 14.12 Potential energy curve of $LaPbAl_4O_3$ model, when displacing oxide ion along diagonal path. *d* is oxide ion conduction distance. Reference [2] by permission from Elsevier

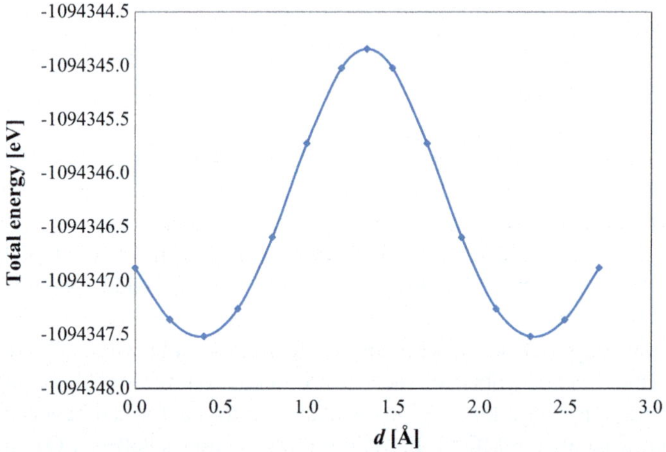

Fig. 14.13 Potential energy curve of $Pb_2Al_4O_3$ model, when displacing oxide ion along diagonal path. *d* is oxide ion conduction distance. Reference [2] by permission from Elsevier

Lead (Pb^{+2}) is another dopant for $LaAlO_3$ perovskite to incorporate oxygen vacancy. Figures 14.12 and 14.13 show the potential energy curves of $LaPbAl_4O_3$ and $Pb_2Al_4O_3$ models, when displacing oxide ion along diagonal path, respectively. Note that one lanthanum cation is replaced by one lead cation in $LaPbAl_4O_3$ model, and two lanthanum cations are replaced by two lead cations in $Pb_2Al_4O_3$ model. In both models, the local maximum is given at the middle of diagonal line, and the local minima are given around 0.4 and 2.3 Å. The highest energy is given at the middle. The activation energies of $LaPbAl_4O_3$ and $Pb_2Al_4O_3$ models for oxide ion conduction are 2.10 and 2.67 eV, respectively. It is found that one lead doping in $La_2Al_4O_3$ decreases the activation energy.

During oxide ion conduction, oxide ion migrates through oxygen vacancy. At the middle of the diagonal path, covalent bonding is formed between oxide ion and counter cations. By the difference of chemical bonding formation with up-and-down counter cations, the activation energy is changeable. For example, when strontium is substituted for lanthanum, the activation energy becomes smaller.

14.3 Proton Conduction in $LaAlO_3$ Perovskite

14.3.1 Introduction of Proton Conduction

It was first observed that $SrCeO_3$ perovskite exhibits high proton conductivity. However, it has disadvantages in structural stability and mechanical strength in practical use. $BaZrO_3$ and $SrTiO_3$ perovskites were considered as next candidate material. It was shown that the activation energy for proton conduction in $BaZrO_3$ perovskite is 2.42 eV. In $SrTiO_3$ perovskite, as titanium 3d orbital is related to chemical bonding formation during proton conduction, it was considered that the stable proton conduction is not expected.

To decrease activation energy for proton conduction, in comparison with $BaZrO_3$ perovskite, $LaAlO_3$ perovskite was proposed. To incorporate proton into $LaAlO_3$ perovskite, dopants are introduced. There are three methods to introduce proton (positive hydrogen atom). One is the direct insertion during synthesis. When divalent counter cation is doped at lanthanum site, one proton is incorporated, due to charge compensation. Proton exists as a part of OH^-. Second is dissolution of H_2 gas. Using Kröger–Vink notation, the formation of OH defect is expressed as,

$$H_2 + O''_O \rightarrow 2OH'_O + 2e' \quad (14.26)$$

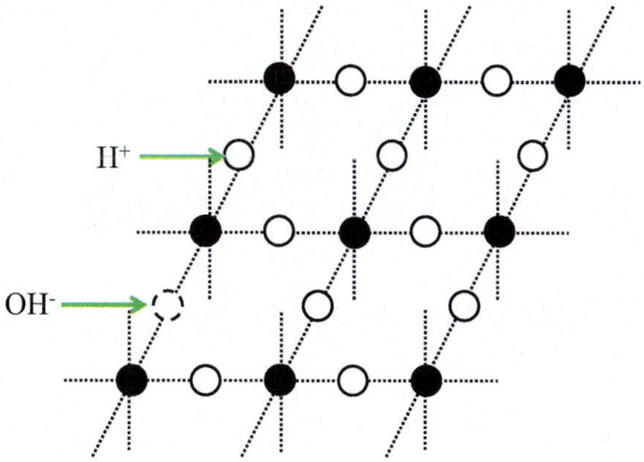

Fig. 14.14 Schematic drawing of proton incorporation on AlO_2 layer of $LaAlO_3$ perovskite, under wet condition. *Black*, *white* and *dotted circles* denote aluminium, oxygen, and oxygen vacancy, respectively

Third is dissolution of water molecule under wet condition (see Fig. 14.14). As same as oxide ion conductor, oxygen vacancy is required as defect. The formation of OH defect is expressed as, in the same manner,

$$H_2O + V_O^{\times} + O_O'' \rightarrow 2OH_O' \tag{14.27}$$

In this case, charge compensation is kept, when OH^- and H^+ are coincidentally incorporated on surface. Note that Kröger–Vink notation is based on the formal charge here.

14.3.2 Proton Conduction Path

In order to take the effect of chemical bonding formation between conducting proton and up-and down counter cations, $La_2Al_4O_4H$ model is constructed (see Fig. 14.15). Figure 14.16 depicts three proton conduction paths within Al_4O_4 square. In A and B paths, proton migrates towards nearest neighbouring and next-nearest neighbouring oxygen, respectively. C path is considered to investigate whether proton directly crosses Al_4O_4 square or not. BHHLYP calculation is performed for $La_2Al_4O_4H$ model. Basis set used for aluminium and oxygen is 6-31G*, combined with MINI(3.3.3.3.3.3/3.3.3.3/3.3/4) for lanthanum.

Figure 14.17 shows the potential energy curves of $La_2Al_4O_4H$ model in A, B and C paths. Figure 14.18 depicts the molecular orbitals of $La_2Al_4O_4H$ model at the local maximum and minimum in A path, and at the local minimum in B path.

In A path, the local maximum is given at the middle, and the local minima are given around 0.95 and 1.75 Å. The activation energy for proton conduction is 0.74 eV. At the local minimum, the obtained wave-functions of MO73, MO83 and MO84 are

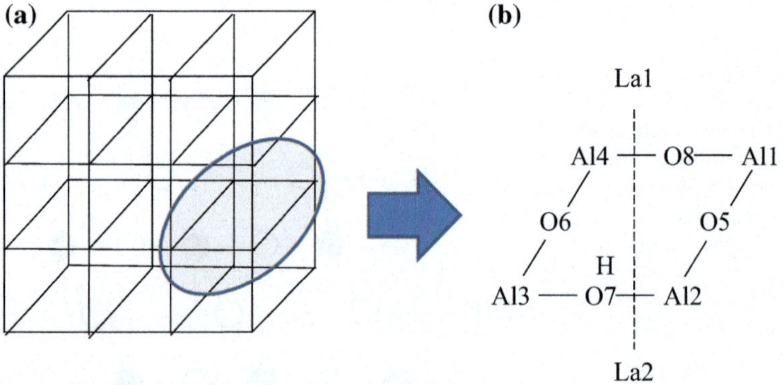

Fig. 14.15 **a** Crystal structure of $LaAlO_3$ perovskite, **b** $La_2Al_4O_4H$ model. Reference [3] by permission from Elsevier

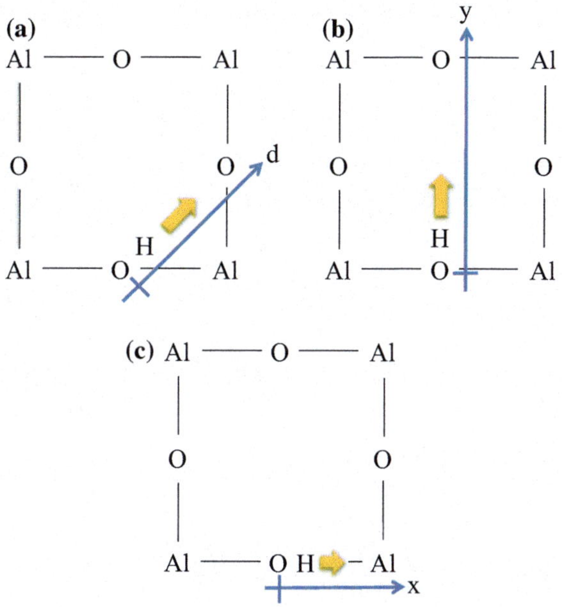

Fig. 14.16 Three proton conduction paths within Al_4O_4 square of $LaAlO_3$ perovskite: **a** A path, **b** B path, **c** C path. Reference [3] by permission from Elsevier

$$\psi_{MO73(min)} = 0.19\phi_{H(1s')} - 0.20\phi_{O7(1s)} + 0.42\phi_{O7(2s')} + 0.47\phi_{O7(2s'')} \quad (14.28)$$

$$\begin{aligned}\psi_{MO83(min)} = {} & 0.27\phi_{H(1s')} + 0.26\phi_{H(1s'')} - 0.13\phi_{O5(2s'')} \\ & - 0.17\phi_{O7(2s')} - 0.39\phi_{O7(2s'')} + 0.40\phi_{O7(2px')} + 0.26\psi_{O7(2px'')} + 0.17\psi_{O7(2py')}\end{aligned} \quad (14.29)$$

$$\begin{aligned}\psi_{MO84(min)} = {} & 0.10\phi_{H(1s')} + 0.37\phi_{H(1s'')} \\ & - 0.10\phi_{O5(2s')} - 0.23\phi_{O5(2s'')} \\ & - 0.26\phi_{O7(2px')} - 0.27\phi_{O7(2px'')} + 0.48\phi_{O7(2py')} + 0.30\phi_{O7(2py'')}\end{aligned} \quad (14.30)$$

In MO73, there is orbital overlap between H 1s and O7 2s orbitals. One hydrogen lobe interacts with one oxygen lobe. From chemical bonding rule, it is found that hydrogen atom forms σ-type covalent bonding with oxygen atom (OH bond). In MO83 and MO84, there are orbital overlaps between H1s and O7 2p orbitals, though O5 2s orbital also participates. One hydrogen lobe interacts with one oxygen lobe. From chemical bonding rule, it is found that hydrogen atom forms σ-type covalent bonding with oxygen atom. Mulliken charge density of hydrogen atom is 0.10. It implies that hydrogen atom exists as not proton but almost neutral hydrogen at the local minimum. It is concluded that hydrogen atom is regarded as a part of OH at the local minimum. At the local maximum, the obtained wave-functions of MO73 and MO83 are

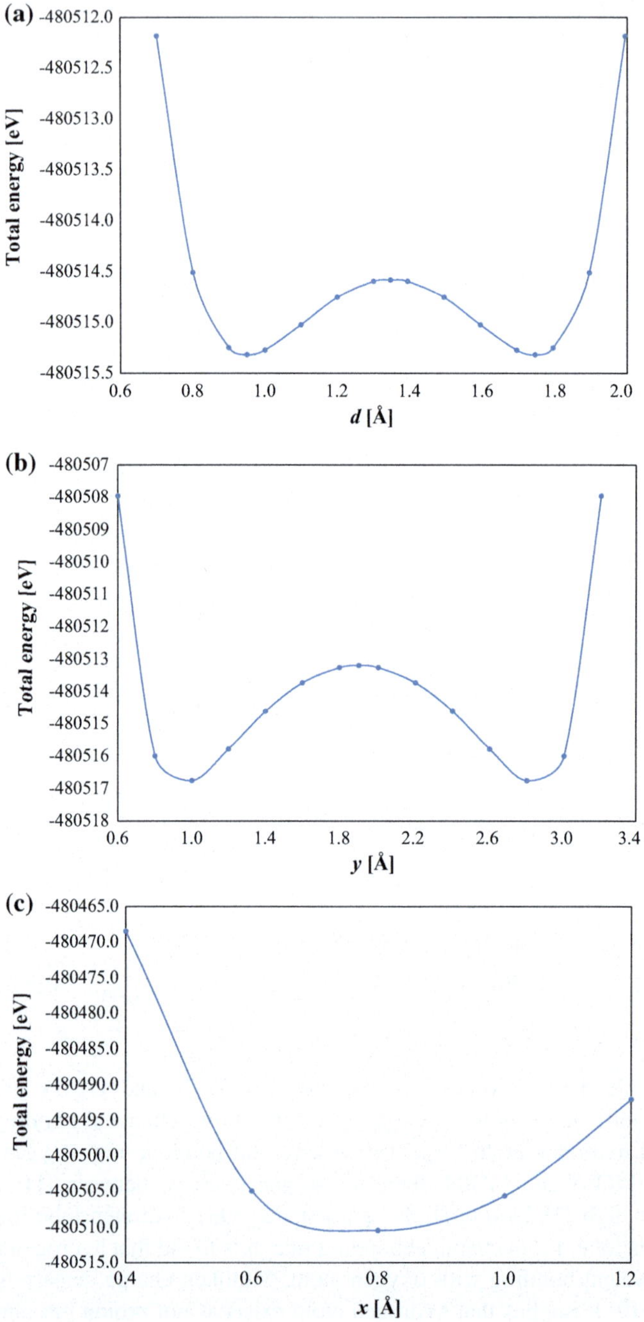

Fig. 14.17 Potential energy curves of $La_2Al_4O_4H$ model: **a** A, **b** B and **c** C paths. Reference [3] by permission from Elsevier

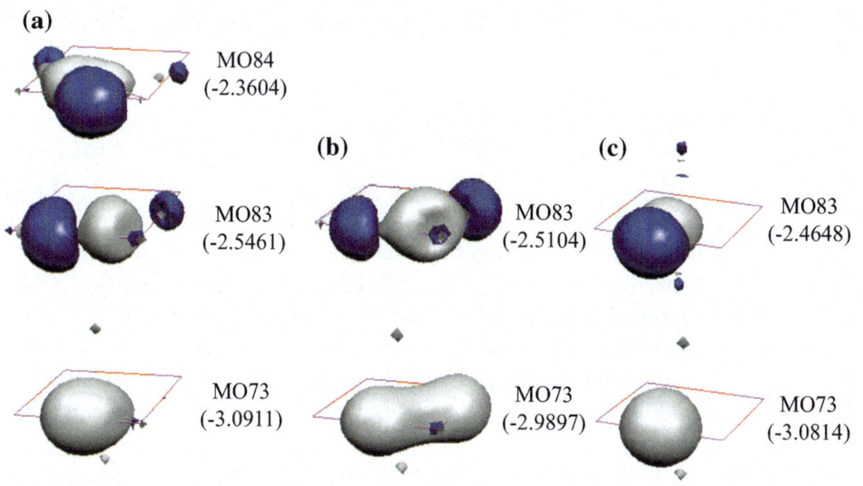

Fig. 14.18 Selected molecular orbitals of $La_2Al_4O_4H$ model: **a** local minimum in A path, **b** local maximum in A path, **c** local minimum in B path. Orbital energy is given in *parenthesis*. Reference [3] by permission from Elsevier

$$\begin{aligned}\psi_{\mathrm{MO73(max)}} &= 0.13\phi_{\mathrm{H(1s')}} - 0.14\phi_{\mathrm{O5(1s)}} + 0.30\phi_{\mathrm{O5(2s')}} + 0.33\phi_{\mathrm{O5(2s'')}} \\ &\quad - 0.14\phi_{\mathrm{O7(1s)}} + 0.30\phi_{\mathrm{O7(2s')}} + 0.33\phi_{\mathrm{O7(2s'')}}\end{aligned} \tag{14.31}$$

$$\begin{aligned}\psi_{\mathrm{MO83(max)}} &= 0.27\phi_{\mathrm{H(1s')}} + 0.37\phi_{\mathrm{H(1s'')}} \\ &\quad - 0.12\phi_{\mathrm{O5(2s')}} - 0.26\phi_{\mathrm{O5(2s'')}} - 0.10\phi_{\mathrm{O5(2px')}} - 0.26\phi_{\mathrm{O5(2py')}} - 0.16\phi_{\mathrm{O5(2py'')}} \\ &\quad - 0.12\phi_{\mathrm{O7(2s')}} - 0.26\phi_{\mathrm{O7(2s'')}} + 0.26\phi_{\mathrm{O7(2px')}} + 0.16\phi_{\mathrm{O7(2px'')}} + 0.10\phi_{\mathrm{O7(2py')}}\end{aligned} \tag{14.32}$$

In MO73, there are orbital overlaps between H 1s and O5 2s orbitals, and between H 1s and O7 2s orbitals. One hydrogen lobe interacts with one oxygen lobe. From chemical bonding rule, it is found that hydrogen atom forms σ-type covalent bonding with two oxygen atoms (OHO bond). In MO83, there are nodes between H and O5, and between H and O7. MO83 is inversion σ-type covalent bonding to MO73. Mulliken charge density of hydrogen atom is 0.00. It implies that hydrogen atom exists as not proton but neutral hydrogen at the local maximum. Finally, it is concluded that OH and OHO covalent bonds are alternately formed during proton conduction in A path.

In B path, the local maximum is given at the centre, and the local minima are given around 1.0 and 2.8 Å. As the activation energy for proton conduction in B path (3.56 eV) is much larger than A path, it is found that A path is dominative. At the local minimum, the obtained wave-functions of MO73 and MO83 are

$$\psi_{\mathrm{MO73(min)}} = 0.16\phi_{\mathrm{H(1s')}} - 0.20\phi_{\mathrm{O7(1s)}} + 0.43\phi_{\mathrm{O7(2s')}} + 0.49\phi_{\mathrm{O7(2s'')}} \quad (14.33)$$

$$\psi_{\mathrm{MO83(min)}} = 0.28\phi_{\mathrm{H(1s')}} + 0.39\phi_{\mathrm{H(1s'')}} - 0.17\phi_{\mathrm{O7(2s')}} - 0.38\phi_{\mathrm{O7(2s'')}} + 0.48\phi_{\mathrm{O7(2py')}} + 0.24\phi_{\mathrm{O7(2py'')}} \quad (14.34)$$

In MO73, there is orbital overlap between H 1s and O7 2s orbitals. In MO83, there are orbital overlaps between H 1s and O7 2s orbitals, and between H 1s and O7 2p orbitals. From chemical bonding rule, it is found that hydrogen atom forms σ-type covalent bonding with oxygen atom (OH bond). Mulliken charge density of hydrogen atom is 0.04. It implies that hydrogen atom exists as not proton but almost neutral hydrogen. Finally, it is concluded that hydrogen atom exists as a part of OH at the local minimum.

In comparison with total energies of three local minima (see Table 14.1), the total energy at the local minimum in C path is 4.15 eV larger than the local minimum in A path. It is found that the direct proton conduction through Al–O–Al bond is impossible. Figure 14.19 depicts the schematic drawing of proton conduction within Al_4O_4 square. Before starting proton conduction, hydrogen exists as a part of OH bond, which is towards a square centre. When proton conduction starts, hydrogen starts to be rotated around oxygen atom, keeping OH bond. When hydrogen reaches the local minimum in diagonal line, hydrogen migrates towards to next-nearest neighbouring oxygen, changing covalent bonding (OH and OHO bonds).

Table 14.1 Total energies at three local minima

Proton conduction path	Total energy (eV)
A	−480515.31
B	−480516.75
C	−480510.43

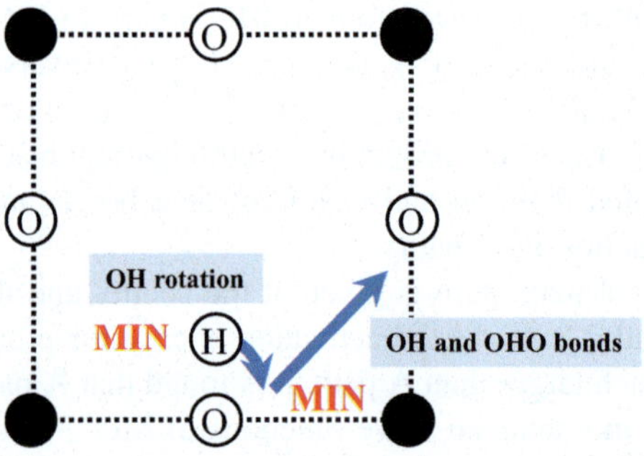

Fig. 14.19 Schematic drawing of proton conduction within Al_4O_4 square. *Black circle* denotes aluminium

Figure 14.20 depicts alternative three-dimensional proton conduction path (D path). In D path, hydrogen is rotated, connecting three different local minima on three different Al_4O_4 squares. Note that hydrogen is three-dimensionally rotated around O5, keeping OH bond. Figure 14.21 shows the potential energy curve of $La_2Al_4O_4H$ model in D path. Hydrogen migration between two local minima is

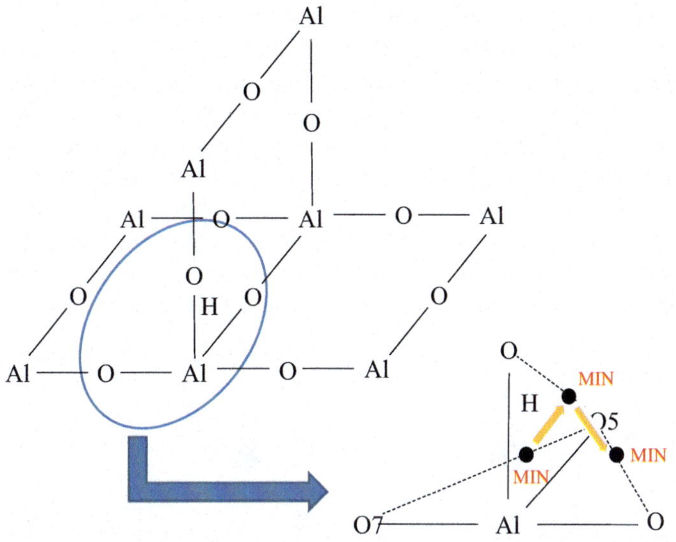

Fig. 14.20 Three-dimensional proton conduction path, crossing Al_4O_4 square (D path). Reference [3] by permission from Elsevier

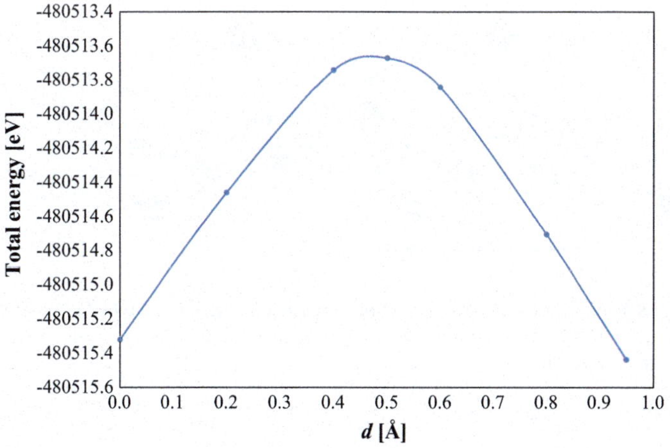

Fig. 14.21 Potential energy curve of $La_2Al_4O_4H$ model in D path. Reference [3] by permission from Elsevier

only shown. It is because the same potential energy curve is given, due to the symmetry. The activation energy of D path is 1.65 eV.

Figure 14.22 depicts the schematic drawing of the whole proton conduction path in $LaAlO_3$ perovskite. Before starting proton conduction, hydrogen is located at the most stable position within Al_4O_4 square, where hydrogen of OH bond is located towards square centre. Hydrogen migrates through OH rotation until the local minimum in diagonal line. Then, hydrogen migrates towards next-nearest neighbouring oxygen site, forming OHO and OH bonds alternately. When crossing Al_4O_4 square, hydrogen migrates through three-dimensional OH rotation, connecting three local minima in the diagonal line. In order to estimate the activation energy for proton conduction in $LaAlO_3$ perovskite, all proton conduction paths within and crossing Al_4O_4 square must be coincidentally considered. Figure 14.23 shows the whole potential energy curve of $La_2Al_4O_4H$ model. Note that proton conduction distance is projected in x axis. The activation energy is estimated to be 2.17 eV, from the total energy difference between minimum (0.0 Å) and local maximum (0.95 Å). Once proton conduction starts, hydrogen can pass the next energy barrier (1.65 eV) through three-dimensional OH rotation.

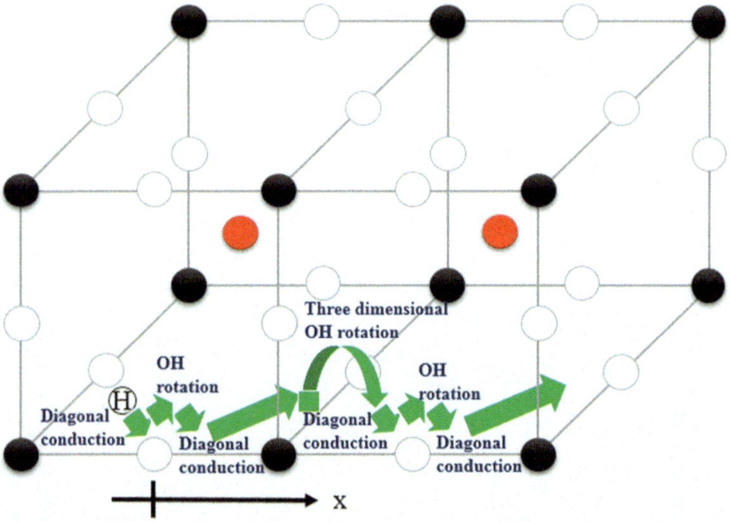

Fig. 14.22 Schematic drawing of the whole proton conduction in $LaAlO_3$ perovskite. *Black*, *white*, and *red circles* denote aluminium, oxygen and lanthanum, respectively. Reference [3] by permission from Elsevier

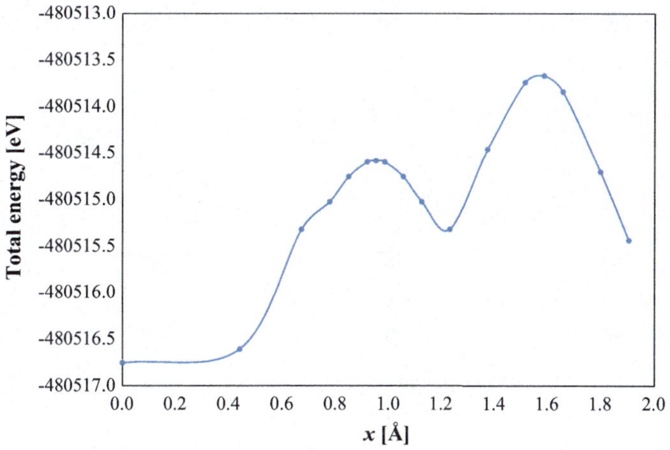

Fig. 14.23 Whole potential energy curve of $La_2Al_4O_4H$ model. The proton conduction distance is projected in *x* axis. Reference [3] by permission from Elsevier

14.3.3 Proton Pumping Effect

Before starting proton conduction, proton exists as a part of OH, at oxygen lattice position. It was theoretically proposed that proton is pumped into the inside of Al_4O_4 square, combined with OH conduction. The effect is called "proton pumping effect". Figure 14.24 depicts the schematic drawing of proton pumping effect. When H_2 molecule is dissolved into $LaAlO_3$ perovskite, proton forms covalent bonding with oxygen. After three-dimensional OH rotation, proton is pumped towards the centre of Al_4O_4 square through OH conduction. Then proton conduction occurs within Al_4O_4 square. When proton reaches second nearest neighbouring oxygen, proton pumping occurs after crossing Al_4O_4 square. On the other hand, when water molecule is dissolved into $LaAlO_3$ perovskite under wet condition, OH and proton are directly captured at oxygen vacancy and oxygen, respectively. In the former case, without three-dimensional OH rotation, proton is pumped towards the centre of Al_4O_4 square through OH conduction. In the latter case, the process is as same as H_2 gas.

Let us confirm proton pumping effect in $LaAlO_3$ perovskite. Figure 14.25 shows the potential energy curve of $La_2Al_4O_4H$ model, when displacing OH towards the centre of Al_4O_4 square. The lattice position of oxygen is defined as origin (see Fig. 14.26). The local minimum is given around 0.1 Å. It implies that proton pumping occurs. Figure 14.27 shows the potential energy curve of $La_2Al_4O_4H$ model, under consideration of proton pumping effect. The activation energy for proton conduction is 1.31 eV, which is much smaller in comparison with no proton pumping (2.17 eV).

It is concluded that high proton conductivity comes from proton pumping effect. Even if no local minimum is given during OH conduction, there may be energetic

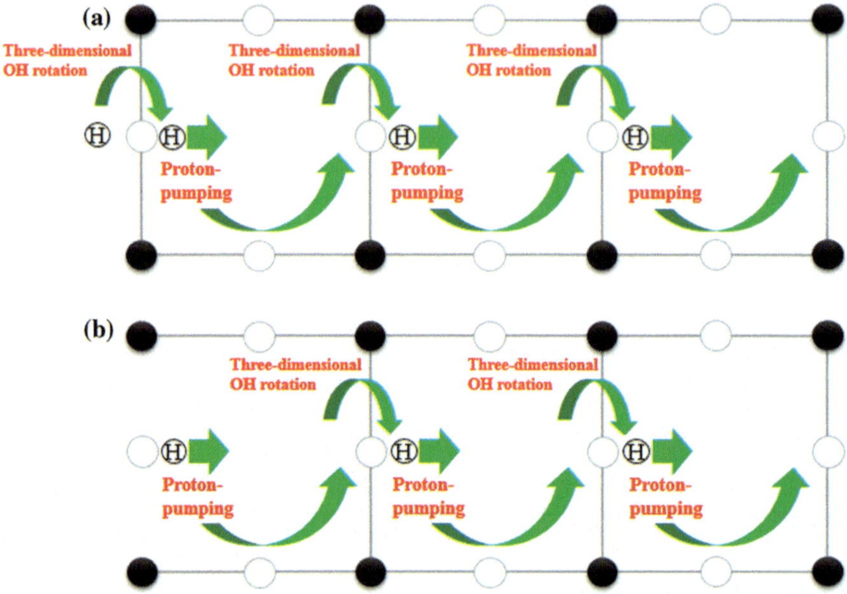

Fig. 14.24 Schematic drawing of proton pumping effect in $LaAlO_3$ perovskite: **a** H_2 gas, **b** water molecule under wet condition. *Black* and *white circles* denote aluminium and oxygen, respectively. Reference [3] by permission from Elsevier

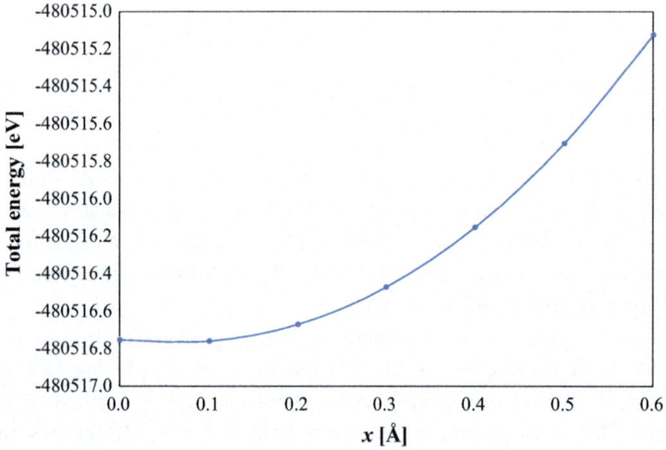

Fig. 14.25 Potential energy curve of $La_2Al_4O_4H$ model, when displacing OH towards the centre of Al_4O_4 square. Reference [3] by permission from Elsevier

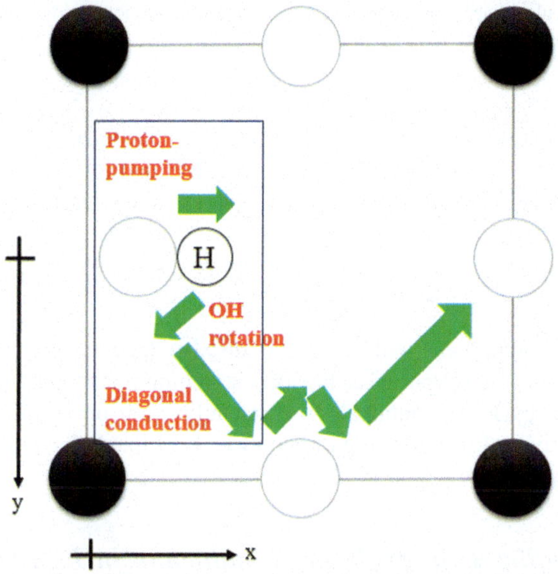

Fig. 14.26 Schematic drawing of proton conduction after proton pumping in $LaAlO_3$ perovskite. Reference [3] by permission from Elsevier

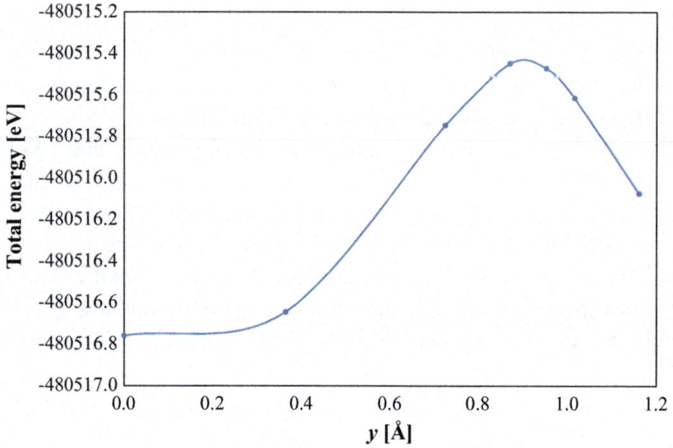

Fig. 14.27 Potential energy curve of $La_2Al_4O_4H$ model, under consideration of proton pumping effect. Reference [3] by permission from Elsevier

Table 14.2 Calculated activation energies for proton and oxide ion conductions in undoped and doped $LaAlO_3$ perovskite (BHHLYP method)

	Undoped case	Sr-doping	Pb-doping	Zn-doping	Pb and Zn co-doping
Proton	$La_2Al_4O_4H$	$LaSrAl_4O_4H$	$LaPbAl_4O_4H$	$La_2Al_3ZnO_4H$	$LaPbAl_3ZnO_4H$
	1.31 eV	1.57 eV	1.78 eV	0.91 eV	1.32 eV
Oxide ion	$La_2Al_4O_3$	$LaSrAl_4O_3$	$LaPbAl_4O_3$	$La_2Al_3ZnO_3$	$LaPbAl_3ZnO_3$
	2.73 eV	2.29 eV	2.10 eV	3.57 eV	4.99 eV

advantage, in comparison with energetic disadvantage of no proton pumping. Proton pumping can be regarded as the local structural relaxation, due to strong OH covalency. However, conventional structural relaxation on surface may suppress proton conductivity. It is because proton may be kept fixed, due to the stabilization.

14.3.4 Conflict with Oxide Ion Conduction in $LaAlO_3$ Perovskite

In $LaAlO_3$ perovskite, oxide ion migrates through oxygen vacancy. When oxygen vacancy exists, the conflict with oxide ion conduction must be taken into consideration. For example, when positively charged proton migrates from left electrode to right electrode through solid electrolyte, negatively charged oxide ion can migrate from right electrode to left electrode through oxygen vacancy (see Fig. 14.1). It implies the coincident oxide ion conduction.

Table 14.2 summarizes the activation energies for proton and oxide ion conductions in $LaAlO_3$ perovskite. In undoped case ($La_2Al_4O_4H$ and $La_2Al_4O_3$ models), the activation energies for proton and oxide ion conductions are 1.31 and 2.73 eV, respectively. When keeping lower operation temperature, only oxide ion can be prevented. As the strategy of materials design, the smaller and larger activation energies are favourable for proton and oxide ion conductions, respectively. For example, Pd and Zn co-doped $LaAlO_3$ perovskite is one of the best candidates.

14.4 Comparison with AC Impedance Measurement

In AC impedance measurement, the real part, which expresses electric resistance, is generally divided into three contributions such as bulk, grain boundary and electrode interface. In experimental analysis, electric resistance is constant in bulk part. However, in real oxide ion and proton conductors, it is changeable, due to change of covalent bonding. In proton conducting $BaZrO_3$ perovskite, BHHLYP calculation predicts that the activation energy for proton conduction is 2.42 eV

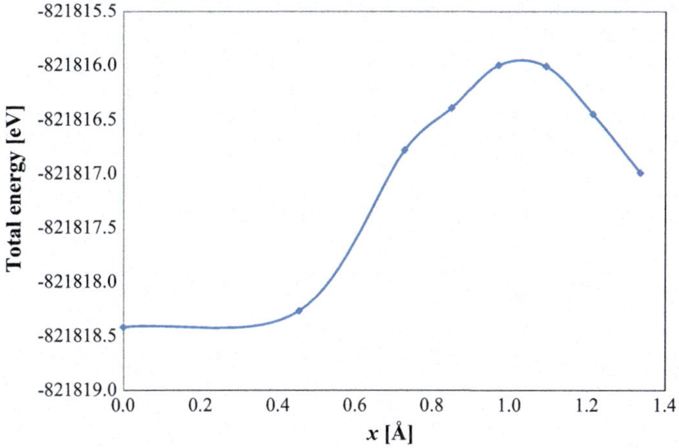

Fig. 14.28 Potential energy curve of $Ba_2Zr_4O_4H$ model. The proton conduction distance is projected in the same manner (*x* axis)

under consideration of proton pumping effect (see Fig. 14.28), though experimental value is 0.44–0.49 eV. The large mismatch comes from neglecting a change of covalency. On the other hand, as lithium and sodium ions form no covalent bonding with other atoms, there is no mismatch between theoretical and experimental values.

References

1. Onishi T (2010) Int J Quant Chem 110:2912–2917
2. Onishi T (2012) Adv Quant Chem 64:59–70
3. Onishi T (2015) Adv Quant Chem 70:31–67

Further Readings

4. Ishihara T, Matsuda H, Takita Y (1994) J Am Chem Soc 116:3801–3803
5. Howard CJ, Kennedy BJ, Chakoumakos BC (2000) J Phys Condens Mat 12:349–365
6. Onishi T, Helgaker T (2012) Int J Quant Chem 112:201–207
7. Onishi T, Helgaker T (2013) Int J Quant Chem 113:599–604
8. Iwahara H, Esaka H, Uchida H, Maeda N (1981) Solid State Ionics 3(4):p359–p363
9. Iwahara H, Yajima T, Hibino T, Ozaki K, Suzuki H (1993) Solid State Ionics 61:p65–p69
10. Schmidt MW, Baldridge KK, Boatz JA, Elbert ST, Gordon MS, Jensen JH, Koseki S, Matsunaga N, Nguyen KA, Su S, Windus TL, Dupuis M, Montgomery JA (1993) J Comput Chem 14:1347–1363
11. Varetto U <MOLEKEL 4.3.>; Swiss National Supercomputing Centre. Manno, Switzerland
12. Huzinaga S, Andzelm J, Radzio-Andzelm E, Sakai Y, Tatewaki H, Klobukowski M (1984) Gaussian basis sets for molecular calculations. Elsevier, Amsterdam

13. Hariharan PC, Pople JA (1973) Theoret Chim Acta 28:213–222
14. Francl MM, Pietro WJ, Hehre WJ, Binkley JS, Gordon MS, DeFrees DJ, Pople JA (1982) J Chem Phys 77(7):3654–3665
15. Rassolov VA, Pople JA, Ratner MA, Windus TL (1998) J Chem Phys 109(4):1223–1229
16. Rassolov VA, Ratner MA, Pople JA, Redfern PC, Curtiss LA (2001) J Comput Chem 22 (9):976–984

Part V
Helium Chemistry and Future

Chapter 15
Helium Chemistry

Abstract Helium is getting rare resource to support recent advanced industry. The electronic structure of helium is introduced, in comparison with hydrogen. In helium dimer, it has been recognized that helium atoms are weakly bound through interatomic interaction, due to zero bond order. In coupled cluster calculation for helium dimer, there is orbital overlap between helium 1s orbitals. One helium lobe interacts with one helium lobe. From chemical bonding rule, it is found that σ-type covalent bonding is formed. Bond order cannot determine whether diatomic molecule is dispersed or not. In He–H system, three formal charges of hydrogen are considered. In He–H$^+$, covalent bonding is formed between helium 1s and hydrogen 1s orbitals. It implies that one electron is shared by helium and hydrogen. However, in He–H and He–H$^-$, no orbital overlap is observed between helium and hydrogen.

Keywords Helium · Hydrogen · Covalent bonding · Bond order · Chemical bonding rule

15.1 Introduction of Helium

The atomic number of helium is 2, and the element symbol is "He". It is known that helium, neon, argon, krypton, xenon and radon are categorized as noble gas, and have closed shell electronic structure. At normal temperature, helium exists as colourless, odourless, non-toxic inert gas. Table 15.1 shows the boiling point of several molecules under normal pressure. The boiling point of helium is 4.2 K, which is much smaller than other molecules.

Helium has been widely utilized as industrial gas, and has been indispensable in manufacturing process of optical fibre and semi-conductor, and coolant for superconductors. Helium is getting to be recognized as rare resource to support recent advanced industry. Industrial production of helium is performed by separation and purification from underground natural gas, since helium scarcely exists in the air.

T. Onishi, *Quantum Computational Chemistry*,
DOI 10.1007/978-981-10-5933-9_15

Table 15.1 Boiling point of several molecules under normal pressure

H_2O	O_2	N_2	H_2	He
373 K	90 K	77 K	20 K	4 K

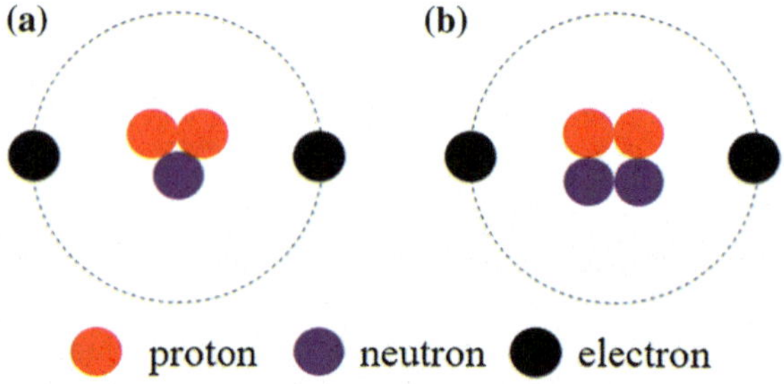

Fig. 15.1 Schematic drawings of particle structures for **a** helium 3 and **b** helium 4

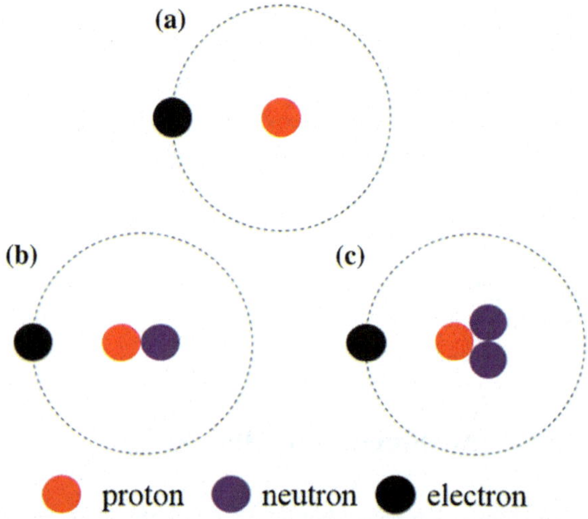

Fig. 15.2 Schematic drawings of particle structures of **a** hydrogen (H), **b** deuterium (D) and tritium (T)

Figure 15.1 depicts the schematic drawings of particle structures of helium 3 and helium 4. In both helium, two electrons occupy 1s orbital. In nature, helium exists as helium 4, where two protons and two neutrons exist within atomic nucleus. Helium 3, where two protons and one neutron in atomic nucleus, is the stable isotope of helium 4.

Figure 15.2 depicts the schematic drawings of particle structures of hydrogen, deuterium and tritium. Hydrogen has only one proton within atomic nucleus, though deuterium and tritium have one and two neutrons, respectively.

As shown in Fig. 15.3, helium 4 is synthesized by nuclear fusion reaction between deuterium and tritium (D–T reaction).

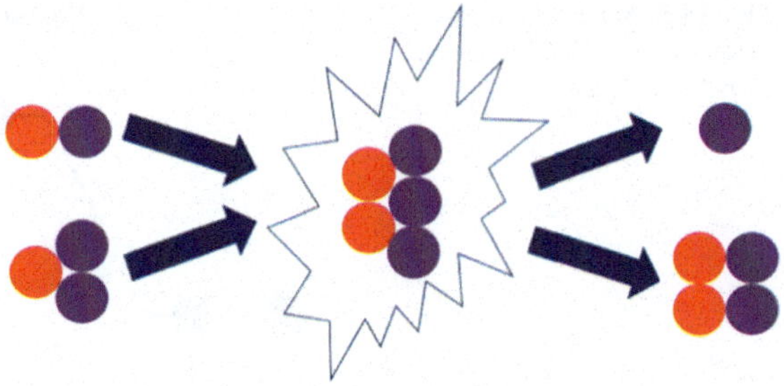

Fig. 15.3 Schematic drawing of D–T reaction. Atomic nucleus is only shown

Fig. 15.4 Schematic drawing of D–D reaction. Atomic nucleus is only shown

$$D + T \rightarrow {}^{4}He + n \tag{15.1}$$

In the reaction, neutron is created, combined with emission of larger energy.

On the other hand, as shown in Fig. 15.4, helium 3 is synthesized by nuclear fusion reaction between deuteriums (D–D reaction).

$$D + D \rightarrow {}^{3}He + n \tag{15.2}$$

The neutron of hydrogen has an important role in the nuclear fusion reaction of helium.

15.2 Helium Dimer

In general, it has been recognized that helium atoms are weakly bound through interatomic interaction, due to the closed shell structure. Let us reconsider chemical bonding between helium atoms, based on molecular orbital (MO) theory. Following empirical manner, it is assumed that two helium 1s orbitals form MO1 and corresponding inversion MO2, as shown in Fig. 15.5. Note that molecular orbital calculation is not performed at this stage.

Bond order (N) is known as an empirical index to judge covalency in diatomic molecule. It is defined by

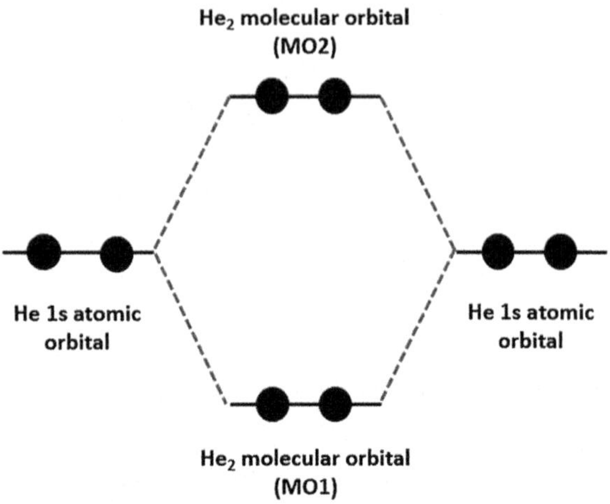

Fig. 15.5 Schematic drawing of molecular orbital diagram of helium dimer

$$N = (N_a - N_b)/2 \tag{15.3}$$

where N_a and N_b denote the numbers of electrons in covalent and corresponding inversion MOs, respectively. In fact, covalent character depends on both orbital overlap pattern and magnitude. In helium dimer, bond order becomes zero, due to $N_a = N_b = 2$. It has been often recognized that helium dimer is dispersed, as bond order is zero. It implies that helium atoms are isolated and there is no orbital overlap between helium 1s orbitals.

Let us discuss again whether helium dimer is dispersed or not. The site numbers of left and right helium atoms are defined as He1 and He2, respectively. CCSD/aug-cc-pVTZ calculation is performed for helium dimer (He1–He2). At geometry optimized structure, He1–He2 distance is 3.04 Å. It is much larger than H–H distance. Four electrons occupy MO1 and MO2. The obtained wave-function of MO1 is

$$\begin{aligned}\psi_{\mathrm{He_2(MO1)}} &= 0.25\phi_{\mathrm{He1(1s')}} + 0.34\phi_{\mathrm{He1(1s'')}} + 0.21\phi_{\mathrm{He1(1s''')}} \\ &\quad + 0.25\phi_{\mathrm{He2(1s')}} + 0.34\phi_{\mathrm{He2(1s'')}} + 0.21\phi_{\mathrm{He2(1s''')}}\end{aligned} \tag{15.4}$$

There is orbital overlap between He1 and He2 1s orbitals. One He1 lobe interacts with one He2 lobe. From chemical bonding rule, it is found that σ-type covalent bonding is formed. The obtained wave-function of MO2 is

$$\begin{aligned}\psi_{\mathrm{He_2(MO2)}} &= 0.25\phi_{\mathrm{He1(1s')}} + 0.34\phi_{\mathrm{He1(1s'')}} + 0.21\phi_{\mathrm{He1(1s''')}} \\ &\quad - 0.25\phi_{\mathrm{He2(1s')}} - 0.34\phi_{\mathrm{He2(1s'')}} - 0.21\phi_{\mathrm{He2(1s''')}}\end{aligned} \tag{15.5}$$

MO2 is corresponding inversion σ-type covalent bonding, due to the different signs of He1 and He2 1s coefficients. It is found that covalent bonding is formed in helium dimer in spite of zero bond order. Finally, it is concluded that bond order

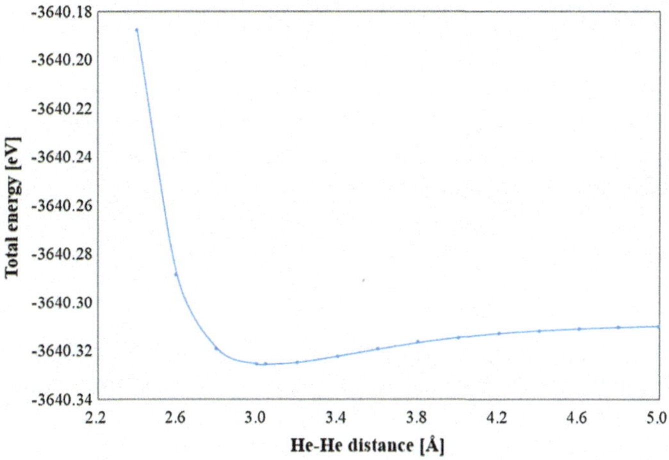

Fig. 15.6 Potential energy curve of helium dimer, changing He–He distance

cannot judge whether covalent bonding is formed or not in diatomic molecule. Note that there is no orbital overlap between them, if two atoms are completely dispersed.

Figure 15.6 shows the potential energy curve of helium dimer, changing He1–He2 distance. The bond dissociation energy can be estimated from the total energy difference between the local minimum and completely dissociated point.

$$E_{\text{dissociation}}(He_2) = E(He) + E(He) - E(He_2) \tag{15.6}$$

The value is 0.017 kcal/mol. It is found that helium dimer can be formed only at very low temperature. At room temperature, the larger energy is given as kinetic energy, in comparison with the bond dissociation energy. Helium dimer will be dispersed.

As the zero point vibration energy (0.038 kcal/mol) is larger than the bond dissociation energy, the effect of quantum vibration is negligible. It is considered that solid state helium aggregation is difficult, due to the quantum fluctuation.

15.3 Helium and Hydrogen

In universe, the abundance ratios of helium and hydrogen are larger than other atoms. It can be expected that helium is interacted with hydrogen at very low temperature or under extreme environment. CCSD/aug-cc-pVTZ calculation is performed for He–H model. Here, we consider three types of hydrogen charges such as positive (+1), neutral (0) and negative (−1).

15.3.1 He–H⁺

Figure 15.7 shows the potential energy curves of He–H^+ model, changing He–H distance. Local minimum is given at 0.776 Å. Mulliken charge densities of helium and hydrogen are 0.315 and 0.685, respectively. Two electrons occupy one MO1. The obtained wave-function of MO1 is

$$\psi_{\mathrm{HeH^+(MO1)}} = 0.13\phi_{\mathrm{H(1s')}} + 0.12\phi_{\mathrm{H(1s'')}} + 0.35\phi_{\mathrm{He(1s')}} + 0.45\phi_{\mathrm{He(1s'')}} + 0.16\phi_{\mathrm{He(1s''')}} \tag{15.7}$$

There is orbital overlap between He and H 1s orbitals. One He lobe interacts with one H lobe. From chemical bonding rule, it is found that σ-type covalent bonding is formed.

Figure 15.8 depicts the schematic drawing of the relationship between atomic orbitals and molecular orbital in He–H^+. As the formal charge of H^+ is +1, no electron exist in atomic orbital of H^+. However, one electron is shared by both helium and hydrogen in He–H^+.

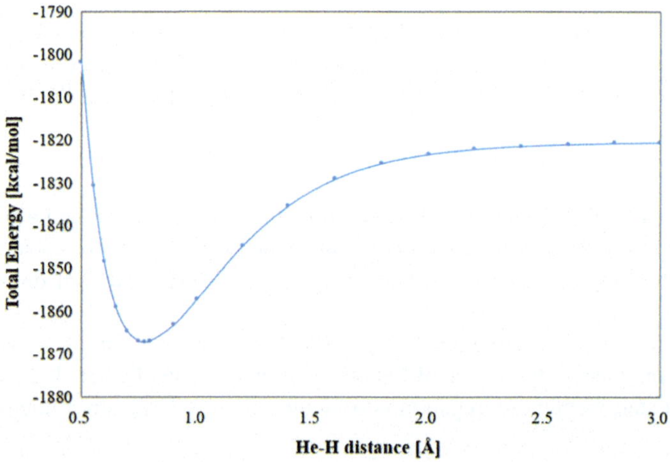

Fig. 15.7 Potential energy curve of He–H^+, changing He–H distance

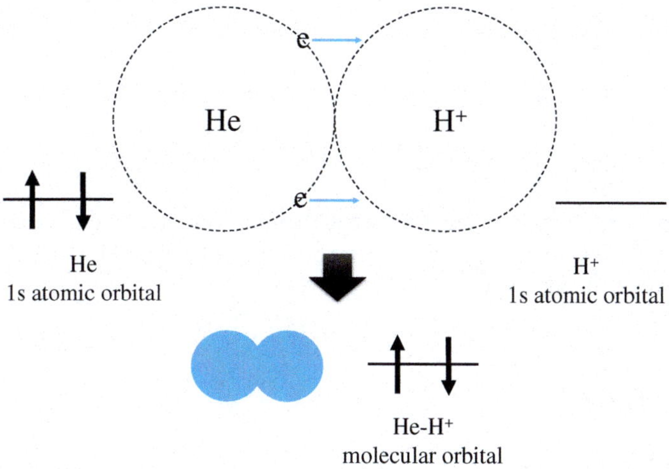

Fig. 15.8 Schematic drawing of the relationship between atomic orbitals and molecular orbital in He–H^+

15.3.2 He–H

Figure 15.9 shows the potential energy curves of He–H, changing He–H distance. Local minimum is given at 3.58 Å. Spin densities of helium and hydrogen are 0.00 and 1.00, respectively. There electrons occupy MO1α, MO1β and MO2α. The obtained wave-functions of MO1α, MO1β and MO2α are

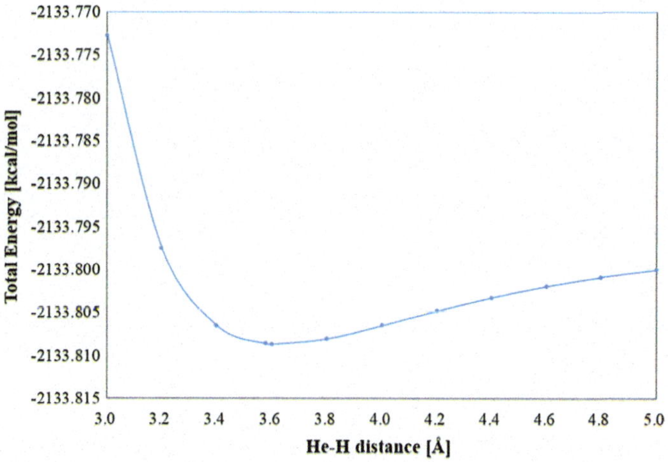

Fig. 15.9 Potential energy curve of He–H, changing He–H distance

$$\psi_{\mathrm{HeH(MO1\alpha)}} = 0.35\phi_{\mathrm{He(1s')}} + 0.48\phi_{\mathrm{He(1s'')}} + 0.30\phi_{\mathrm{He(1s''')}} \tag{15.8}$$

$$\psi_{\mathrm{HeH(MO1\beta)}} = 0.35\phi_{\mathrm{He(1s')}} + 0.48\phi_{\mathrm{He(1s'')}} + 0.30\phi_{\mathrm{He(1s''')}} \tag{15.9}$$

$$\psi_{\mathrm{HeH(MO2\alpha)}} = 0.24\phi_{\mathrm{H(1s')}} + 0.51\phi_{\mathrm{H(1s'')}} + 0.38\phi_{\mathrm{H(1s''')}} \tag{15.10}$$

There is no orbital overlap between helium and hydrogen. From chemical bonding rule, it is found that no covalent bonding is formed. MO1α and MO1β are paired. MO2α is unpaired.

15.3.3 He–H$^-$

Figure 15.10 shows the potential energy curves of He–H$^-$, changing He–H distance. Local minimum is given at 6.45 Å. Four electrons occupy MO1 and MO2. The obtained wave-functions of MO1 and MO2 are

$$\psi_{\psi_{\mathrm{HeH^-(MO1)}}} = 0.35\phi_{\mathrm{He(1s')}} + 0.48\phi_{\mathrm{He(1s'')}} + 0.30\phi_{\mathrm{He(1s''')}} \tag{15.11}$$

$$\psi_{\psi_{\mathrm{HeH^-(MO2)}}} = 0.16\phi_{\mathrm{H(1s')}} + 0.27\phi_{\mathrm{H(1s'')}} + 0.41\phi_{\mathrm{H(1s''')}} + 0.37\phi_{\mathrm{H(2s')}} \tag{15.12}$$

As same as He–H, there is no orbital overlap between helium and hydrogen. From chemical bonding rule, it is found that no covalent bonding is formed. He–H$^-$ distance is larger than He–H. It is because Coulomb repulsion between helium and hydrogen is larger.

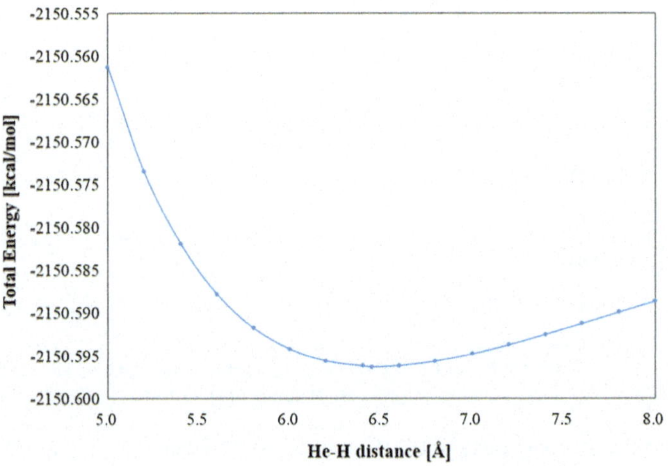

Fig. 15.10 Potential energy curve of He–H$^-$, changing He–H distance

15.3.4 Comparison with Three Cases

The dissociation energies of He–H^{+}, He–H and He–H^{-} are 46.9, 0.0117 and 0.0191 kcal/mol, respectively. The large dissociation energy of He–H^{+} is due to covalent bonding formation. Zero point vibration energies for He–H^{+}, He–H and He–H^{-} are 4.58, 0.0467 and 0.0208 kcal/mol, respectively. In He–H^{+}, zero point energy is smaller than dissociation energy. It implies that the effect of quantum vibration hydrogen is negligible in He–H^{+}.

Further Readings

1. Atkins P, de Paula J (2006) Physical chemistry, 8th edn, Chap. 11 (in Japanese)
2. Atkins P, de Paula J, Friedman R (2009) Quanta, matter, and change a molecular approach to physical chemistry, Chap. 5 (in Japanese)
3. Gaussian 09, Frisch MJ, Trucks GW, Schlegel HB, Scuseria GE, Robb MA, Cheeseman JR, Scalmani G, Barone V, Mennucci B, Petersson GA, Nakatsuji H, Caricato M, Li X, Hratchian HP, Izmaylov AF, Bloino J, Zheng G, Sonnenberg JL, Hada M, Ehara M, Toyota K, Fukuda R, Hasegawa J, Ishida M, Nakajima T, Honda Y, Kitao O, Nakai H, Vreven T, Montgomery JA Jr, Peralta JE, Ogliaro F, Bearpark M, Heyd JJ, Brothers E, Kudin KN, Staroverov VN, Kobayashi R, Normand J, Raghavachari K, Rendell A, Burant JC, Iyengar SS, Tomasi J, Cossi M, Rega N, Millam JM, Klene M, Knox JE, Cross JB, Bakken V, Adamo C, Jaramillo J, Gomperts R, Stratmann RE, Yazyev O, Austin AJ, Cammi R, Pomelli C, Ochterski JW, Martin RL, Morokuma K, Zakrzewski VG, Voth GA, Salvador P, Dannenberg JJ, Dapprich S, Daniels AD, Farkas Ö, Foresman JB, Ortiz JV, Cioslowski J, Fox DJ, Gaussian, Inc., Wallingford CT, 2009
4. Dunning TH Jr (1989) J Chem Phys 90(2):1007–1023
5. Woon DE, Dunning TH Jr (1994) J Chem Phys 100(4):2975–2988
6. Woon DE, Dunning TH Jr (1994) J Chem Phys 98(2):1358–1371
7. Dunning TH Jr, Peterson KA, Woon DE (1999) Encyclopedia of computational chemistry, pp 88–115
8. Onishi T (2016) J Chin Chem Soc 63:p83–p86
9. Onishi T (2016) AIP Conf Proc 1790:02002
10. Onishi T, Prog Theor Chem Phys (in press)

Chapter 16
Summary and Future

Abstract At the beginning of last century, it was difficult to extend quantum theory to many-electron system. In Bohr model, the electron structure of hydrogen atom was reproduced by introducing the concept of matter wave. However, it could not be applicable for many-electron system. To represent electron as quantum particle, the wave-function was proposed. One electron is allocated in one wave-function. It is not split into two wave-functions. Though the basic equation of electrons is Schrödinger equation, it cannot be analytically solved. Hartree-Fock method was developed to solve it numerically. It is important to analyse the obtained wave-function, which stands for molecular orbital. By using chemical bonding rule, chemical bonding character is specified. Though natural orbital is often utilized in molecular orbital analysis, it is completely different from molecular orbital. DFT is the scientifically reasonable method, due to incorporation of electron correlation effect. It makes it possible to design advanced materials. Finally, the future research in collaboration with particle physics is explained.

Keywords Quantum particle · Wave-function · Chemical bonding rule · Molecular orbital · Chemistry of the universe

16.1 From Quantum Theory to Molecular Orbital

16.1.1 Quantum Electron and Schrödinger Equation

Electron is categorized as quantum particle. In Bohr model, wave-particle duality is incorporated by the introduction of the concept of matter wave. However, Bohr model could not be applicable for many-electron system. To represent electron as quantum particle, quantum wave-function was proposed. Though the wave-function represents no figure, the square of wave-function represents electron density. The basic equation of electron is Schrödinger equation. Operating wave-function to Hamiltonian, discrete energy is given as eigenvalue. One electron is allocated in one wave-function. It is not split into two wave-functions.

T. Onishi, *Quantum Computational Chemistry*,
DOI 10.1007/978-981-10-5933-9_16

16.1.2 Orbital and Hartree-Fock Equation

In principal, the electronic structure of many-electron system can be obtained by solving Schrödinger equation. In hydrogen atom and hydrogenic atom, the exact wave-function is analytically obtained. It corresponds to orbital, which is expressed in real space. The kind of orbital is determined by the difference of two quantum numbers such as principal quantum number and quantum number of orbital angular momentum. In many-electron system, spin orbital is introduced to include the effect of spin angular momentum. The total wave-function is expressed by Slater determinant, to satisfy inverse principle.

Hartree-Fock equation is derived by minimizing the total energy of Schrödinger equation, under Born–Oppenheimer approximation. Hartree-Fock equation is one-electron equation. The eigenfunction and eigenvalue correspond to spin orbital and orbital energy, respectively.

16.1.3 Wave-Function Analysis

To solve Hartree-Fock equation numerically, basis function is introduced to spatial wave-function. In Hartree-Fock matrix equation, the coefficients and orbital energy are calculated under self-consistent-field (SCF) procedure. Initial atomic orbital is defined by basis function designated for calculation. Molecular orbital is represented by the combination of initial atomic orbitals.

In molecule and solid, atoms are bound through chemical bonding formation. In covalent bonding, outer shell electron is shared between different atoms. On the other hand, in ionic bonding, different atoms are bound through Coulomb interaction. In chemical bonding rule, chemical bonding character is specified by checking molecular orbitals including outer shell electrons: With orbital overlap, it is covalent; without orbital overlap, it is ionic. The essence of covalency is sharing electron at the same energy level. Note that ionic bonding is essentially included, due to the existence of positively charged atomic nucleus and negatively charged electron. Mulliken charge density is also utilized to estimate a net electron density.

In principal, from the communication relation of Hamiltonian operator and square of total spin angular momentum operator ($\boldsymbol{S}^2$), the total wave-function must be the eigenfunction of $\boldsymbol{S}^2$ operator. However, the relation is not satisfied in open shell system. It is much expected that spin function is prepared for not isolated electron but electron in atom, under consideration of spin-orbital interaction.

Natural orbital is completely different from molecular orbital. It loses the information of energy by the diagonalization of reduced charge density matrix. Hence, molecular orbitals with different orbital energies are mixed. In addition, spin information disappears, because spatial orbitals of α and β spins are mixed. Natural orbital is not eigenfunction of Hartree-Fock equation. It is apart from quantum mechanics. Natural orbital analysis sometimes misleads wave-function analysis.

16.2 Electron Correlation

In Hartree-Fock method, the electron–electron interaction is represented in an average manner. The exact total energy is different from Hartree-Fock total energy. Electron correlation energy is defined as the difference between the exact and Hartree-Fock total energies. In configuration interaction method, the electron correlation effect is incorporated by the combination of wave-functions with excited electron configurations. In coupled cluster method, the expansion of wave-function is performed by using cluster operator. However, they have the scientific contradiction that several Hartree-Fock equations are taken into account at the same time. CI and CI based methods sometimes predict the wrong electronic structure, especially in transition metal compounds.

In density functional theory, the electron correlation effect is incorporated by revising the electron correlation energy directly. Though it is scientifically reasonable, universal exchange-correlation functional has not been developed. At present, the best functional must be selected for considering system. The electron–electron interaction differs according to material (combination of atoms). For example, in transition metal compounds, hybrid-density functional theory is applied. It is because localization and delocalization properties are incorporated by Hartree-Fock exchange functional and LDA or GGA functionals, respectively. To perform very accurate calculation, the essence is to reproduce the electron–electron interaction precisely. Note that the calculation accuracy must be discussed by energetics and bonding. The latter is often missing in many research.

16.3 Solid State Calculation

In boundary system, molecular orbital calculation for minimum cluster model is applicable under three conditions: (1) no neutral condition, (2) no geometry optimization and (3) experimental distance. If using large cluster model consisting of many minimum cluster models, the electronic structure of nanoparticle will be reproduced, due to the breakdown of boundary condition within large cluster model.

The antiferromagnetic interaction of MXM system, where M and X denote transition metal and ligand anion, respectively, is explained by superexchange rule. When α and β molecular orbitals represent up and down spins of transition metals, respectively, α and β spins of ligand anion are cancelled out. As the result, the antiferromagnetic interaction occurs between transition metals via ligand anion.

In transition metal compounds, the energy splitting of 3d orbitals is explained by ligand bonding effect. The 3d electron configuration of transition metal is affected by not only Coulomb interaction but also charge transfer and orbital overlap. In some cases, the structural distortion is combined.

16.4 Materials Design

It is desired that molecular orbital calculation is performed, in parallel to experimental study. It is because much useful chemical information is provided from the experimental results. For example, much experimental information makes it faster to construct the scientific reasonable calculation model. The calculation results can be compared with the experimental results. Contrarily, the validity of experimental results is also judged. Now, it is possible to store much calculation data for several materials. By using much data effectively, we can design new material theoretically.

The major targets of materials design

(1) **Battery materials**
(2) **Catalysts materials**
(3) **Medicine, Biomolecules**

16.5 Chemistry of the Universe

It is known that most of elements in the universe are hydrogen and helium. It is because they are produced by nuclear fusion reaction of proton and neutron. Other heavy elements are synthesized from hydrogen and helium. To investigate nucleosynthesis, the collaboration of particle physics is indispensable. It is because quantum particles such as proton and neutron participate in nucleosynthesis. In addition, orbital theory under extreme condition must be explored. Previously, molecular orbital theory has succeeded in reproducing the real electronic structure and real chemical reaction for many materials on earth. At next stage, it is much expected that nucleosynthesis in the universe is clarified, based on the next molecular orbital theory, combined with particle physics.

Further Readings

1. Onishi T (2016) J Chin Chem Soc 63:83–86
2. Onishi T (2016) AIP Conf Proc 1790:02002
3. Onishi T, Prog Theor Chem Phys, in press

The manufacturer's authorised representative in the EU is Springer
Nature Customer Service Centre GmbH, Europaplatz 3, 69115 Heidelberg,
Germany. If you have any concerns regarding our products, please
contact ProductSafety@springernature.com

Printed and bound by CPI Group (UK) Ltd, Croydon, CR0 4YY
01/10/2026
02225871-0004